International Encyclopaedia of

NANO TECHNOLOGY, SCIENCE AND CHEMISTRY

International Encyclopaedia of

NANO TECHNOLOGY, SCIENCE AND CHEMISTRY

Volume 2

R.N. Parmar

ANMOL PUBLICATIONS PVT. LTD.

NEW DELHI - 110 002 (INDIA)

ANMOL PUBLICATIONS PVT. LTD.
H.O.: 4374/4B, Ansari Road, Daryaganj,
New Delhi-110 002 (India)
Ph.: 23278000, 23261597
B.O.: No. 1015, Ist Main Road, BSK IIIrd Stage
IIIrd Phase, IIIrd Block,
Bangalore - 560 085 (India)
Visit us at: www.anmolpublications.com

International Encyclopaedia of Nano Technology, Science & Chemistry

First Edition, 2008
ISBN 978-81-261-3681-0 (Set)

PRINTED IN INDIA

Printed at Mehra Offset Press, Delhi.

Contents

Preface

Nano Chemistry is the branch of science used for analyzing ultra small systems and for this one needs ultra small tools. Whether dealing with single molecules or single cells, the precision and sensitivity of the manipulation and detection schemes used are absolutely critical. Nano Chemistry scientists have been interested in confining single molecules in solution to perform chemistry at the lowest level possible, as well as in delivering small volumes of chemicals to single sub-cellular organelles. By mimicking what nature has developed over billions of years to conduct chemistry in ultra small packets (cells and organelles), Nano Chemistry scientists have arrived at what is to date the smallest test tubes known - synthetic liposomes which are manipulated and electrofused to mix their contents. These liposomes have volumes of a few femtoliters or less, matching the probe volumes of standard single molecule detection techniques, and are made in minutes by a protocol developed in this lab. In sum, Nano Chemistry is the science of tools, technologies, and methodologies for chemical synthesis, analysis, and biochemical diagnostics, performed in nanolitre to femtolitre domains. Nano Chemistry is the use of synthetic chemistry to make nanoscale building blocks of desired shape, size, composition and surface structure, charge and functionality with an optional target to control self-assembly of these building blocks at various scale-lengths.

This book entitled “International Encyclopaedia of Nano Technology, Science & Chemistry” provides readers with an elaborate introduction to molecular nanotechnology, Nanostructured materials, Nanoparticles and Nanodevices. An overview of relationship between nanotechnology and Nano-velcro is presented herein. The modern global issues regarding Nanoparticles, Nanostructured materials and polymer nanotechnology are discussed in detail. The subject areas of Nanocomposites and Nanostructured materials are covered detailing about their respective properties and applications. The inter-relationship among today’s Nanoworld, Nano Chemistry and the all important process of self-assembly is highlighted. An overview of nanotech products is presented. There are so many other applications of Nano Chemistry ranging from mundane to incredibly profound. Other applications include energy storage materials, photovoltaic cells, fuel cells, polymer Nanocomposites, super-tough Nanocoatings, landmine detectors, and thousands more. Although it is a field that has already seen its share of hype, there is a general consensus that over the long term, Nano Chemistry is likely to reconfigure everything. Nanostructured materials do not represent a new phenomenon. However, the ability to probe, manipulate, understand and engineer matter at atomic scales has only recently come within our grasp. Developments such as the invention of the Scanning Tunneling Microscope have since made nanoscale science a reality. Today, the term nanotechnology is used to describe specific construction and manipulation performed on an atomic and molecular scale. It could be one's strong opinion that Nano Chemistry will revolutionize many areas of chemistry, and is of strategic importance. In the first instance, the goals are focused around the future needs in pharmaceutical chemistry and clinical diagnostics, where paradigm shifts are very close. Novel approaches towards combinatorial Nano Chemistry are studied and developed. Tools like chip-based Nanovials, flow reaction channels, dispensing devices, mechanical and chemical microactuators, optical detectors etc. are being developed by joint expertise in micro structure technology, chemistry as well as specialists in surface and coating technology. Applications which are being studied are e.g. optimization of catalysis, asymmetric synthesis, combinatorial chemistry, novel photochemistry etc. In this context, new strategies for high throughput handling of very small amounts of materials of major concern, which will have important applications in biochemistry, molecular biology, clinical diagnostics, drug development and material science.

Generally, systems with a maximum size of less than a hundred nanometers are classified as nanotechnology. Nanotechnology is a completely new production process in a molecular world, opening up possibilities far beyond what we have known before. Nature provides the inspiration for this new technology – from time immemorial, nanomachines have been at work in all organic cells, building our world of plants, animals and people from a store of chemical elements. Experts world over believe nanotechnology will have a more far-reaching impact than the silicon integrated chip with many more applications than electronics. Nanotechnology is expected to revolutionize certain areas, such as semiconductors, pharmaceuticals and materials. The total societal impact of nanotechnology is expected to be much greater than that of the silicon integrated circuit because it is applicable in many more fields than electronics. Nanotechnology experts believe that because of Nano Chemistry, we would see more changes in the next 30 years than we saw in the entire last century. The main focus of research work in Nano Chemistry is on the synthesis and characterization of metal and semiconductor Nanoparticles. These particles have interesting size-dependent properties and can be used as building blocks for new materials. We use dynamic light scattering to study the formation of superstructures of Nanoparticles.

This book tends to cover the aforementioned subject areas in their broadest possible terms. All useful supplementary research and reference tools have been included for readers' further investigations in the subject area of Nano Chemistry. This book mainly focuses on the chemical aspects of Nanotechnology; Nanoparticles,

Nanomaterials and Nanostructures. This book "International Encyclopaedia of Nano Technology, Science & Chemistry" seeks to explain this complex field of studying its totality. It also seeks to provide an understanding of what is happening now in the field of Nano Chemistry, as well as what to expect in future by generations to come. Keeping in mind the applications and utilities of Nano Chemistry for meeting the contemporary global medical requirements, it seems that this subject will become the core area of future nanoscience and nanotechnology research. Developments such as the invention of the Scanning Tunneling Microscope have since made nanoscale science a reality. Today, the term nanotechnology is used to describe specific construction and manipulation performed on an atomic and molecular scale. Generally, systems with a maximum size of less than a hundred nanometers are classified as nanotechnology. This book tends to cover the aforementioned subject areas in their broadest possible terms. All useful supplementary research and reference tools have been included for readers' further investigations in the subject area of Nano Chemistry. This "International Encyclopaedia of Nano Technology, Science & Chemistry" is user friendly and future research oriented.

—EDITOR

5

Multiple Applications of Nanocomposites and Nano Chemistry with Special Focus on Industries

Applications of Nanocomposites and Plastics

The definition of nanocomposite material has broadened significantly to encompass a large variety of systems such as one-dimensional, two-dimensional, three-dimensional and amorphous materials, made of distinctly dissimilar components and mixed at the nanometer scale. The general class of nanocomposite organic or inorganic materials is a fast growing area of research. Significant effort is focused on the ability to obtain control of the nanoscale structures via innovative synthetic approaches. The properties of nanocomposite materials depend not only on the properties of their individual parents but also on their morphology and interfacial characteristics. This rapidly expanding field is generating many exciting new materials with novel properties. The latter can derive by combining properties from the parent constituents into a single material. There is also the possibility of new properties which are unknown in the parent constituent materials. The inorganic components can be three-dimensional framework systems such as zeolites, two-dimensional layered materials such as clays, metal oxides, metal phosphates, chalcogenides, and even one-dimensional and zero-dimensional materials such as $(Mo_3Se_3\text{-})_n$ chains and clusters. Experimental work has generally shown that virtually all types and classes of nanocomposite materials lead to new and improved properties when compared to their macrocomposite counterparts. Therefore, nanocomposites promise new applications in many fields such as mechanically reinforced lightweight components, non-linear optics, battery cathodes and ionics, nanowires, sensors and other systems.

The general class of organic/inorganic nanocomposites may also be of relevance to issues of bio-ceramics and biomineralisation in which *in-situ* growth and polymerisation of biopolymer and inorganic matrix is occurring. Finally, lamellar nanocomposites represent an extreme case of a composite in which interface interactions between the two phases are maximised. Since the remarkable properties of conventional composites are mainly due to interface interactions, the materials dealt with here could provide good model systems in which such interactions can be studied in detail using conventional bulk sample (as opposed to surface) techniques. By judiciously engineering the polymer-host interactions, nanocomposites may be produced with a broad range of properties. Inorganic layered materials exist in great variety. They possess well defined, ordered intralamellar space potentially accessible by foreign species. This ability enables them to act as matrices or hosts for polymers, yielding interesting hybrid nano-composite materials. Lamellar nanocomposites can be divided into two distinct classes, intercalated and exfoliated. In the former, the polymer chains alternate with the inorganic layers in a fixed compositional ratio and have a well defined number of polymer layers in the intralamellar space. In exfoliated nano-composites the number of polymer chains between the layers is almost continuously variable and the layers stand >100 Å apart. The intercalated nanocomposites are also more compound-like because of the fixed polymer/layer ratio, and they are interesting for their electronic and charge transport properties. On the other hand, exfoliated nanocomposites are more interesting for their superior mechanical properties. A subclass of nanocomposites which is the lamellar class of intercalated organic/ inorganic nanocomposites and namely those systems that exhibit electronic properties in at least one of the components offers the possibility of obtaining well ordered systems some of which may lead to unusual electrical and mechanical properties. Selected members of this class may be amenable to direct structural characterisation by standard crystallographic methods. An important issue in this area is the few structural details that are available, therefore, any system which can be subjected to such analysis is of interest. Nanocomposites also offer the possibility to combine diverse properties which are impossible within a single material, e.g. flexible mechanical properties and superconducting properties. This work is now in its infancy and extensive investigations are underway. Another exciting aspect is the possibility of creating heterostructures composed of different kinds of inorganic layers, which could lead to new multifuntional materials. Several general synthetic routes have been developed for inserting polymer chains into host structures and many novel nanocomposites have been designed.

These include:

- *In-situ* intercalative polymerisation (ISIP) of a monomer using the host itself as the oxidant. The rationale behind intercalative polymerisation is that host matrices with high electron affinity can oxidatively polymerize appropriate monomers in their interior.
- Monomer intercalation followed by topotactic intralamellar solid state polymerisation. This route creates conjugated polymers inside non-oxidising hosts.

- Direct precipitative encapsulation of polymer chains by colloidally dispersed single layers of a host. This approach gives access to a large variety of nanocomposites with many kinds of polymers and hosts.

Materials with features on the scale of nanometers (10^{-9} meter) often have properties dramatically different from their bulk-scale counterparts. For example, nanocrystalline copper is five times harder than ordinary copper with its micrometer-sized crystalline structure. The development of such materials is currently a research area of great interest. Important among these nanoscale materials are nanocomposites, in which the constituents are mixed on a nanometer-length scale. They often have properties that are superior to conventional microscale composites and can be synthesised using surprisingly simple and inexpensive techniques. A nanoscale dispersion of sheet-like inorganic silicate particles in a polymer matrix, for example, is superior to either constituent in such properties as optical clarity, strength, stiffness, thermal stability, reduced permeability, and flame retardancy. The key to the synthesis of nanocomposites is how the silicates are made to disperse in the polymer. Since the organic and inorganic components are typically immiscible (like water and oil), the silicate surface is modified by attaching surfactant molecules (like adding soap to water). Polymer-based nanocomposites are also being developed for electronics applications such as thin-film capacitors in integrated circuits and solid polymer electrolyes for batteries. Nature also makes fabulous nanocomposites, and scientists are trying to emulate such processes. The abalone shell, for example, has alternating layers of calcium carbonate and a rubbery biopolymer; it is twice as hard as and a thousand times tougher than its components. The study of nanocomposite materials requires a multidisciplinary approach, involving novel chemical techniques and an understanding of physics and surface science. It is a field of broad scientific interest with impressive technological promise. Studies have revealed clearly the property advantages that nanomaterial additives can provide in comparison to both their conventional filler counterparts and base polymer. Properties which have been shown to undergo substantial improvements include:

- Mechanical properties e.g. strength, modulus and dimensional stability.
- Decreased permeability to gases, water and hydrocarbons.
- Thermal stability and heat distortion temperature.
- Flame retardancy and reduced smoke emissions.
- Chemical resistance.
- Surface appearance.
- Electrical conductivity.
- Optical clarity in comparison to conventionally filled polymers.

To date one of the few disadvantages associated with nanoparticle incorporation has concerned toughness and impact performance. Some of the data presented has suggested that nanoclay modification of polymers such as polyamides, could reduce impact performance. Clearly this is an issue which would require consideration for applications where impact loading events are likely. In addition, further research will be necessary to, for example, develop a better understanding of formulation/ structure/property relationships, better routes to platelet exfoliation and dispersion etc. In addition it is important to recognise that nanoparticulate/fibrous loading confers significant property improvements with very low loading levels, traditional microparticle additives requiring much higher loading levels to achieve similar performance. This in turn can result in significant weight reductions (of obvious importance for various military and aerospace applications) for similar performance, greater strength for similar structural dimensions and, for barrier applications, increased barrier performance for similar material thickness.

Data provided by Hartmut Fischer of TNO in the Netherlands relating to polyamide – montmorillonite nanocomposites indicates tensile strength improvements of approximately 40 and 20 per cent at temperatures of 23ºC and 120ºC respectively and modulus improvements of 70 per cent and a very impressive 220 per cent at the same temperatures. In addition Heat Distortion Temperature was shown to increase from 65ºC for the unmodified polyamide to 152ºC for the nanoclay-modified material, all the above being achieved with just a 5 per cent loading of montmorillonite clay. Similar mechanical property improvements were presented for polymethyl methacrylate—clay hybrids.

Further data provided by Akkepeddi of Honeywell relating to polyamide-6 polymers confirms these property trends. In addition, the further benefits of short/long glass fibre incorporation, together with nanoclay incorporation, are clearly revealed.

Nanocomposite materials with an inorganic glass and an organic polymer constitute a relatively new and unique area in material science. The term "ormocers", "ormosils" and "ceramers" are often utilised to describe this class of nanocomposite. By combining at the molecular level inorganic and organic polymeric material a blending of unique physical properties can be achieved. The value in these materials is apparent, from fiber optics to paints these materials may provide the requisite physical properties to achieve the next technological advance. There are several different ways of synthesising this class of nanocomposite; therefore a means of classification is necessary. Most developed nomenclature is based on synthetic techniques; Wilkes has a relatively recent and exhaustive categorisation. However, we chose to classify these materials upon a simpler system first suggested by Novak. Five categories cover the majority of composites synthesised with more recent techniques being modifications or combinations from this list.

Type I: Organic polymer embedded in an inorganic matrix without covalent bonding between the components.

Type II: Organic polymer embedded in an inorganic matrix with sites of covalent bonding between the components

Type III: Co-formed interpenetrating networks of inorganic and organic polymers without covalent bonds between phases.

Type IV: Co-formed interpenetrating networks of inorganic and organic polymers with covalent bonds between phases.

Type V: Non-shrinking simultaneous polymerisation of inorganic and organic polymers.

Before discussing different kinds of nanocomposites it is appropriate to discuss the sol-gel process. Although zirconium, titanium, aluminum and boron oxides have been utilised as the inorganic component, the great majority of nanocomposites incorporate silica from tetraethoxysilane (TEOS). The formation of the inorganic component involves two steps, hydrolysis and condensation. The important insight in the formation of this part of the composite is based upon the relative kinetic rates of each step. For example, if the rate of hydrolysis is fast compared to condensation, then simple particles or highly branched silicate matrices are formed. Conversely, if the condensation step is quicker than hydrolysis, then string-like filaments are formed. These changes in morphology of the silicate matrix can be manipulated by the choice of sol-gel catalyst with dramatic effect on the physical properties of the nanocomposite.

Perhaps the most common and straightforward nanocomposites found in the literature are the Type I composites. Typically, TEOS is used to form an inorganic network around an organic polymer component. The goal in the process is to form a completely interpenetrating network (IPN) of both inorganic and organic phases. Homogeneous nanocomposites with good IPNs are often stronger, more resilient, and optically transparent, whereas heterogeneous composites are often mechanically weaker and opaque. Choice of co-solvents is critical throughout the formation of the inorganic component. As the hydrolysis and/or condensation reaction occurs, changes in the polarity of the solvent mixture can result in undesired phase separation of the inorganic and organic components. Phase separations can produce a range of results from small zones of one component dominance to total inorganic/organic separations. However, the result is that the same sites of instability are created. The timing and physical conditions in which phase separations occur can produce isolated silicate formations within composite formulations.

Type II composites utilize modified organic polymers with sites capable of covalently bonding directly to the forming inorganic phase of the nanocomposite. Most often an organosilane with a reactive pendant group is used to form the covalent bond to the organic polymer component. These pendant groups can be isocyanates, which easily react with polymers containing alcohols or amines. Another option that has been explored is hydrosilation reactions with polymers containing a terminal alkene. Once bound, the silane's remaining pendant ethoxide or chloride groups will be available to bond with such sites on other silane species, forming the inorganic component.

Type V composites are unique examples of these materials because of the physical conditions in which they are formed. Novak and Grubbs developa method of simultaneously polymerising both inorganic and organic components without shrinkage of the material. The insight was that by replacing the ethoxide found in TEOS with an organic monomer oxide, solvent loss created during the hydrolysis step is eliminated. It is the loss of ethanol (or other solvents) in the sol-gel process, which creates material shrinkage and leads to cracking from capillary pressures. Although extremely innovative, unfortunately a limited number of composites have been formed successfully with this process. This is presumably because of the difficulties in precisely matching the rates of polymerisation of the organic component with both the rates of hydrolysis and condensation of the ceramic component to prevent disastrous phase separations.

A general synthesis for a base, acid, or salt catalysed polyphosphazene, polyethylene oxide (PEO), and polyethylene oxide/polypropylene oxide (PPO/PEO) block nanocomposite is as follows: 300 mg of polymer is dissolved into 10 mL of a 50/50 by volume tetrahydrofuran (THF)/ethanol mixed solvent in a capped vial. To this solution is added TEOS (336 mg). A catalyst is then introduced as an aqueous solution (150 μl) and the mixture is capped and sonicated at 50°C for 30 minutes. The solution is aged from hours to days depending upon the catalyst used in a sealed vial and poured into a teflon mold and loosely covered at room temperature. The nanocomposite self assembles as the volatile solvent slowly escapes during the condensation process. The synthesis of polyvinyl acetate (PVAc)/silicate nanocomposites requires a different approach from the other nanocomposites. PVAc (300 mg) is dissolved into an 50/50 by volume acetic acid/methanol (10 mL) mixed solvent in a capped vial. To this solution is added TEOS (373 mg). The solution is then sonicated for 5 minutes in a sealed vial at room temperature and poured into a teflon mould and loosely covered at room temperature. The nanocomposite self assembles during the curing process, which typically lasts up to 24 hours. Additional heating at 100°C for 30 minutes aids in removing lingering acetic acid from the nanocomposite.

Various membrane samples were tested on a Dynamic Mechanical Analyzer (TA Instruments, Model 2980, New Castle, DE) fitted with a tension (film) or penetration (compression) clamp. A simple linear ramp-force protocol was developed to evaluate the membrane strength in tension and compression. Both tension and compression protocols consist of the establishment of 0.03 Newton (N) static force and equilibration to 25°C. A force is then applied to the specimen at a specific rate, 0.5 N per minute for tension, and 1.0 N per minute for compression (18 N maximum) or until the maximum limit of travel is reached (24.7 mm). The membrane specimens are cut with a laboratory fixture consisting of two razor blades mounted on a parallel plate yielding a 6.20 mm by 39.0 mm sample. The specimen's thickness is measured with a caliper (Mitutoyo, Model ID-C112EBS, Japan) in several places along the length to obtain an average membrane thickness.

Various thin films were imaged as an unstressed, swelled, or stretched nanocomposite with a Philips XL30ESEM. A 10 to 20 kV electron beam was utilised. Wet mode analyses employed a 500 μm pressure-limiting aperture (PLA). To image the topography either a gaseous secondary electron detector (wet mode) or secondary electron detector (HiVac) was used.

The polyphosphazene family of polymers was chosen for the initial investigations into nanocomposites due to their diverse array of physical and solubility properties. The researcher was specifically interested in some heretofore inaccessible phosphazenes that lacked the mechanical stability to be useful for membrane and barrier applications. The polymer poly[*bis*-(2-(2-methoxyethoxy)ethoxy)phosphazene] (MEEP) is such a polymer - one where the chemical properties, such as selective transport of ions are extremely interesting, yet the polymer alone lacks the requisite mechanical integrity to form a stable membrane for evaluation and subsequent use. This water-soluble polyphosphazene is a viscous liquid with two phosphorus bound polyether ligands on an alternating nitrogen phosphorus backbone. In all of the nanocomposites TEOS was employed as the ceramic precursor. The condensation of the ceramic component was catalysed in a solvent composed of various ratios of water, ethanol and THF. Depending upon the nature of the catalyst employed, differing properties were observed in the resulting nanocomposites. These catalysts are best separated into three categories; acid, base or ionic salt; and each will be discussed separately below. Acid catalysed nanocomposites that were examined in this study were found to have the highest tensile strength of any of the catalyst types studied; yet they were also found to be glassy and brittle when compared to salt or base catalysed analogues. While providing a maximum of physical stabilisation for the liquid-like MEEP polymer, this translates directly into extremely poor durability when wetted, as exhibited in aqueous swelling tests. An ESEM image of an aqueous swelled/deswelled membrane, which has fractured from the internal hydrostatic pressure. The dominant reason for this lack of durability when wet is the nature of the ceramic morphology of these acid catalysed nanocomposites. It is known from the ceramic literature that the general morphology of acid catalysed sol-gel condensations forms long filament structures instead of star-like or particulate structures. This more rigid morphology imparts a high tensile strength to these nanocomposites, yet leaves them less elastic and highly susceptible to damage upon contact with liquids. Base catalysed nanocomposites have the best elastic and swelling characteristics of all the nanocomposites studied. As seen in the acid catalysed nanocomposites, the morphology of the condensing ceramic component dominates the physical properties of the material. An ESEM image of a MEEP nanocomposite after being soaked to equilibrium in water. Before swelling, the nanocomposite has a featureless uniform appearance, as it does after drying. During wetting, the hydrophilic polymer component expands as it absorbs the water yet the ceramic is not capable of expansion. This behaviour produces the creases seen in the micrograph. The villae-like features reveal the underlying structure of the ceramic component of the nanocomposite, i.e. areas of higher silicate composition. This is consistent with the morphology expected for a base catalysed nanocomposite—one that contains interlocking star-like structures. In regions between the ceramic star structures, the polymer is capable of absorbing water and expanding, while regions within the ceramic star structures maintain their structural rigidity. Because a single material contains both of these types of different regions, this renders base catalysed nanocomposites simultaneously both tough and flexible. Salt catalysed nanocomposites exhibit properties that are between those found for acid and base catalysed nanocomposites. Again, the nature of the ceramic dominates the physical properties of the nanocomposite. There is considerable variation in the properties of this class of nanocomposites as there are a variety of simple salts that have proven to be effective catalysts. To investigate the nature of ionic salt catalysts in MEEP nanocomposites, a course of fluoride salts with various alkali counter-cations was examined, as it is known that fluoride anion salts are superior ceramic condensation catalysts over other halide salts. As was expected, the mechanical properties of fluoride catalysed nanocomposites more closely resembled their base catalysed analogues because the condensation was driven by an anionic catalyst, similar to OH^-.

The initial results indicated salt catalysts that contain large cations like cesium performed better than small ones such as lithium. Intuitively, hard ions such as lithium, the closest of the alkali metal ions to the proton (an extremely effective catalyst) should outperform the softer ionic catalysts, yet exactly the opposite trend is observed. The reason behind the observed trend becomes apparent when one considers that the strongly electronegative fluoride ion has a poorer affinity towards large polarizable cations (like cesium) than harder cations (like lithium). Catalyst availability in these polar, protic organic solvent systems becomes the dominant factor in the total catalytic effectiveness of these systems. There is a linear correlation between the lattice energy of the alkali fluoride salt employed and the final strength of the resulting nanocomposite, with the lowest energy salt yielding the strongest material. This is due to higher availability of the "naked" fluoride anion in systems possessing the lower lattice energies. In these systems, the catalyst ions spend less time during synthesis associated as intimate ion pairs in solution and are consequently more effective in driving the ceramic condensation.

A better understanding of the formation, properties, and morphology of Type I nanocomposites was gained during studies with polyphosphazene polymers. This information was used to try to better understand the properties of nanocomposites employing more conventional organic polymers. Commercially available polymers were used in the preparation of these polymer/silicate materials. A block polymer consisting of PEO/PPO (M_w = 2K), PEO (2K), polyvinylalcohol (PVA, 30–70K), PVAc (83K and 167K) and poly (acrylonitrile) (PAN, 150K) were the primary organic polymers of interest. Drawing upon previous experience, a 50/50 (w/v) polymer to silicate ratio was used to form these nanocomposites. Acid catalysed composites were the focus of most of these studies. Numerous attempts were made with low molecular weight PEO and PEO/PPO block polymers to prepare a nanocomposite with acceptable mechanical integrity. Unfortunately, all attempts resulted in materials that were too brittle to be lifted as a single unit suitable for mechanical analyses. This was surprising in light of the fact that acid catalysed composites were previously found to yield the strongest nanocomposites. Explanation for this behaviour lies in the low molecular weight of the polymer component. Typical molecular weights for the MEEP polymer were in the 10^6-10^7 range, providing a directing effect for the condensing ceramic. The PEO polymer was sufficiently light that the ceramic exhibits a directing effect on the orientation of the polymer leading to a significantly different ceramic morphology, more base-like in nature. Polyvinyl alcohol (PVA) and polyvinyl acetate (PVAc) were found to readily form tractable nanocomposites with good mechanical integrity, good flexibility and interesting wetting characteristics. In general, the PVA nanocomposites yielded a significantly higher storage modulus then the PVAc nanocomposites. However, results of a simple destructive force ramp test, at the individual material's glass transition temperature (PVA 85°C, PVAc 30°C), suggests that the PVAc nanocomposite actually is mechanically stronger. Another interesting feature of these two nanocomposites that was noted is that they tend to behave somewhat as a thermoplastic. That is, once heated above the polymer component's glass transition temperature, the material can be easily deformed or bent, and the material will retain its shape once cooled. This is a rare example of where the polymer dictates the physical properties of the entire nanocomposite.

The behaviour of PAN nanocomposites was also explored. However, PAN has limited solubility in solvents commonly used to form this type of nanocomposite due to polarity differences between the polymer (lower polarity) and the solvents or TEOS (higher polarity). Attempts to form PAN nanocomposites through our usual synthetic protocols resulted in an unusual material unlike the other homogeneous nanocomposites previously produced. The polymer and ceramic components retreated into separate phase domains, forming an inhomogeneous composite material. As was expected, the mechanical properties of this material were poor due the poor interfacial characteristics between the polymer and ceramic phases. Additional attempts were made at producing a true homogeneous nanocomposite by using N-methylpyrrolidone as the solvent for the condensation. Though a homogeneous nanocomposite was formed with this method, difficulties in the condensation/curing process resulted in a highly fractured material such that a mechanically acceptable material was never obtained.

Applications

There are a host of applications of nanocomposites. One of the most interesting of these applications is as solid polymer electrolytes (SPE) for lithium batteries. The polyphosphazene MEEP is a well-known SPE with very high room temperature conductivity, however it lacks the mechanical stability to be used in a practical device. Traditional stabilisation methods, such as deep UV or electron beam crosslinking methods do improve the physical stability of SPEs, however this crosslinking lowers ionic conductivity – tests performed in the laboratory revealed this to be a factor of 30-45 for MEEP-like phosphazene polymers. This reduction is due to the additional covalent linkages formed during the crosslinking process that inhibit chain segmental motion and ion transfer. Since the nanocomposites formed by the ceramic condensation process do not form bonds to the polymer component, (Type I nanocomposites) mechanical stabilisation is achieved without a great loss of ionic conductivity. AThe ceramic component forms a "scaffolding" around the polymer providing the needed stability, while not interfering with the polymer's ability to easily transport ions. It is hoped that these nanocomposites will afford the development of all-solid lithium batteries that will be better, safer, and more environmentally friendly than those available today. Another set of applications for the nanocomposites is in separations membranes and other selective barriers. Polymers that are used in selective gas separations, and liquid pervaporation separations must usually be mechanically stabilised, typically by crosslinking, before they become practically usable. Since the nanocomposite is able to physically stabilize the polymer component in a manner very different than these traditional methods, a real advantage may be gained through nanocompositing rather than crosslinking. One final separation application that is being explored is using nanocomposites as subsurface permeable reactive barriers. The nanocomposite matrix is being used to provide a platform for selective chelating agents that remove contaminants from subsurface water. The design employs reproducibly swellable hydrophilic materials (such as the base catalysed materials discussed above) that allow subsurface water to flow through the barrier, allowing the entrained chelators to selectively sequester the contaminants of concern. This type of *in-situ* remediation technology is of particular interest to the U.S. Department of Energy, but is also applicable to other contaminated sites. From auto parts to barrier packaging, the race is on to commercialize nanoclay thermoplastic composites. Just a pinch of these infinitesimally small particles can dramatically raise mechanical, thermal, barrier, and flame-retardant properties.

Why all the excitement about the new class of mineral-reinforced plastics known as nanocomposites? Because relatively small amounts (2-5 per cent) of nanometer-sized clay particles can provide large improvements in mechanical and thermal properties, as well as gas barrier and flame resistance. They also reduce shrinkage and warpage. And all of these benefits are available without significantly raising the density of the compound or reducing light transmission (nanoclays are in the same size range as visible-light wavelengths). Most of the current R&D work is focused on automotive parts and packaging. The goals are physical and thermal property enhancements and reduced permeability to gases, moisture, and hydrocarbons. Nylon-based nanocomposites are the first commercial materials to emerge, and there is now a frenzy of activity aimed at nano-reinforcing commodity thermoplastics such as polypropylene and PET. Two research consortia on nanocomposites have been formed in recent months. The Edison Polymer Innovation Corp. (EPIC) of Akron, Ohio, got one under way in March. Headed by staff scientist Jon Collister, this three-year, $3-million programme will coordinate investigations by industry and academic researchers on thermoplastics (PP, PET, PE, and acrylic), advanced composites, coatings, and elastomers. Meanwhile, the National Institute of Standards and Technology (NIST) in Gaithersburg, Md., has formed a consortium of government and industry scientists to explore nanocomposites' potential for reducing the flammability of thermoplastics.

The current favourite nanoclay is montmorillonite, a layered alumino-silicate whose individual platelets measure on the order of one micron diameter, giving them an aspect ratio of about 1000:1. Because it is hydrophilic, montmorillonite is not compatible with most polymers and must be chemically modified to render its surface more hydrophobic. The most popular surface treatments are organic ammonium cations, which can be exchanged for the inorganic cations existing on the silicate surface. The treated clay is then incorporated into the resin matrix either during polymerisation or by melt compounding. As an example of what can result, commercial nylon 6 nanocomposite materials provide tensile modulus, tensile strength, and HDT equivalent to 40 per cent mineral-filled nylon, while specific gravity remains at the level of unfilled nylon 6. With lower-modulus polymers such as polyolefins, nanocomposites have the potential to upgrade physical properties to levels competitive with costlier engineering resins. Nanocomposites already look attractive for molded car parts such as body panels and under-hood components, as well as electrical/electronic parts, power-tool housings, lawnmowers, aircraft interiors, and appliance components. On the packaging side, nanocomposites can slow transmission of gases and moisture vapor through plastics by creating a "tortuous path" for gas molecules to thread their way among the obstructing platelets. Bottles and food packaging aren't the only areas of interest here. According to EPIC, nanocomposites hold commercial benefits for reducing hydrocarbon emissions from hoses, seals, and auto fuel-system components. Nanocomposites of polypropylene might permit blow molding auto fuel tanks at lower cost than current multi-layer barrier structures.

The first nanocomposites on the market are based on nylon 6 and produced in the reactor. Toyota Central R& D Laboratories in Japan pioneered the work on nanocomposites a decade ago. Within the last few years, Toyota has licensed its patented in-reactor nylon nanocomposite technology to companies such as Nanocor and Japan's Ube Industries. Nanocor, in turn, sublicenses the technology to firms that want to make nylon nanocomposites. Ube was among the first to come out with a commercial nanocomposite—its "nylonclay hybrids" (NCH) of nylon 6 and nylon 6/66 copolymer for film and structural applications. Commercial applications of NCH include nylon 6 barrier film for food packaging and a timing-belt cover for Toyota Motors. Compared with unfilled nylon 6, Ube's NCH is said to have 68 per cent higher tensile modulus and 126 per cent higher flexural modulus. Ube also reports that cast films of NCH-2 grade (2 per cent clay) have half the oxygen permeability of straight nylon 6. At about the same time, Unitika Co. of Japan introduced a nylon 6 nanocomposite for injection molding, which is marketed and distributed in North America by Toyota Tsusho. Nylon M2350 is produced with Unitika's proprietary technology, which incorporates a synthetic silicate during polymerisation. Nylon M2350 has been used by Mitsubishi Motors for an engine cover on its GDI models, where the nanocomposite is said to offer a 20 per cent weight reduction and excellent surface finish. Lighting covers and knife handles are other applications.

Sometime back, Bayer AG in Germany announced the development of nylon 6 nanocomposites for transparent barrier film packaging. The compounds are made in the reactor using Nanocor's nanoclay. Bayer is currently marketing nanocomposites only in Europe. It has two nylon 6-based nanocomposite grades for cast film: Durethan LPDU 601-1 and LPDU 601-2, which offer different degrees of barrier improvement. Relative to neat nylon 6, they reportedly cut oxygen transmission rate (OTR) by up to 50 per cent, plus offer increased barrier to other gases, chemicals, and flavors/odors. The nanofilms are also said to be stiffer than unfilled film. AlliedSignal Corp. has been exploring nylon 6 nanocomposites for potential applications in automotive, consumer, and packaging markets. According to Kris Akkapeddi, team leader in plastics R&D, the company is currently evaluating this technology for some niche applications with specific customers. Developmental products at AlliedSignal include toughened nylon nanocomposites for injection molding and high-viscosity nanocomposites for blow molding. Compared with unfilled nylon, Akkapeddi says the nanocomposites have 50-80 per cent higher stiffness and up to 126°F higher HDT without much increase in specific gravity.

Nylon nanocomposites produced by melt compounding that reportedly approach the performance of their in-reactor counterparts are now emerging. The trick is to exfoliate (delaminate) the clay particles sufficiently so that the ultimate level of reinforcement is achieved. Nanocor has introduced a new nanoclay grade, Nanomer I.24TC, which reportedly allows compounders to produce a nylon 6 compound with flexural modulus and HDT about 90 per cent of the levels achieved with in-reactor blends. RTP Co., a specialty thermoplastics compounder, announced in late April that it has produced what it believed to be the first commercially available compounded nylon 6 nanocomposite. The material is offered in injection molding grades and film grades with 3-5 per cent loadings. As shown in the table, RTP's injection grade shows 53 per cent higher HDT than unfilled resin—comparable to 20-30 per cent mineral-filled nylon—but with little change in specific gravity. Tensile strength is much better than either unfilled or mineral-filled nylon 6. Reduced flammability is also indicated by a 50 per cent reduction in rate of heat release. Toyota Central R&D Labs. recently succeeded in producing nylon 6 clay nanocomposites by melt processing in a twin-screw extruder. Dr. Arimitsu Usuki reports that the compounding approach improves productivity and opens the way for making nanocomposites of nylons 12, 66, and 6/66 copolymer. Showa Denko of Japan recently started commercial production of nylon 66 nanocomposites using a kneading extruder. The materials are said to show improved flame retardancy and rigidity. Two special FR grades contain no halogen or phosphorus. Systemer FE 30600 and 30602 reportedly provide UL 94V-0 and V-2 performance, respectively, at only 1/64-in. thickness. Their flex moduli are 30-80 per cent higher, and HDTs are 54-144°F higher, than those of standard nylon 66. Melt-compounded nylon 6 and 66 nanocomposites are also reported to be forthcoming from BASF AG and from Solutia, respectively. (Solutia markets through Dow Plastics.)

An agreement announced sometime ago by Eastman Chemical and Nanocor to jointly develop nanocomposites of PET and other packaging materials (e.g., polyethylene) is the first indication that commercialisation of such materials may not be far off. Jim Lewis, v.p. of container plastics at Eastman, says his firm has worked with Nanocor since 1995, but several breakthroughs have occurred over the last year and a commercial product introduction is likely to come fairly soon. Eastman is developing its first PET nanocomposites via the in-reactor approach. Its initial focus is on rigid containers. "Nanocomposites are one of the strategic technology components we are developing to enable emerging markets in food packaging that require superior barrier, including beer bottles and single-serve juice bottles, " Lewis says. He hints that these applications are likely to utilize a coinjected multi-layer preform with the nanocomposite barrier layer on the inside. Active projects on other barrier resins are also under way. Nanocor says it has made significant progress in joint development of EVOH nanocomposites for multi-layer packaging. Nanocor even offers a new nanoclay, Nanomer I.35L, designed for compounding with EVOH. Within the range of 50-80 per cent R.H., it reduces oxygen permeability of EVOH by 66-80 per cent. Besides packaging, EVOH nanocomposites are being investigated as a hydrocarbon barrier for automotive fuel tanks and fuel-system components.Nanocor has another new nanoclay, Nanomer I.28MC, for compounding with MXD6 aromatic nylon from Mitsubishi Chemical. Applications could include packaging films for moisture- and oxygen-sensitive foods and electronics.

Automotived-driven development of PP and TPO nanocomposites is one of the hottest areas of current interest. Because PP is a strongly nonpolar resin, it has been especially challenging to come up with an appropriate surface treatment for the clay. Nonetheless, late last year, three development partners—Montell North America, General Motors R& D in Warren, Mich., and Southern Clay—jointly announced the world's first TPO-based nanocomposite. According to Jim Keeler, advanced programme manager at Montell's Automotive Business Group, the partners are testing several TPO nanocomposites for exterior and interior applications. The initial focus is on exterior door and rear quarter panels. Although the nanocomposites are produced by melt compounding, Montell is using its Catalloy in-reactor TPOs in order to simplify the overall compounding process. According to Keeler, previous PP/nanoclay compounds achieved only 10-15 per cent exfoliation of the layered clay particles, whereas the current work has achieved much greater exfoliation and thus enhanced property development. TPO nanocomposites with specific gravity as low as 0.91 have been produced with better dimensional stability than unfilled TPOs and comparable to talc-filled grades (CLTE of 5 × 10-5 mm/mm/°C). These developmental TPO nancomposites also have flexural moduli of 150, 000 to 250, 000 psi. In a 5-mph impact test, the TPO remains ductile down to –30°C. The partners also report that the nanocomposite has improved surface appearance due to the low filler loading. Work on nanocomposites is also well under way at Ford Motor Co., according to Jeff Helms, director of research and materials science. "We have been focusing on anything polypropylene-based, including TPOs, impact-modified PP, and PP alloys, " he says. "We have explored several different methods via compounding with which we can make PP nanocomposites."

Ford's initial target is interior trim—instrument panels, A and B pillars, door panels, and consoles. The chief challenge, Helms says, is to maximize coupling between the silicate and the polymer matrix. He indicates that this could prove simpler with surface-treated synthetic clays. "With natural clays, you may need a surface treatment plus a compatibilizer." He says his group recently has achieved very good results in terms of stiffness/impact balance. The next step is to team up with material suppliers. Meanwhile, Toyota Central R&D Labs recently produced PP nanocomposites in a twin-screw extruder using 5 per cent clay and maleic anhydride-modified PP oligomer as a compatibilizer. According to Toyota, the "polypropylene clay hybrid" (PPCH) provides higher stiffness than the same loading of talc or glass.

Dow Plastics and automotive processor Magna International, Southfield, Mich., are in their second year of a joint effort to develop nanocomposites of PP and TPO. The five-year project, which is partially funded by the U.S. Commerce Dept.'s Advanced Technology Programme, is aimed at developing methods for dispersing the nanoclay in such resins and demonstrating cost-effectiveness of nanocomposites for exterior front and rear car parts. For auto under-hood and electronics uses, Showa Denko just commercialised the first acetal nanocomposite. It reportedly shows low warpage and few sinks. Compared with unfilled acetal copolymer, it has 40 per cent higher flexural modulus and 45° F higher heat resistance.

Flame-retardant properties of nanocomposites are of interest on many fronts, as indicated by ongoing work at NIST and its new consortium. Over the last three years, NIST fire researchers demonstrated that the rate of heat release for nylon 6 nanocomposites with 5 per cent clay is 63 per cent lower than the RHR of pure nylon 6. There was also no increase in carbon monoxide or soot generated during combustion. Though the mechanism of nanocomposites' FR action is not known for sure, a durable carbon-silicate char was observed to form very rapidly in the burn tests. NIST's work last year showed that the reduced flammability of nanocomposites extends to other thermoplastics, including PP and PS. In one study, PP samples were grafted with 0.4 per cent maleic anhydride to make them more compatible with the clay. Fire researcher Jeffrey Gilman thinks that nanoclay in combination with reduced amounts of conventional FR fillers such as ATH and MgOH at lower use levels are likely to meet FR standards such as 94V-0 and V-2. Gilman notes that Showa-Denko has reported V-0 performance of nylon 6 nanocomposites containing melamine.

Flame-retardant properties of nanocomposites are of interest on many fronts, as indicated by ongoing work at NIST and its new consortium. Over the last three years, NIST fire researchers demonstrated that the rate of heat release for nylon 6 nanocomposites with 5 per cent clay is 63 per cent lower than the RHR of pure nylon 6. There was also no increase in carbon monoxide or soot generated during combustion. Though the mechanism of nanocomposites' FR action is not known for sure, a durable carbon-silicate char was observed to form very rapidly in the burn tests. NIST's work last year showed that the reduced flammability of nanocomposites extends to other thermoplastics, including PP and PS. In one study, PP samples were grafted with 0.4 per cent maleic anhydride to make them more compatible with the clay. Fire researcher Jeffrey Gilman thinks that nanoclay in combination with reduced amounts of conventional FR fillers such as ATH and MgOH at lower use levels are likely to meet FR standards such as 94V-0 and V-2. Gilman notes that Showa-Denko has reported V-0 performance of nylon 6 nanocomposites containing melamine.

Nanosized particles have mega-potential in plastics because just a pinch does so much more than heavy loadings of other additives. Three recent conferences presented almost 200 papers on the feverish pace of 'nano' R&D on boosting plastics' mechanical and barrier properties, flame retardancy, and electrical conductivity. They are still in their infancy, but if the forecasts are right, nanocomposites could turn out to be the biggest little thing to hit plastics in decades. Polymers reinforced with as little as 2 per cent to 5 per cent of these particles via melt compounding or *in-situ* polymerisation exhibit dramatic improvements in thermo-mechanical properties, barrier properties, and flame retardancy. They also can outperform standard fillers and fibers in raising heat resistance, dimensional stability, and electrical conductivity. Dispersions of nanoscale reinforcements in polymers are already entering the marketplace in automotive and packaging applications, albeit in a low-profile manner and slower than had been anticipated. But that pace is expected to speed up dramatically, as indicated by the enthusiasm of researchers and marketers shown in roughly 200 papers delivered at three technical conferences. These were Nanocomposites 2004 in San Francisco, the SPE Antec 2004 in Chicago, and Nanocomposites 2004 in Brussels, Belgium. A report from market-research firm pegs the total worldwide market for polymer nanocomposites at 24.5 million lb in 2003, valued at $90.8 million. It also projects the market to grow at an average annual rate of 18.4 per cent to reach $211.1 million by 2008. Even if nano developments hit some snags, BCC says some applications will grow faster than 20 per cent per year. The leading nanoscale fillers in R&D and commercial projects are layered silicate nanoclays and nanotalcs, plus carbon nanotubes and graphite platelets. But other candidates are being actively investigated, such as synthetic clays, polyhedral oligomeric silsesquioxane (POSS), and even natural fibers like flax and hemp.

The two types of nanofillers that have been most widely discussed and the first to break into commercial use are nanoclays and carbon nanotubes. Both must be chemically modified with surface treatments in order to achieve the fine dispersion and resin coupling that are required to derive maximum benefit. Both of these nanofillers have demonstrated improvements in structural, thermal, barrier, and flame-retardant properties of plastics. Carbon nanotubes also enhance electrical conductivity. So far, nanoclays haveshown the broadest commercial viability due to their lower cost—$2.25 to $3.25/lb—and their utility in common thermoplastics like PP, TPO, PET, PE, PS, and nylon. The leading nanoclay is montmorillonite, a layered alumino-silicate whose individual platelets measure around 1 micron diam., giving them an aspect ratio of 1000:1. The two major US producers are Nanocor with its Nanomer line and Southern Clay Products with its Cloisite line. Both companies have formed alliances with suppliers of resins and surfactants, plus compounders and automotive OEMs and packaging firms. While much of their work is proprietary, they have disclosed several commercial successes.

General Motors has taken the lead in putting nanocomposites on the road. GM launched the first commercial auto exterior use of a nanocomposite in the step assist on the 2002 GMC Safari and Chevrolet Astro van. The part also appears on 2003 and 2004 models. More recently, a PP/nanoclay composite appeared on the body side molding of General Motors' highest-volume car, the 2004 Chevrolet Impala. The compound was developed by GM's R&D Center in Warren, Mich., in cooperation with Basell North America and Southern Clay Products. The latest application is on the 2005 GM Hummer H2 SUT. The vehicle's cargo bed uses about seven pounds of molded-in-colour nanocomposite parts for its center bridge, sail panel, and box-rail protector. The material is Basell's Profax CX-284 reactor TPO with nanoclay. While nanoclay adds muscle to plastics, carbon nanotubes impart electrical and thermal conductivity. Nanotubes' commercial potential has been limited by their high price tags—reportedly in the range of $100/gram, although they are available in masterbatches for $50/lb and up. Still, nearly every car produced in the U.S. since the late 1990s contains some carbon nanotubes, typically blended into nylon to protect against static electricity in the fuel system. Static-dissipative compounds containing nanotubes are also protecting computer read/write heads. Carbon nanotubes include both single-and multi-walled structures. The former have a typical outside diameter of 1 to 2 nm while the latter have an OD of 8 to 12 nm. They can range in length from the typical 10 microns to as much as 100 microns and have at least a 1000:1 aspect ratio. Carbon nanotubes have 50 times the tensile strength of stainless steel (100 GPa vs. 2 GPa) and five times the thermal conductivity of copper. When incorporated into a polymer matrix, they have the potential to boost electrical or thermal conductivity by orders of magnitude over the performance possible with traditional fillers such as carbon black or metal powders. US suppliers of nanotubes include Hyperion Catalysis with its Fibril multi-walled nanotubes and newcomer Zyvex Corp. with its Nanosolve single- or multi-walled tubes. Both suppliers now offer their products in masterbatches that typically contain 15 to 20 per cent nanotubes. A different but related category is vapor-grown carbon nanofibers from Pyrograf Products, a spin-off from Applied Sciences. Its Pyrograf III nanofibers reportedly can compete with nanotubes in providing thermal and electrical conductivity and dramatically enhancing mechanical properties and fire resistance (char formation). What's more, nanofibers cost significantly less—around $100 to $150/lb. Evaluations are under way in nylon, PP, and polyurethanes.

Alliances between Nanocor and two specialty compounders, have resulted in commercial-scale nanocomposite concentrates and compounds for structural and barrier applications. Noble Polymers new Forte PP nanocomposite compound made its commercial debut in the seat backs of the Honda Acura TL 2004 car. Forte replaced glass-filled PP, which caused processing difficulties, visual defects, and warping. Forte has a low density of 0.928 g/cc, superior mechanical properties, and improved surface quality and recyclability. Noble reports that the Forte nanocomposite will also be used to produce the center console for a 2006 model light truck. Other applications in the works include office furniture—replacing 20 per cent glass-filled PP—and appliance parts, where Forte reduces weight and possibly material cost.

PolyOne recently introduced the Maxxam LST line of PP homopolymer/nanoclay compounds that boast high stiffness and impact resistance. Through a patent-pending process, PolyOne reports that it has been able to overcome previous problems of incomplete exfoliation and dispersion of the nanoclay, resulting in performance that meets or exceeds many engineering thermoplastics. Lighter weight, aesthetic and processing advantages, and lower cost are also claimed. PolyOne also offers Nanoblend concentrates of up to 40 per cent nanoclay in homopolymer PP, modified PP, LLDPE, LDPE, HDPE, or an ethylene copolymer. Some grades are tailored specifically for barrier enhancement. PolyOne reports that applications nearing commercialisation include pallets and dunnage, where Maxxam LST compounds are specified as alternatives to engineering resins due to their improved dimensional control, which is critical for robotic assembly. In addition, they boast good impact strength and lighter weight. Maxxam LST is also being considered for consumer disposable applications due to a combination of chemical resistance and stiffness, as well as dramatic cycle-time improvements. Meanwhile, the Nanoblend concentrates are being considered for auto interior and exterior TPO parts. Key drivers are dimensional stability, lighter weight, and stiffness without loss of impact. Nanoblend concentrates are being evaluated in films for enhancing barrier, stiffness, HDT, and controlled release or migration of additives such as biocides and dyes. In blow molded packaging, Nanoblend is being considered for improved barrier and the potential for thinwalling and faster cycles. Thin-walling and faster cycles are also attractions in injection molded containers and totes. Some industry sectors are evaluating the concentrates for improving flame retardancy.

Papers presented by General Motors and Southern Clay Products discussed numerous enhancements to automotive TPOs obtained with nanoclays. Those advances did not come easily: Early processing problems caused by clay agglomeration were ultimately resolved by optimising the clay infeed position at the extruder, the screw design, screw speed, temperature, and pressure. Once processing issues were resolved, nanocomposite TPOs outperformed conventional talc-filled TPOs in consistency of properties, retention of low-temperature ductility, elimination of "tiger striping," reduced paint delamination, and improved knit-line appearance, colourability, grain patterns, scratch and mar resistance, and recyclability. What's more, the lower filler level means 3 per cent to 21 per cent lower density (0.92 vs. 0.96 to 1.13 g/cc). Lighter weight requires less adhesive for attachment, which cuts cost. Among the many auto exterior, interior, and under-hood applications for which nanocomposites appear suited are fascias, rocker covers, side trim, grilles, hood louvers, instrument panels, seat/IP foams, door inners, pillar covers, vertical and horizontal body and closure panels, engine shrouds, fan shrouds, air intakes, fuel tanks, and fuel lines. In addition to TPO/nanoclay composites, GM has explored using carbon nanotubes to replace current thermoset structural composites. GM is interested in reducing reinforcement levels in Class A applications by replacing continuous carbon fibers with nanotubes or short nanofibers. nanotubes also have the potential to reduce plastics' coefficient of thermal expansion more effectively.

Polymer barrier technology is also getting a boost from nanoclays. Mitsubishi Gas Chemical (MGC) and Honeywell Specialty Polymers both are using Nanocor's nanoclays in nylons as barrier layers in multi-layer PET bottles and films for food packaging. MGC's MXD6 nylon nanocomposite, called Imperm N, is used commercially in Europe in multi-layer PET bottles for beer and other alcoholic beverages. It is also being evaluated for small carbonated soft-drink bottles. Other Imperm applications that will debut in the next six months are multi-layer thermoformed containers for deli meats and cheeses and flexible multi-layer films for potato chips and ketchup. Honeywell has aimed its Aegis nylon 6 nanocomposites initially at PET beer bottles. In late 2003, a version containing an oxygen scavenger made a commercial splash with the introduction of the 1.6-liter Hite Pitcher beer bottle from Hite Brewery Co. in South Korea. Aegis is the barrier layer in this three-layer structure, which is said to provide a 26-week shelf life. Honeywell is aiming other Aegis nanocomposite grades (without oxygen scavenger) as replacements for EVOH in films and pouches. Such grades reportedly are lower in cost than EVOH, provide a better barrier that allows for lightweighting, and also have better puncture resistance and good clarity. (Because of their size, nanoparticles do not interfere with light transmission.) The U.S. military and NASA, in conjunction with Triton Systems, Inc., Chelmsford, Mass., are looking into nanoclay as a barrier enhancer for EVOH in long-shelflife packaging. An experimental thermoformed food tray was made from EVOH plus 3 per cent of Southern Clay's Cloisite in a layer sandwiched between two PP layers. It reportedly imparts three- to five-year shelf life without refrigeration, plus good clarity, processability, and recyclability. Alcoa CSI, Crawfordsville, Ind., is seeking a patent on coextruded barrier liners for plastic bottle caps for beer, juice, or carbonated soft drinks. The liners include a layer of nylon 6/nanoclay composite plus one or two EVA layers with oxygen scavengers. This liner is said to outperform other barrier materials at very high humidity (95 per cent to 96 per cent RH). LG Chem Ltd. of South Korea has developed high-barrier, monolayer blow molded containers of HDPE with 3 per cent to 5 per cent nanoclay for handling toluene and light hydrocarbon fluids. LG reports that permeation of the hydrocarbon solvents is cut by a factor of 40 to 200 compared with neat HDPE.

Since the early 1990s, automotive fuel-line components such as quick connectors and filters have used inner barrier layers consisting of nylon 12 and carbon nanotubes.

Hyperion Catalysis now aims to introduce nanotubes into other resins used in auto fuel systems, such modified nylons and fluoropolymers. A new fluoropolymer/nanotube compound is being used to make O-rings for automotive fuel connectors. In electronics, polycarbonate and polyetherimide (GE's Ultem) components of computer hard drives have been reinforced with nanotubes to render them conductive and very smooth. Over the last three years, a major automotive OEM in Europe has been using carbon nanotubes in GE's Noryl GTX nylon/PPO alloy to mold exterior fenders. This conductive nanocomposite allows for electrostatic painting.

Michigan State University's Composite Materials and Structures Center in East Lansing developed a new surface-treated graphite nano-platelet. Graphite has a modulus several times that of clay and also has excellent electrical and thermal properties. When incorporated into an epoxy, it results in superior mechanical properties and excellent electrical conductivity compared with standard carbon fibers and nanosized carbon black. MSU sees potential in ESD protection and EMI shielding. Plastic nano-graphite compounds are expected to sell for up to $5/lb, significantly less than compounds based on nanotubes or vapor-grown carbon fibers. Carbon nanotubes have more going for them than just conductivity. Researchers at the National Institute of Standards and Technology (NIST), Gaithersburg, Md., report that carbon nanotubes in PP not only enhance the material's strength and properties, but also dramatically change how the molten polymer flows, virtually eliminating die swell.

Extensive research at NIST has established nanoclays' effectiveness as flame-retardant synergists. Nanoclay levels of 2 per cent and 5 per cent in nylon 6 reduced the rate of heat release by 32 per cent and 63 per cent, respectively, NIST found. Specialty compounder Foster Corp. recently demonstrated that higher levels (13.9 per cent) of nanoclay can be added to nylon 12 elastomers to achieve UL 94V-0 rating at 1/8-in. thickness. Used as a char former, the nanoclay allows the typical 50 per cent loading of halogen/antimony oxide flame-retardant system to be cut in half, which significantly reduces detrimental effects on physical properties. The company first introduced nylon 12/nanoclay compounds for tubing and film in 2001. Germany's Sud-Chemie (U.S. office in Louisville, Ky.) offers modified nanoclays called Nanofil as flame retardants. It recently developed halogen-free EVA/PE wire and cable compounds containing 3 per cent to 5 per cent of new Nanofil SE 3000 plus 52 per cent to 55 per cent alumina trihydrate or magnesium hydroxide (typically used at 65 per cent levels). The result is said to be improved mechanical properties, smoother cable, and higher extrusion speeds. According to Hyperion Catalysis, two recent studies show that multi-walled carbon nanotubes may act as a flame retardant without use of halogen. In both EVA and maleic-anhydride-modified PP, 2.4 per cent to 4.8 per cent loadings of nanotubes show heat-release rates comparable to or better than those obtained with nanoclays.

Among its many virtues, nanoclay can work as a nucleating agent to control foam cell structure and enhance properties of polymeric foams for applications from insulation to packaging. The University of Toronto's Dept. of Mechanical and Industrial Engineering studied extrusion of chemically foamed LDPE/wood-fiber compounds. Addition of 5 per cent nanoclay to the mix decreased the cell size, increased the cell density and facilitated foam expansion. When burned, the foam showed good char formation. Similar results wer_2 gas. Researchers at Ohio State University's Dept. of Chemical Engineering (Columbus) in USA found that small amounts of nanoclay surface-grafted with PMMA can reduce cell size and increase cell density in microcellular PS foamed with CO_2. Another OSU study showed that smaller cell size and higher density can be achieved with 5 per cent nanoclay in polyurethane foams blown with pentane or water. Louisiana State University's Mechanical Engineering Dept. (Baton Rouge) reports that 4 per cent to 5 per cent nanoclay increases the flexural strength and elongation of epoxy syntactic foams used as core materials for sandwich composites in structural applications.

Automotive and Barrier Packaging Properties: Future Application

Nanocomposites are gradually gaining acceptance in the mainstream of global plastics processing. These polymer compounds, containing relatively low loadings (under 6 per cent by weight) of nanometer-sized mineral particles, are beginning to show up in polypropylene and TPO-based automotive exterior claddings (see sidebar), barrier beer bottles, nylon packaging films, polyethylene pipe and wire/cable coatings, and more. Optimism surrounding these novel materials has increased since they burst into industry consciousness two or three years ago. Exploratory effort has intensified as a growing body of data substantiates the potential of established nylon/clay nanocomposites, emerging polyolefin versions, and a range of other resin matrixes and nanofillers. Real-world applications are coming more slowly, in part because of the need to validate cost-effectiveness in the face of high price tags for nano-particle ingredients. Some early application-development programmes have lapsed for cost reasons. Such casualties include an automotive timing-belt cover based on a nylon 6 nanocomposite from Japan's Unitika and an automotive mirror housing of conductive PPO/nylon alloy from GE Plastics. Yet the promise of nanocomposites is undiminished. They can improve polymers' stiffness, HDT, dimensional stability, gas barrier, electrical conductivity, and flame retardancy. Nanoparticles are so small and their aspect ratio (L/D) so high that properties improve with lower loadings and fewer penalties (such as higher density, brittleness, or loss of clarity) than with conventional reinforcers like talc or glass. Nanoclays are believed to increase barrier properties by creating a maze or "tortuous path" that slows the progress of gas molecules through the matrix resin. At the same time, these nanoplatelets are only 1 nm thick, less than the wavelength of light, so they do not impede light's passage.

Nanocomposites now draw on a wider menu of resin matrices, including PP, TPO, EVA, acetal, polycarbonate, biodegradable polylactic acid (PLA), and inherently conductive polyaniline. The nano-particles most widely used so far in these compounds are clays supplied by Nanocor and Southern Clay Products. But a new generation of emerging nanomaterials—including nanostructured silicas, carbon nano-tubes, and ceramic nanofibers—suggest that impressive gains in nanocomposite performance lie just a few years ahead.

"Progress has moved beyond nylon 6/clay composites to include products based on PP and PE, " declares Nanocor president Peter Maul. Nanocor's data show up to 98 per cent stiffness improvement in PP and up to 52° F higher HDT. The company says the nanocomposite has virtually the same impact strength as unfilled PP homopolymer.

General Motors recently announced the first-ever automotive production part in an olefinic nanocomposite. It's an exterior step assist for 2002 vans, made of a nanoclay/TPO compound from Basell. William Windscheif, Basell's global business v.p. for advanced polyolefins, calls this application "a small step, but a giant one for nanocomposites, " adding that it heralds a broader shift to nano-PP in automotive. Initially, the auto industry expressed most interest in nylon 6 nanocomposites for use under the hood, where higher HDT and lightweighting were the goals. Ingolf Buethe, senior v.p. for polymer research at BASF in Germany, says nylon nanocomposites show considerable promise in terms of enhanced stiffness, heat resistance, and gloss. But a serious downside for a 5 per cent nanoclay nylon compound tested by BASF was a loss of toughness more pronounced than with standard fillers. More recently, automotive OEMs and molders have turned their attention to PP and TPO nanocomposites. These polyolefin materials potentially offer engineering-thermoplastic properties at 20 per cent lower density and 50 per cent lower cost per pound. Consultant Kenneth Sinclair, head of STA Research in Snohomish, Wash., says the auto makers find that combination difficult to ignore. He estimates that about 30 per cent of nano-PP usage by 2004 will be in autos, mostly cannibalising existing PP applications. But replacement of metals and engineering thermoplastics will follow. Sinclair says nano-PPs are stiffer and process better than standard PPs, so thin-walling of parts by around 40 per cent is feasible. In turn, thin-walling permits around 25 per cent reduction in cycle times, for 60-80 per cent total savings per part. Nonetheless, an official at one of GM's domestic competitors insists that car companies will balk at paying a premium for new materials. Nanocor, concluding that automotive OEMs want lighter parts at no extra cost, has focused its attention on non-automotive uses, including pallets, electronics, and appliance housings. Volvo Corp. of Sweden has studied 5 per cent-nanoclay composites based on a Basell TPO modified with maleic anhydride as a coupling agent. Volvo observed 32 per cent to 50 per cent higher stiffness than 20 per cent talc-filled PP. Impact strength was lower than unmodified TPO but higher than 20 per cent talc-filled TPO. Volvo found that nano-TPO still has 68 per cent lower stiffness than aluminum sheet. A long-term goal for Dow Plastics is in-reactor compounding of nano-PP by using nanoclays as the catalyst support for in-situ polymerisation of PP homopolymer. Dow's effort is focused on highly loaded (up to 10 per cent clay) nano-PPs for semi-structural automotive uses. Dow sources say preliminary findings show "quite promising" performance of these composites.

Nanocor has developed a 40-50 per cent nanoclay concentrate in PP. One potential use is in heavy-duty electrical enclosures that must meet various fire ratings plus demanding specs for low-temperature toughness and weatherability. Switching to nano-PPs could bring 18 per cent weight savings and permit use of less halogenated FR additive to reach a given UL rating. In other polyolefins, Kabelwerk Eupen of Belgium says melt blending of nanoclay into EVA shows promise for wire and cable compounds. Calorimeter tests reveal a dramatic decline in heat release at relatively low (3-5 per cent) loadings. Nano-EVAs also exhibit superior mechanical properties, chemical resistance, and thermal stability. Meanwhile, not all automotive work in nanocomposites involves polyolefins. A role for nanocomposites in polycarbonate automotive glazing is being explored by Exatec of Wixom, Mich., the joint venture of Bayer and GE Plastics that is dedicated to PC auto-glazing development. Exatec marketing director Fritz Stein says nanotechnology is being considered for the exterior coating needed to achieve weatherability and abrasion-resistance without reducing clarity. A Bayer coating containing nano-particles is one of several promising approaches being pursued, Stein reports.

The major application focus for nylon 6 nanocomposites today is in high-barrier packaging. Much of the attention is on PET bottles, where nanocomposites demonstrate improved oxygen and carbon dioxide barrier. Honeywell offers a 2 per cent nanoclay nylon 6 for bottles and has a 4 per cent nanoclay version in development. An early commercial use is a pasteurizable beer bottle that will be introduced in China later this year. Nanoclays also enhance the oxygen barrier and stiffness of nylon 6 films. That could permit downgauging of packaging of oxygen-sensitive products—pet foods, boil-in bags, vacuum packs, and stand-up pouches. No modification of cast film equipment is needed. "Nanoclays significantly boost the barrier performance of nylon 6 while retaining most of its existing favourable characteristics, " states Lance Altizer, Honeywell's market-development manager. He notes that nanocomposites retain nylon's toughness, clarity, hot-fill heat resistance, and oil/grease resistance. Honeywell claims that nylon 6 with 2 per cent nanoclay has three times the oxygen barrier of straight nylon 6, and 4 per cent nanoclay confers a six-fold improvement. That makes Honeywell's Aegis NC nano-nylon a candidate for medium-barrier bottles and films—those offering around 0.5 to 1 cc/mil/day O_2 transmission rate (OTR). Data also show a doubling of stiffness, higher HDT, and improved clarity for nano-nylon 6 packaging. Honeywell has turned its attention to creating nano-nylon materials that can beat the cost of high-barrier plastics or even glass. Its current contender is an active-passive barrier system called Aegis OX, which synergizes nanoclays as the passive barrier and a proprietary, nylon-specific oxygen scavenger as the active agent. Honeywell claims this one-two punch brings a 100-fold reduction in OTR versus nylon 6, taking oxygen ingress to near-zero levels. It also addresses a drawback of existing O_2 scavengers: In the Honeywell system, the passive barrier protects the scavenger from premature depletion. Efficiency of the system is also improved by uniform dispersion of the nano-platelets and by ensuring that the scavenger is positioned "to easily find the oxygen, " as a Honeywell source puts it. Aegis materials are being tested by major PET bottle makers. Honeywell says nano-nylon 6 tends to stretch and orient in ways compatible with stretch-blow molding processes. Current barrier requirements for beer bottles (in which Aegis OX would be the core of a three-layer structure) set a maximum limit on oxygen ingress over 120 days, as well as a limit of 10 per cent CO_2 escape in that time. The beer industry appears headed toward a 180-day barrier standard for both gases. "We feel Aegis OX will compete against any existing barrier system for beer, " says Honeywell's Altizer. He claims that Aegis OX meets both 120-day requirements and that the 180-day standard is achievable with processing refinements. Nanocor has come up with a different high-barrier option. Its Imperm compound supplements the inherent gas barrier (0.35 cc/mil/day) of amorphous MDX6 nylon from Mitsubishi Gas Chemical with the addition of a nanoclay. Used as the core of a three-layer PET bottle, Imperm is said to have 100-fold lower OTR than that of straight PET. It is being used in a 16-oz, non-pasteurised beer bottle in which the Imperm core (10 per cent of bottle thickness) reportedly ensures a 28.5-week shelf life. Imperm is said to adhere to PET without tie layers. Sufficient clarity is retained to meet requirements for the amber bottle. Bayer is aiming nylon 6 nanocomposites at cast film for multi-layer packaging, protective films for medical and corrosion-prone items, and more. Bayer's pre-commercial Durethan KU2-2601 compound uses Nanocor's clay to reduce OTR by around 50 per cent versus nucleated nylon 6. Stiffness of the nanocomposite is doubled, and its gloss and clarity rival those of a costly high-clarity copolyamide film, Bayer reports. Anti-blocking properties are also improved. Meanwhile, nanocomposites also limit emissions of gasoline, methanol, and organic solvents. Ube America is developing nanocomposite barriers for automotive fuel systems. It uses up to 5 per cent nanoclay in nylon 6 and 6/66 blends. Nylon 6 with 2 per cent nanoclay is said to be five times more resistant to gasoline permeation than unmodified nylon 6. Ube has developed a coextruded barrier fuel line, trade-named Ecobesta, using nylon 6/66 nanocomposite as the core layer.

Nanocomposite technology has been described by some as the next great frontier of material science in packaging. This technology was developed to improve barrier performance pertaining to gases such as oxygen and carbon dioxide. It also enhances the barrier performance to ultraviolet rays, as well as adding strength, stiffness, dimensional stability, and heat resistance. New plastics created with this technology demonstrate an increased shelf life and are less likely to shatter. Once perfected, these plastics will offer these improved characteristics at competitive prices. It will also make them attractive for use in food and beverage packaging and pharmaceutical packaging applications. In 2001, nearly 100, 000 pounds of nanocomposite materials were used in the packaging industry; by 2011, the market in the United States is expected to be around 100 million pounds for rigid and flexible packaging. Polymer nanocomposites consist of resins [either thermoset or thermoplastics] that have fillers added with a least one dimension measured in nanometers. These fillers fall into three categories: nanotubes, nanoscale oxides and metals, and nanoclays. Nanotube-based nanocomposites are used for electrostatic dissipation applications; nanoscale oxides and metals are used for abrasion-resistant films; and nanoclay-based nanocomposites are used for barrier packaging applications. Nanocomposites have received considerable attention over the past decade because of their potential to achieve property enhancements significantly greater than that attainable using conventional fillers or polymer blends. The use of nanoclay-based nanocomposites in food and beverage packaging applications, and the properties that make them superior to conventional plastics, will be focused on. Nanocomposites were first referenced in the 1950s and polyamide nanocomposites were reported around 1976. In the early 1980s Toyota's Central Research and Development Labs began working with polymer-layered silicate-clay mineral composites and this is when the technology began to be studied more widely. The clay mineral that is generating the most interest for use in nanocomposites is montmorillonite, generically referred to as nanoclay, and sometimes referred to as bentonite. It is natural clay that is most commonly formed by the in situ alteration of volcanic ash or by the hydrothermal alteration of volcanic rocks. This clay is widely available and relatively inexpensive, thus becoming the most widely used clay in nanocomposite applications. The first step in the preparation of the nanoclay involves purifying approximately 99 per cent of the montmorillonite. The second step involves surface treatment of the clay. Montmorillonite is hydrophilic and relatively incompatible with most hydrophobic polymers, so it must be chemically modified to make its surface more receptive to dispersion. After the clays are chemically treated, they are dispersed in the polymer. The clays are incorporated into the polymer matrix by one of two approaches: during polymerisation or by melt compounding. This is the difficult part of the technology and may limit the use of nanocomposites. The dispersion process requires a custom solution for each polymer used, so developing polymer nanocomposites becomes a capitol intensive research and development project. Very few compounding firms have this kind of capability, and this leaves resin producers or well-funded startup companies to develop these materials ,

A nanocomposite behaviour is entirely dependent on the size of the clay fibers added. The smaller and thinner the fiber, the more surface area is available to interact with the polymer matrix and an improved plastic. The size of the clay fibers has grown smaller, which means that even tiny amounts of filler (two per cent of total volume) can be added to achieve the same effects. The nanoclays used in plastic composites are extremely small, irregular platelets. They are approximately one nanometer thick and 100 nanometers in diameter. A single human hair is 80, 000 nanometers thick, so one can understand how a little goes a long way. Once these tiny flat platelets are dispersed into the plastic, they create a path that gases must follow to move through the material, greatly slowing their transmission.

Today's nanocomposites typically demonstrate unique improvements in materials properties, including rigidity, strength, and barrier characteristics, while maintaining a level of transparency and offering the potential for recyclability. Polymer-clay promise a veritable explosion of thermoplastic polymer applications because they are lightweight materials that rival metals in stiffness and strength while providing enhanced gas-barrier characteristics, superior dimensional stability, and high heat distortion temperatures—all with low mineral loading and virtually no loss in impact resistance. Making this technology even more impressive is the fact that the properties improve with few penalties (such as higher density, brittleness, or loss of clarity) than with conventional reinforcements such as talc or glass.

Nylon is often touted as one of the most promising resins for nanocomposite additives. In packaging, the major application focus for nylon 6 nanocomposites is in high-barrier packaging. Much of the attention is on PET bottles, where nanocomposites demonstrate an improved oxygen and carbon dioxide barrier.

Honeywell offers three grades of nanoclay nylon 6 resins; Aegis OX, HFX, and CSD. Aegis OX resin is an oxygen-scavenging nylon formulated for the high oxygen barrier demands of plastic beer bottles. Honeywell has turned its attention to creating nano-nylon materials that can beat the cost of higher-barrier plastics or even glass. The nanoclays act as the passive barrier and the nylon specific oxygen scavenger acts as the active agent Honeywell claims this one-two punch brings a 100 fold reduction in OTR versus nylon 6, taking oxygen ingress to near-zero levels. Aegis OX demonstrates excellent resistance to delamination, it can be processed easily, has excellent clarity, is recyclable, and cost competitive. Aegis OX is currently used in a three-layer PET bottle where it is the core layer for a 12-ounce premium beer bottle for Anchor Brewing of San Francisco

Aegis HFX is specifically formulated to address the growing requirement for improved delamination resistance in hot-fill bottle applications. It is a passive-active system that adds more oxygen resistance for foods. Customer trials have demonstrated that Aegis HFX meets shelf life requirements for hot-fill applications including juice, tea, and condiments.

For carbonated soft drinks, Aegis CSD provides delamination resistance during production. It also provides a high level of a passive carbon dioxide barrier compared with monolayer PET. At 3.5 per cent of bottle weight, Aegis CSD extends the shelf life of 0.5 liter bottles from 9 to 14 weeks. New data extends the shelf life to 16 weeks.

Films produced from Honeywell Specialty Polymers' provide excellent barrier properties to oxygen, flavors, and aromas. Each resin provides the extra performance needed for suitable films for a wide range of flexible packaging applications. They also provide toughness, strength, tear, and puncture resistance, and resistance to grease and gas penetration. Applications for this film include processed meat, fresh red meat, poultry, fish, cheese, dried food, and chilled fruit juices.

Nanocor has a different high barrier option. Through an alliance with the Mitsubishi Gas Chemical Company, Inc., Nanocor has rights to melt-compound its nanoclay additives with Mitsubishi's MXD6 nylon for use in barrier PET bottles and films under the trade name of "M9". Mitsubishi Gas Chemical Company, Inc. has supplied MXD6 plastics to the packaging market for 15 years. MXD6 features an excellent barrier for foods and beverages against harmful gases. It is preferred over competitive barrier plastics because it is easy to process and yields packages with high transparency. M9 boosts the CO_2 and O_2 barrier of standard MXD6 respectively by 50 per cent and 75 per cent. The material also retains high clarity and delamination resistance equal to standard MXD6. It is currently being used as the core of a three-layer 16-ounce non-pasteurised PET beer bottle and is said to have 100-fold lower OTR than that of straight PET. It also adheres to PET without tie layers and retains sufficient clarity to meet requirements for the amber bottle. M9's superior gas barrier provides flexibility in bottle design with a range of layer thickness to meet specific shelf life needs. Layer adhesion is virtually the same as standard N-MXD6, so it can be separated from PET by normal recycling methods. Because M9 is miscible with PET at layer thickness of five percent or lower, recyclers have the option of reprocessing thin-layer bottles in total, eliminating the separation step. Rancor-Mitsubishi is able to extend the shelf life of a 28 gram weight PET bottle from 14 weeks to 21 weeks. They use the standard N-MXD6 with a 5 per cent barrier layer of M9 due to its high carbon dioxide barrier. The cut-off is at 90 per cent CO_2 retention.

Triton Systems, Inc. also has a class of commodity polymers that have been enhanced with organically modified nanoclays. These materials are sold under the trade name, ORMLAS, and have demonstrated improved barrier performance, impact resistance, as well as flame resistance. ORMLAS have been incorporated into a long-life food tray that offers high barrier performance and impact resistance. They also include high clarity, ease of manufacture, and recyclability.

In addition to its alliance with Mitsubishi, Nanocor offers its nanoclay resins to other converters for use in various consumer applications. The dominant property improvement seen is a higher quality shelf life. This improvement can lead to lower weight packages because less material is needed to obtain the same, and if not better, barrier properties. This, in turn, can lead to reduced package cost. Improved shelf life and lower package cost are the reasons why nanotechnology is being pursued in consumer packaging.

Bayer's research team is aiming nylon 6 nanocomposites at cast film. It is referred to as a hybrid system that uses Nanocor's clay to reduce OTR by around 50 per cent versus nucleated nylon. Stiffness of the nanocomposite is doubled, and its gloss and clarity rival those of a costly high-clarity polyamide film. Anti-blocking properties are also improved. The new film material with nanoparticles unites the advantages of EVOH and polyamide 6. The embedded particles prevent gases from penetrating the film and moisture from escaping. Durethan KU 2-2601 is suitable for applications where conventional polyamides are too permeable and EVOH is too expensive. Bayer's Durethan KU 2-2601 makes an ideal candidate as a plastic coating for paperboard juice containers. It can protect highly oxygen-sensitive package contents, such as orange juice, at a much lower cost.

Real world applications for nanocomposites are coming slowly. There appears to be a reluctance to embrace this new technology due to cost and variability in the quality of some of the products. John Jones, a market development specialist at Honeywell, believes that manufacturers of these materials have to prove to the market that these new materials can meet their performance expectations. Another challenge nanocomposite producers face is the production of the nanocomposite itself. Both methods, pre-polymerisation and post polymerisation, for preparing nanocomposites have drawbacks. Pre-polymerisation production can disrupt the polymerisation process, which is often critical and requires much developmental time and expense to achieve good yields and controllability, and post polymerisation often requires a lot of time to achieve a good dispersion of the nanoparticles in the composite. This then becomes an expensive and low cost-competitive initiative. Another concern deals with equipment conversion that accepts new material through recalibration; this is a big investment for converters to make. It's a complicated process to go from plastic pellets to a blown bottle. It requires heating and blowing that form to the shape of the bottle. This is expensive equipment, very high-speed equipment, designed for the material that you're going to run. You can't just take another material with different flow characteristics, crystallisation rate, and those kinds of things, throw that in there and make it run.

Today, nanocomposite research is widespread and is conducted by companies and universities across the globe. Plastic suppliers who have already commercialised nanocomposite materials include Basell USA, Bayer, Dow Chemical, Eastman Chemical, Mitsubishi Gas Chemical, Nanocor, Triton Systems, Honeywell, and RTP Co. Most of these efforts are currently focused on either polyolefins or nylons, but, in theory, the clay nanoparticles could be used in any resin family. Optimism surrounding these novel materials has increased since they burst into industry consciousness two or three years ago, and exploratory effort has intensified as a growing body of data substantiates the potential established by nylon/clay nanocomposites, emerging polyolefin versions, and a range of other resin matrixes and nanofillers....the promise of nanocomposites is undiminished. It appears that the momentum is building. The success of Honeywell, Mitsubishi Gas and Chemical, Bayer, Triton Systems and Nanocor will lead to other successes. As production reaches a sufficient scale, incorporating the clay into polymers will become more cost-effective as well. Equipment recalibration used for conventional plastic resins will also show success. The popularity of nanocomposites comes from the fact that a little goes a long way. It provides a marked increase in oxygen, carbon dioxide, moisture and odor barrier properties, increased stiffness, strength and heat resistance, and maintains film clarity and impact strength. For the packaging industry, these new materials and their commercial applications are coming into focus.

Suppliers of various nano-chemicals, nanofibers, and nanotubes argue that the potential of their technologies exceeds that of current nanoclay materials. But the prices are prohibitively high, and practical impacts probably lie five years ahead. Static-dissipative applications are emerging as one potentially large market. Applied Sciences has developed a vapor-grown carbon nano-tube made by pyrolysis of coal. Pyrograf-III comes in 100- and 200-nm diam. and has potential as an electrically conductive additive and modifier of plastics' coefficient of thermal expansion. Max Lake, president of Applied Sciences, says the nanotubes enhance electrical conductivity over a broad resistivity range and boost mechanical properties. Just 0.5 per cent loading provides volume resistivity in the 104 ohm-cm range. Another class of nanotube is a graphitic carbon type that is designed primarily to enhance electrical conductivity. These nano-tubes or "fibrils" are made from a hydrocarbon gas by Hyperion Catalysis International. They offer surface resistivity of 103 to 105 ohm/sq at 4 per cent to 7 per cent loadings. "Fibrils are more efficient at building electrical conductivity into plastics than carbon black or PAN carbon fibers, " states John Hagerstrom, Hyperion's technical service manager. He says the fibrils' small diameter (10 nm average) and high aspect ratio (1000:1) mean a given level of conductivity is achieved at lower loadings than with conventional carbon particles or fibers. That reportedly means less sacrifice of matrix properties, lower warpage, and better surface smoothness. Fibrils have been used to enhance electrostatic painting of automotive mirrors, and as a static-dissipative additive in semiconductor components and disk drives, where the non-sloughing feature of nanotubes helps retain purity.

Hybrid Plastics offers "POSS" nanochemicals, named for their polyhedral oligomeric silsesquioxane molecular building blocks.

These molecular silicas are hybrid organic-inorganic materials said to bridge conventional differences between minerals and monomers. POSS molecules are cage-like structures typically measuring 1.5 nm along each axis. The single molecule particles are truly dispersable and have no affinity for one another. Even loadings of 50 per cent or more by weight reportedly disperse without agglomeration. POSS molecules dissolve in a plastic melt, then recrystallize on cooling into a network that enhances mechanical and thermal properties, as well as flame-retardancy. These chemical tools allow the creation of superior reinforced, crosslinked, and chemically coupled alloys. A 10 per cent POSS loading elevates PP's flexural modulus by 12 per cent, HDT by 21 per cent, and impact strength by 36 per cent.

A 50 per cent POSS loading in crystal PS reportedly has no effect on optical clarity. Hybrid cites optical disks, micro-electronics, and medical products as target niches. At present, POSS costs around \$200/lb. A larger plant for making POSS nanochemicals, due on-line in early 2002, could push down the cost of some POSS products to about \$15/lb. Hybrid Plastics' sales and marketing arm is Divex, Inc. Argonide has a technology for producing alumina ceramic nano-whiskers by electro-explosion of metal wire. These NanoCeram whiskers offer potential for reinforcement because of their small size (2 nm diam.) and high aspect ratio (50:1 average). Applications for NanoCeram are being pursued as reinforcements and thermally conductive additives. Current price of pilot-plant materials is \$280/lb, but that is likely to fall as usage develops.

· For the first time, a PP-based nanocomposite has been adopted for use in an automotive production part. A step-assist for 2002 General Motors Safari and Astro vans is molded by Blackhawk Automotive Plastics, Troy, Mich., from a TPO matrix reinforced with 2.5 per cent nanoclay supplied by Southern Clay Products. The part is a low-volume, dealer-installed option. The nano-TPO replaces a 15 per cent talc-filled PP but is molded in the same tool. This is a beginning. Parties to this effort are investigating other exterior and interior automotive parts. Nano-TPO delivers stiffness equivalent to the talc-filled PP it replaced, translating into 7-8 per cent weight savings. Further, nano-TPO improves low-temperature impact strength and part surface quality. GM officials termed the project "cost-neutral, " explaining that the higher price of Basell's nano-TPO is offset by lightweighting and other cost benefits. The nano-TPO processes at a somewhat lower temperature than talc-filled PP, decreasing cycle time, while also decreasing knit-line creation and other surface imperfections. Reports also show molding other prototype exterior claddings, including rocker and sail panels and body side moldings. Newer versions of Basell materials are being investigated, including some containing up to 5 per cent nanoclay and offering stiffness comparable to 30-40 per cent talc-filled PP. It is also noted that the much lower filler levels in nano-TPOs result in smoother surfaces while enhancing scratch resistance. When GM's R&D unit first announced the nano-TPO project in 1999, emphasis then was on vertical body panels and interior parts. Instead, the first commercial application is a relatively simple, low-volume part. But GM declares that there is no retreat from the applications initially targeted. Work is well under way on a rocker-panel prototype that is 20 per cent lighter than one of talc-filled PP. Interior trim is also being actively explored. A significant advance in chemistry by GM is critical to success of the van step. It allows more complete exfoliation (layer separation) and better dispersion of nanoclay particles into TPO. GM has granted Basell exclusive rights to use its advances in nano-PP technology. In turn, GM has been guaranteed exclusive rights to use Basell's nano-TPOs in the automotive field.

Future Application

Such mechanical property improvements have resulted in major interest in nanocomposite materials in numerous automotive and general/industrial applications. These include potential for utilisation as mirror housings on various vehicle types, door handles, engine covers and intake manifolds and timing belt covers. More general applications currently being considered include usage as impellers and blades for vacuum cleaners, power tool housings, mower hoods and covers for portable electronic equipment such as mobile phones, pagers etc.

The gaseous barrier property improvement that can result from incorporation of relatively small quantities of nanoclay materials is shown to be substantial. Data provided from various sources indicates oxygen transmission rates for polyamide-organoclay composites which are usually less than half that of the unmodified polymer. Further data reveals the extent to which both the amount of clay incorporated in the polymer, and the aspect ratio of the filler contributes to overall barrier performance. In particular, aspect ratio is shown to have a major effect, with high ratios (and hence tendencies towards filler incorporation at the nano-level) quite dramatically enhancing gaseous barrier properties. Such excellent barrier characteristics have resulted in considerable interest in nanoclay composites in food packaging applications, both flexible and rigid. Specific examples include packaging for processed meats, cheese, confectionery, cereals and boil-in-the-bag foods, also extrusion-coating applications in association with paperboard for fruit juice and dairy products, together with co-extrusion processes for the manufacture of beer and carbonated drinks bottles. The use of nanocomposite formulations would be expected to enhance considerably the shelf life of many types of food.

Honeywell have also been active in developing a combined active/passive oxygen barrier system for polyamide-6 materials. Passive barrier characteristics are provided by nanoclay particles incorporated via melt processing techniques whilst the active contribution comes from an oxygen scavenging ingredient (undisclosed). Oxygen transmission results reveal substantial benefits provided by nanoclay incorporation in comparison to the base polymer (rates approximately 15-20 per cent of the bulk polymer value, with further benefits provided by the combined active/passive system). Akkapeddi suggests that the increased tortuosity provided by the nanoclay particles essentially slows transmission of oxygen through the composite and drives molecules to the active scavenging species resulting in near zero oxygen transmission for a considerable period of time.

Triton Systems and the US Army are conducting further work on barrier performance in a joint investigation. The requirement here is for a non-refrigerated packaging system capable of maintaining food freshness for three years. Nanoclay polymer composites are currently showing considerable promise for this application. It is likely that excellent gaseous barrier properties exhibited by nanocomposite polymer systems will result in their substantial use as packaging materials in future years. A somewhat more esoteric possibility arising from enhanced barrier performance recently suggested has been blown–films for artificial intestines!

The ability of nanoclay incorporation to reduce solvent transmission through polymers such as polyamides has been demonstrated. Data provided by De Bievre and Nakamura of UBE Industries reveals significant reductions in fuel transmission through polyamide–6/66 polymers by incorporation of a nanoclay filler. As a result, considerable interest is now being shown in these materials as both fuel tank and fuel line components for cars. Of further interest for this type of application, the reduced fuel transmission characteristics are accompanied by significant material cost reductions.

The presence of filler incorporation at nano-levels has also been shown to have significant effects on the transparency and haze characteristics of films. In comparison to conventionally filled polymers, nanoclay incorporation has been shown to significantly enhance transparency and reduce haze. With polyamide based composites, this effect has been shown to be due to modifications in the crystallisation behaviour brought about by the nanoclay particles; spherilitic domain dimensions being considerably smaller. Similarly, nanomodified polymers have been shown, when employed to coat polymeric transparency materials, to enhance both toughness and hardness of these materials without interfering with light transmission characteristics. An ability to resist high velocity impact combined with substantially improved abrasion resistance was demonstrated by Haghighat of Triton Systems.

Water laden atmospheres have long been regarded as one of the most damaging environments which polymeric materials can encounter. Thus an ability to minimise the extent to which water is absorbed can be a major advantage. Data provided by

Beall from Missouri Baptist College indicates the significant extent to which nanoclay incorporation can reduce the extent of water absorption in a polymer. Similar effects have been observed by van Es of DSM with polyamide based nanocomposites. In addition, van Es noted a significant effect of nanoclay aspect ratio on water diffusion characteristics in a polyimide nanocomposite. Specifically, increasing aspect ratio was found to diminish substantially the amount of water absorbed, thus indicating the beneficial effects likely from nanoparticle incorporation in comparison to conventional microparticle loading. Hydrophobic enhancement would clearly promote both improved nanocomposite properties and diminish the extent to which water would be transmitted through to an underlying substrate. Thus, applications in which contact with water or moist environments is likely could clearly benefit from materials incorporating nanoclay particles.

The ability of nanoclay incorporation to reduce the flammability of polymeric materials was a major theme of the paper presented by Gilman of the National Institute of Standards and Technology in the US. In his work Gilman demonstrated the extent to which flammability behaviour could be restricted in polymers such as polypropylene with as little as 2 per cent nanoclay loading. In particular heat release rates, as obtained from cone calorimetry experiments, were found to diminish substantially by nanoclay incorporation. Although conventional microparticle filler incorporation, together with the use of flame retardant and intumescent agents would also minimise flammability behaviour, this is usually accompanied by reductions in various other important properties. With the nanoclay approach, this is usually achieved whilst maintaining or enhancing other properties and characteristics.

Nanostructured Materials and Nanoscale Interfacial Material

Over the past decade nanostructured materials have been the subject of enormous interest, with the potential for wide-ranging industrial, biomedical, and electronic applications. These can include metals, ceramics, polymeric materials, or nanocomposites. The nanomaterials field has seen a huge increase in funding from private enterprises and government, and academic researchers within the field have formed many partnerships.

Nanomaterials have actually been produced and used by humans for hundreds of years - the ruby red colour of some stained glass is due to gold nanoparticles trapped in the glass matrix. The decorative glaze known as lustre, found on some medieval pottery, contains metallic spherical nanoparticles dispersed in a complex way in the glaze, which gives rise to its special optical properties.

The variety of nanomaterials is great, and their range of properties and possible applications enormous, from miniature batteries, to biomedical implants, as packaging films, super absorbants, components of army, and automobile parts.

Nanocrystalline materials are polycrystalline with grain sizes in the 1-100nm range that are found to exhibit improved hardness, increased ductility and toughness, superior magnetic and optoelectronic properties. Nanocrystals are aggregates of atoms from a few hundred to tens of thousands that combine into a crystalline form of matter known as a 'cluster'. Typically around 10nm in diameter, nanocrystals are larger than molecules but smaller than bulk solids and therefore frequently exhibit physical and chemical properties somewhere in-between.

Given that a nanocrystal is virtually all surface and no interior, its properties can vary considerably as the crystal grows in size. Nanoscale crystals are often harder, stronger and more wear resistant than the same materials in bulk form. Nanocrystals might be used to make super-strong and long-lasting metal parts. The crystals also might be added to plastics and other metals to make new types of composite structures for everything from cars to electronics.

These are particles that are less than 100nm in size. At this level the percentage of exposed area in proportion to its total volume changes dramatically when compared to bulk materials. In particles of around three to 5nm in diameter, about a third of the atoms are actually surface atoms; this means they behave a lot more like gas particles rather than solids. This gives them new properties, they float where bigger particles would sink, and dissolve more easily than the same material in bulk form. The extra exposed surface area also changes depending on how the elements and particles interact with each other.

Recently clay nanoparticles have made their way into composites in cars and packaging materials, where they offer transparency, reduction in permeability and increased strength, the latter being already attractive to the automotive industry, but potentially much more so to the aerospace industry. Sunscreens utilize nanoparticulate zinc oxide. Nanoparticles are also being used in antiseptics, as abrasives, in paints, in new coatings for spectacles (making them scratchproof and unbreakable), for tiles, and in electrochromic or self-cleaning coatings for windows.

Nanoparticles are the basis for anti-graffiti coatings for walls and improved ski waxes and ceramic coatings for solar cells. Glues containing nanoparticles have optical properties that give rise to uses in optoelectronics. Casings, containing nanoparticles used in for electronic devices, such as computers, offer improved shielding against electromagnetic interference.

Nanoscale Interfacial Material

In the development of nanometer-scale materials, ever since the early 1980s, scientists began discovering ultra-small crystal grains. The important point is that small is not only beautiful but also eminently useful. Therefore, the virtues of working in the nanodomain are increasingly recognised by the scientific community, the technological world, and the popular press. These small groups of atoms go by several different names: nanoparticles, nanocrystals, atomic clusters, and quantum dots. Much recent research has gone into finding ways to makesmall and uniform-size clusters that possess unusual optical, electrical, magnetic, and mechanical properties.

This effort has already paid several commercial dividends, for example, ceramics and chemical catalysts with increased efficiency due to high surface-to-volume ratios. Moreover, advances with semi- conducting nanoclusters are bringing several commercial applications into view as well. However, in order to develop new kinds of high-level functional materials, one should not only consider the proper- ties of a single nanoparticle (e.g., high surface ratios, quantum confinement effect), but also the cooperative effect among nanostructures and the macroscopic properties of their organised states. In this report, we attempt to introduce a new concept for the design and creation of new types of interfacial materials, i.e., binary cooperative complementary *nanoscale* materials.

In fact, from ancient times in China, the binary cooperative complementary concept has been used as "yin" and "yang" to explain the phenomenon and catalog the materials of nature and the universe. This initial philosophical principle has had a great influence on classical and modern science. Following the development of science, people have realised that binary cooperative complementary character is a universal feature of the material world. For example, the constitution of the chemical elements and the universe, as well as the physical and chemical properties of materials, are all involved in this binary concept.

In the development of materials science, people have also used the binary cooperative complementary effect to create unique functional materials. So far, binary atomic materials (e.g., binary alloys) and binary molecular materials have been developed that demonstrate tremendous potential in both basic research and practical applications. Well-known examples of binary atomic materials exist, such as the combination of Fe and Cr, which creates a type of stainless steel (also, cobalt is added to iron to make cobalt steels, among the hardest alloys known). Furthermore, many other binary alloys show a number of unique properties. Binary molecular, amphiphilic molecules, with both hydrophilic and hydrophobic groups on the same molecule, have been used in different fields, such as surfactants, in which a huge variety of industrial products have been developed. When electron donors and acceptors are incorporated into a single molecule, nonlinear optical materials can result. Also, binary molecular solids provide another good example. Through the synthesis of donor-acceptor complexes, organic conductors and even superconductors may be generated. The quintessential functional binary molecules (multi-binary molecules) are DNA and RNA, which precisely control the heredity of all living things, animal and plant. Still many unknowns remain, waiting to be elucidated in further research.

Above all, binary cooperative complementary atomic and molecular systems have shown us unique material words. Here we extend this concept from the molecular level to the nanostructural level and introduce a new material system, i.e., binary nanoscale materials. Before going into this topic, we would first like to start from the binary interfacial material, since understanding interfacial materials is key to the design of high-level functional materials. The motivation to develop binary cooperative complementary nanoscale interfacial materials is the desire to create unique functional macroscopic properties. By applying the binary coordinating complementary concept on the nanoscale on a solid surface, we expect to generate nanostructures with mutually compensating properties, e.g., hydrophilic and hydrophobic; conducting and insulating; convex and concave; p-type and n-type; oxidising and reducing; ferromagnetic and antiferromagnetic, and so on. Whenever an interface (or a surface) is formed by the mixture of nanostructures with a pair of these mutually complementary properties, under certain cooperative conditions, unusual interfacial properties can be created, i.e., binary cooperative complementary nanoscale interfacial materials. Our recent results indicate that such a concept is useful for designing new types of interfacial materials. In the following sections we will introduce several examples for the unique character of binary cooperative complementary nanoscale interfacial materials.

These kinds of surfaces or interfaces can be generated by outside stimuli, e.g., light, electric fields, magnetic fields, heat, plasma etching, or chemical synthesis on the surface of the solid. The surface nanostructures were characterised by atomic force microscopy (AFM) in air (SPI300 Seiko Instruments) or in high vacuum (SPI300HV Seiko Instruments).

Recent research indicates that the binary cooperative complementary concept is truly useful for the design and creation of *nanoscale* functional interfacial materials. Here, we will simply mention these as follows.

Binary complementary hydrophilic and hydrophobic nanoscale structures on the surface of TiO thin films and single crystals appear to create super-amphiphilic properties, with contact angles of 0° for both water and oil. AFM images on the face of rutile TiO_2 single crystals and suggests the binary TiO_2 surface can be induced by UV light illumination, heating at 300 °C, or argon ion sputtering, respectively. The key transformation involves loss of oxygen atoms from Ti_{4+} sites to Ti_{3+} sites, which can adsorb hydroxyl groups and thus produce hydrophilic regions. In all of the images, hydrophilic domains appear as rectangular shapes, aligned along the direction of the face which was confirmed by friction force microscopy (FFM). This unique amphiphilic TiO surface character is ascribed to a composite of hydrophilic and hydrophobic (oleophilic) phases on the scale of several tens of nanometers, with the hydrophilic phases formed, for dissociative water adsorption. Microscopically, these areas may be hydrophilic and oleophilic; however macroscopically, the surface is highly amphiphilic. Furthermore, no visible water drops can form on the surface, and the material appears to have an antifogging property. The polycrystalline TiO_2 thin film, which has same character, can be coated on various substrates, e.g., glass, ceramics, plastics and metals. This highly amphiphilic surface has self-cleaning, antifogging properties. Furthermore, this basic idea can be also used to design and synthesize other types of super-amphiphilic materials, e.g., super-amphiphilic polymers.

Pt nanograins on the surface of TiO_2 thin films, i.e., a binary (with both oxidising and reducing sites) coordinating complementary catalyst surface, can be created through synthesis processes. The effect is useful for the prevention of icing on windows and electrical cables in winter. The important advantage of this technique is that the anti-icing process is carried out through a photo-electrochemical reaction, but not a physical heating process. The necessary reaction for removing ice blocks on the substrate is just the decomposition of single molecular layer of H_2O contacting with the substrate. Furthermore, this kind of structure can be also applied to other catalysis systems for various industrial applications.

Through a phase-separation process, polyporous with hydrophobic convex and hydrophilic concave surfaces are generated in copolymer film which appears as an “amphiphobic” surface (i.e., the surface appears to have both hydrophobic and oileophobic properties). It was known that the roughness can enhance the wettability of a surface. However, a pure hydrophilic or hydrophobic surface of polyporcus may only enhance the hydrophilic or hydrophobic property of a solid surface, but “amphiphobic” behaviour can not be created. Interestingly, as-grown copolymer films show an “amphiphobic” property.

6

Towards Understanding the Global Contemporary Issues Regarding Nanoparticles, Nanostructured Materials' Chemistry and Polymer Nanotechnology

Nanoparticles & Nanostructured Materials

During the course of this study, the WTEC panelists were privileged to hear of and to observe a wide variety of current work on nanostructured dispersions and coatings in various laboratories around the world. Descriptions follow of a number of research projects and the preparation issues being addressed in the United States and in some of the foreign laboratories that panelists visited. The nanoparticle preparation techniques that are currently in use, many of which will be discussed in the paragraphs that follow.

Synthesis and Application of Polymer Nanotechnology

Polymer nanoparticles have been produced for decades for use in a variety of high performance materials such as high impact resistant polymers and specialty coatings long before it was fashionable to use the "nano" label. The extraordinarily large surface area on the nanoparticles presents diverse opportunities to place functional groups on the surface. Particles can be created that can expand/contract with changes in pH, or interact with anti-bodies in special ways to provide rapid ex-vivo medical diagnostic tests. Important extensions have been made in combining inorganic materials with polymers and in combining different classes of polymers together in nanoparticle form.

More recent advances in chemistry, processing techniques, and analytical instrumentation allow a whole host of new types of polymer particles to be made. We now have particles that are hollow, multi-lobed, magnetic, functionalised with reactive groups on the surface, conductive, etc. Our ability to devise new process control strategies have led to the ability to control the shape, chemical composition, internal structure, and morphology of the nanoparticles so as to develop new levels of product performance. Advanced analytical techniques allow us to measure structure at ever-decreasing length scales. Computer simulations of the events occurring during particle formation have also benefited us in developing control strategies to produce structured particles

The first part of the course focuses on the wet synthesis methods and chemistries used to produce polymer particles in the 10-1000 nanometer range. Careful attention is paid to the analytical techniques that allow confident measurement of phase-separated structure within these composite particles and the varied applications of such polymer particles. The objectives are to:

- Introduce the diversity of polymer particle sizes, shapes, and chemical composition with structural control at the nano scale
- Describe how particles can be made from either reactive or non-reactive processing techniques, including emulsion polymerisation and self assembly

The second part focuses on characterisation (analysis and verification) and the wide range of application areas of economic importance. The objectives are to:

- Demonstrate how particles can be characterised by modern instrumental techniques
- Discuss the multitude of applications for polymer nanoparticles
- Provide an interactive learning experience through the use of individual and group exercises and case studies

Polymeric nanoparticles are predominantly prepared presently, by wet synthetic routes. Several industrial processes will be described. Emphasis will be placed on the type of polymers and morphology structures that can be synthesised using each process. Controlled radical polymerisation will be explored for their ability to provide structural control of polymer chains.

A brief overview of suspension and dispersion-precipitation polymerisation relevant to nanoparticles is included. Solution polymerisation methods are provided to allow a more detailed presentation of a non-reactive pathway to nanoparticle formation. The third day is dedicated to polymeric nanoparticles characterisation. Of critical importance is the measurement of particle size, bulk and surface chemical composition. Importantly, the internal structure of the particles, can be measured by combinations of microscopy and thermal analysis. All techniques will be presented for their utility toward polymeric nanoparticles characterisation.

The fourth day of the course is focused on the applications of polymeric nanoparticles. The most traditional field of application is waterborne paints, adhesives, and coatings. The details of some formulation and application issues will be presented. Of more recent emergence is the field of redispersible lattices, powder coatings, and pressure sensitive adhesives. A recent boom in the range of application of polymeric nanoparticles is the sector of biotechnology, and more specifically biomedical products. These include the critical delivery of sensitive drugs and medical diagnostics. Last but not least, polymeric nanoparticles have found their ways in EMO devices.

All through the course practical problems will be solved both individually and in groups.

Synthesis

- Polycondensation polymerisation
- Radical polymerisation
- Living polymerisation
- SFRP, ATRP, RAFT for controlled radical polymerisation
- Problem solving
- Suspension and dispersion-precipitation polymerisation
- Emulsion polymerisation
- Semi-batch and batch processes
- Single stage and multistage products.
- Problem solving
- Micro-emulsion and mini-emulsion
- Nanoencapsulation in direct and reverse phases.
- Self-assembly
- Directed assembly
- Patterned substrate effects
- Solvent removal processes
- Problem solving

Characterisation

- Size and size distribution.
- CHDF, Capillary Hydro Dynamic Fractionation
- MALS, Multi Angle Light Scattering
- DLS, Dynamic Light Scattering

- Field flow fractionation
- UAC, Analytical Ultra Centrifugation
- SEM, TEM, AFM, microscopy techniques.
- Problem solving
- Bulk chemical composition
- NMR, Nuclear Magnetic Resonance
- Spectroscopic techniques, FTIR, Raman, UV, and fluorescence,
- Problem solving
- Pyrolysis-gas chromatography-mass spectroscopy,
- Mass-spectroscopy
- Surface composition
- Surfactant titration
- XPS, X-ray Photon Spectroscopy
- EDS, Electron Dispersion Spectroscopy
- SERS, Surfaced Enhanced Raman Spectroscopy
- ITC, Isothermal Titration Calorimetry
- Internal structure, TEM, STXM, thermal analysis
- Problem solving

Applications

- Waterborne paints
- Adhesives
- Coatings
- Redispersible latices
- Powder coatings
- Pressure sensitive adhesives
- Hot melts
- Biotechnology
- Biomedical products
- Drug delivery
- Medical diagnostics
- Problem solving
- Electronics

- Magnetics
- Optoelectronics

High Surface Area Materials

The trend to smaller and smaller structures, that is, miniaturisation, is well known in the manufacturing and microelectronics industries, as evidenced by the rapid increase in computing power through reduction on chips of the area and volume needed per transistor (Roher 1993). In the materials area this same trend towards miniaturisation also is occurring, but for different reasons. Smallness in itself is not the goal. Instead, it is the realisation, or now possibly even the expectation, that new properties intrinsic to novel structures will enable breakthroughs in a multitude of technologically important areas (Siegel 1991; Gleiter 1989). Of particular interest to materials scientists is the fact that nanostructures have higher surface areas than do conventional materials. The impact of nanostructure on the properties of high surface area materials is an area of increasing importance to understanding, creating, and improving materials for diverse applications. High surface areas can be attained either by fabricating small particles or clusters where the surface-to-volume ratio of each particle is high, or by creating materials where the void surface area (pores) is high compared to the amount of bulk support material. Materials such as highly dispersed supported metal catalysts and gas phase clusters fall into the former category, and microporous (nanometer-pored) materials such as zeolites, high surface area inorganic oxides, porous carbons, and amorphous silicas fall into the latter category. There are many areas of current academic and industrial activity where the use of the nanostructure approach to high surface area materials may have significant impact:

- microporous materials for energy storage and separations technologies, including nanostructured materials for highly selective adsorption/separation processes such as H2O, H2S, or CO2 removal; high capacity, low volume gas storage of H2 and CH4 for fuel cell applications and high selectivity; high permeance gas separations such as O2 enrichment; and H2 separation and recovery
- thermal barrier materials for use in high temperature engines
- understanding certain atmospheric reactions
- incorporation into construction industry materials for improved strength or for fault diagnostics
- battery or capacitor elements for new or improved operation
- biochemical and pharmaceutical separations
- product-specific catalysts for almost every petrochemical process In catalysis

the key goal is to promote reactions that have high selectivity with high yield.

It is anticipated that this goal will be more closely approached through tailoring a catalyst particle via nanoparticle synthesis and assembly so that it performs only specific chemical conversions, performs these at high yield, and does so with greater energy efficiency. In the electronics area one may anticipate manufacture of single electron devices on a grand scale. Manufacture of materials with greatly improved properties in one or more areas such as strength, toughness, or ductility may become commonplace. In separations science new materials with well defined pore sizes and high surface areas are already being fabricated and tested in the laboratory for potential use in energy storage and separations technologies.

In addition, many laboratories around the world are actively pursuing the potential to create novel thermal barrier materials, highly selective sensors, and novel construction materials whose bonding and strength depend upon the surface area and morphology of the nanoscale constituents. Many are also engaged in developing molecular replication technologies for rapid scaleup and manufacturing. The nanoscale revolution in high surface area materials comes about for several reasons. First, since the late 1970s the scientific community has experienced enormous progress in the synthesis, characterisation, and basic theoretical and experimental understanding of materials with nanoscale dimensions, i.e., small particles and clusters and their very high surface-tovolume ratios.

Second, the properties of such materials have opened a third dimension to the periodic table, that is, the number of atoms, N (for a recent example see Rosen 1998). N now becomes a critical parameter by which the properties for "small" systems are defined. As a simple example, for metals we have known for decades that the atomic ionisation potential (IP) is typically about twice the value of the bulk work function (Lide 1993).

It is only relatively recently that experiments have shown that the IP (and electron affinity) for clusters containing a specific number N of (metal) atoms varies dramatically and non-monotonically with N for clusters containing less than 100-200 atoms. Other properties such as chemical reactivity, magnetic moment, polarizability, and geometric structure, where they have been investigated, are also found to exhibit a strong dependence on N. Expectations for new materials with properties different from the atom or the bulk material have been realised (e.g., see Jena 1996 and reports therein). The opportunity is now open to precisely tailor new materials through atom-by-atom control of the composition (controlling the types as well as the numbers of atoms) in order to generate the clusters or particles of precision design for use in their own right or as building blocks of larger-scale materials or devices—that is, nanotechnology and fabrication at its ultimate. Such precision engineering or tailoring of materials is the goal of much of the effort driving nanoscale technology. Scientists and engineers typically have approached the synthesis and fabrication of high surface area nanostructures from one of two directions:

1. The "bottom up" approach in which the nanostructures are built up from individual atoms or molecules. This is the basis of most "cluster science" as well as crystal materials synthesis, usually via chemical means. Both high surface area particles and micro- and mesoporous crystalline materials with high void volume (pore volume) are included in this "bottom up" approach.

2. The "top down" approach in which nanostructures are generated from breaking up bulk materials.

This is the basis for techniques such as mechanical milling, lithography, precision engineering, and similar techniques that are commonly used to fabricate nanoscale materials, which in turn can be used directly or as building blocks for macroscopic structures. A fundamental driving force towards efforts to exploit the nanoscale or nanostructure is based upon two concepts or realisations:

(1) that the macroscopic bulk behaviour with which we are most familiar is significantly different from quantum behaviour, and

(2) that materials with some aspect of quantum behaviour can now be synthesised and studied in the laboratory. Obviously, quantum behaviour becomes increasingly important as the controlling parameter gets smaller and smaller. There are numerous examples of quantum behaviour showing up in high surface materials: the fact introduced above that clusters are found to exhibit novel (compared to the bulk) electronic, magnetic, chemical, and structural properties; the fact that the diffusivity of molecules through molecular sieving materials cannot be predicted or modeled by hard sphere molecule properties or fixed wall apertures; and the fact that catalysts with one, two, or three spatial dimensions in the nanometer size range exhibit unique (compared to the bulk) catalytic or chemical activity.

Polymeric Nanocomposites

Since the predominant feature of nanoparticles lies in their ultra-fine dimension, a large fraction of the filler atoms can reside at the interface and can lead to a strong interfacial interaction (Roy, 1986), but only if the nanoparticles are well dispersed on a nanometer level in the surrounding polymer matrix. As the interfacial structure plays a critical role in determining the composites' properties, nanocomposites coupled with a great number of interfaces could be expected to provide unusual properties, and the shortcomings induced by the heterogeneity of conventional particle filled composites would also be avoided (Rong et al., 2001). Recent progress indicated that polymer-based nanocomposites acquired reduced frictional coefficients and wear rates. For example, Wang et al. (1998) reported that the incorporation of nanosized ZrO_2 into polyetheretherketone caused a considerable improvement in the tribological characteristics. It was found that the dominant wear mechanism changed from melting adhesive transfer wear to slight transfer wear, and finally to abrasive wear with increasing nano-ZrO_2 content. Wang and his co-workers (1998) prepared nanoscale ceramic (TiC, Si_3N_4, B_4C)/polytetrafluoroethylene multilayers by ion beam sputtering deposition at room temperature. Ball-on-disk tribological tests showed that the multilayers with optimised layer thickness arrangement experienced good performance in wear resistance. There was no obvious periodic variation in the friction coefficient. Schadler et al. (1997) produced silica/polyamide nanocomposite coatings using high-speed oxyfuel thermal spray processing; they found that the surface chemistry of the nano-silica affected the final coating properties. Silica particles with a hydrophobic surface resulted in higher scratch resistance than those with a hydrophilic surface. However, very few works have been presented to date concerning the relationship between microstructural details and tribological properties of bulk, polymer-based nanocomposites. This is very surprising, since this is a topic which has both a fundamental and an applied significance for the development of nanocomposites for new technical applications. Therefore, the objective of the present work is focused on the tribological performance of composites with identical species of the reinforcing components, but with different ways of compounding and different particle sizes. In particular, the effects of the dispersion status and the size of nanoparticles on the friction and wear behaviour of such nanocomposites are studied, so as to provide knowledge for an optimum material preparation.

Commercially available epoxy (E-51) and tetraethylenepentamine (TEPA), provided by Guangzhou Adhesive Co. and Shanghai Dye Chemical Engineering Co., China, were used as matrix polymer and curing agent, respectively. Its reinforcement was carried out by the use of home-synthesised TiO_2 particles, having average diameters of 10nm and 40mm, respectively. To prepare nanocomposites with different filler arrangements, three approaches were employed (Liu, 1997):

1. (1) Nanosized TiO_2 particles were directly mixed into the epoxy resin, which was put into an oven, preset at 80°C. After half an hour, the mixture was taken out and stirred mechanically for another half an hour. To improve the homogeneity of nanoparticles dispersion, the nano-TiO_2/epoxy compounds were further treated under ultrasonic environment for 10min. Finally, the curing agent was added, and the solid nanocomposites were obtained after holding the mixture for 2hr at 80°C.

2. (2) At first, nanosized TiO_2 particles were dispersed in chloroform, aided by sonication. This suspension was then mixed with the epoxy resin by mechanical

stirring for half an hour and subsequent sonication for 10min, respectively. When the solvent was evaporated through vacuum drying, the mixture of nano-TiO_2 and epoxy was again stirred and ultrasonically vibrated for half an hour and 10min, respectively. Then the curing agent was added to the mixture, and the above curing process was followed.

3. (3) The nano-TiO_2 suspension in chloroform made according to the last approach was mixed with epoxy. The mixture was then stirred mechanically for 2.5hr, and subsequently ultrasonically vibrated for 10min. In the next step, the mixture was heated to 80°C to remove the solvent while being stirred once more for 2hr. Eventually, the TiO_2/epoxy nanocomposite was cured as described in the last part of approach (1).

When the epoxy was filled with coarse TiO_2 particles (diameter: 40ìm), the composites were manufactured in accordance with the third approach mentioned above. Unlubricated sliding wear tests were carried out on a pin-on-ring apparatus under a pressure of 1MPa and a constant velocity of 0.4m/s. The carbon steel ring (0.42~0.45wt% C, 0.17~0.37wt% Si and 0.5~0.80wt% Mn, HRC 50) had a diameter of 40mm and an initial surface roughness of 0.1ìm. Prior to wear testing, all the samples were pre-worn so as to average surface conditions and reduce the running-in period. The actual steady-state test period was set to 2hr. After that, a weight measurement of the specimens was carried out. Microindentation Vickers hardness values of the composites were determined by a HDX-1000 digitised microhardness tester. The reported hardness data are the average of five successful measurements. Dynamic mechanical tests were conducted through a DMTA BMARK-III analyzer at a frequency of 1Hz. The internal friction, expressed by the tangential loss angle, tan ä, of the materials was measured under three-point bending configuration at a heating rate of 2úC/min.

The specific wear rates ($wÿ_s$) and the frictional coefficients (ì) of nanocomposites which were prepared through a direct compounding process. In general, it can be stated that especially the wear rates are very high, which is, however, mainly due to the low stiffness nature at elevated temperature of the epoxy system used for this study. With a rise in the nano-TiO_2 concentration, $wÿ_s$ keeps almost unchanged at the beginning, but increases furthermore when the filler content exceeds 3phr (phr = amount of filler by weight per 100 parts matrix by weight). At the same time, ì exhibits a fluctuating mode of variation. As can be imagined, approach (1) of composite manufacturing, as described in the experimental part, is bound to generate a rather heterogeneous dispersion of the nanoparticles. This is due to two facts, a strong tendency of the particles to agglomerate, and a high viscosity of the epoxy matrix. As a result, an extensive material loss due to disintegration and crumbling of the nanoparticle agglomerates under tribological conditions occurs (Bely et al.,

1982). Besides, the unevenly distributed nanoparticles also provide the pin surface with an inhomogeneous modification of the mechanical properties, leading to the irregular dependence of ì on the TiO_2 loading. It is generally believed that solvent blending helps to improve the homogeneity of composites' microstructure. Therefore, TiO_2/epoxy nanocomposites were produced by using a nano-TiO_2/chloroform suspension. The tribological properties of the nanocomposites made in accordance with approach (2). In comparison to the values of $w\ddot{y}_s$ were slightly reduced in the TiO_2 content range, between 1 and 7phr. In the case of 3phr of nano-TiO_2 particles, for example, the wear rate of the composite prepared with the aid of chloroform is more than two times lower than both the wear rate of the neat epoxy ($2.67 \times 10^{-2}mm^3/Nm$) and that of the composite prepared through a direct mixing of the components. This should be attributed to the improved dispersion status of the nanoparticles. However, it can also be seen that the profiles of the $w\ddot{y}_s$ vs TiO_2 content-curves in both resemble each other in general, implying that the two types of composites have a similar microstructure. When the TiO_2 loading is higher, splitting of the agglomerated nanoparticles likely occurs and increases the wear loss.

When nano-TiO_2 particles are incorporated into the epoxy, the microhardness of the composites is increased. This corresponds to an increase in the wear resistance of the composites, especially if a rather uniform dispersion of the nanoparticles exists. On the other hand, the deviation of Hv of the composites (prepared according to approaches (1) and (2)) is not remarkably higher than that of the neat epoxy matrix. This means that the effect of reinforcement imparted by the additives, which is closely related to the dispersion state of the particles, was probably not fully utilised. In other words, it can be concluded that approach (2) is not effective enough to produce a homogeneous dispersion of the particles, and as a result, the expected, positive influence of the nano-TiO_2 particles on the wear performance of the epoxy resin could not be brought into full play. With respect to the frictional coefficient, its dependence on the TiO_2 content is characterised by a "U"-shape, which manifests a stable friction process different from that shown. It suggests that the addition of nano-TiO_2 is able to provide certain lubricating effects, but the more severe removal of material under higher filler loading might lead to a three-body abrasive wear situation. Hence, ì increases with increasing TiO_2 content above 5phr. The friction and wear properties of the nanocomposites manufactured through the third approach. The distinct step in approach (3), which is different from approach (2), lies in the fact that chloroform was removed while the mixture of epoxy and TiO_2 particles was being stirred continuously. It is believed that such a way of solvent removal results in a more uniform distribution of the particulate fillers, which is substantiated by the greatly increased microhardness and a rather small deviation (±0.4MPa) close to that of neat epoxy measurements (0.6MPa). The wear resistance of the epoxy resin can be increased by factors of four to 16 for the observed range of nano-TiO_2 loading, while the frictional coefficient shows a decreasing tendency when the particle loading exceeds 5phr. This means the tribological behaviour is completely different from that reflected.

Clearly indicates that much more chain motion is restricted by the nanoparticles when the composite is produced by approach (3). This means the mobility of the macromolecules at filler/matrix interfaces becomes clearly lower, as characterised by the unusual increase in the á transition temperature, implying increased contacts between the fillers and the matrix. It further demonstrates the relatively uniform microstructure of the nanocomposites tested. It is interesting to find that the decrease in frictional coefficient corresponds to a rise in wear rate when the nano-TiO_2 content is higher than 5phr. This is somewhat like the case of polytetrafluoroethylene, where the extremely low frictional coefficient wear resistance due to the spec wear resistance due to the specific lamellae arrangement. In the case of lower filler content, however, the composites and the unfilled matrix exhibit almost the same frictional coefficient. This can be attributed to the size effect of the additives. That is, when the nano-fillers are well dispersed, they can have the same size as the segments of the surrounding polymer chains so that the polymer filler phase can make a predominant contribution to friction due to its structural integrity. When the content of the nanoparticles is increased to a certain threshold (e.g. 5phr for the current work), at which approach (3) is no longer effective to prevent agglomeration of the particles, the increasing amount of detached nano-TiO_2 particles at the interface between the steel counterpart and the composite pin can significantly reduce the friction. Further investigation of the micromechanisms involved in the present systems are, however, needed.

It is worth noting that approach (3) is not universally applicable. When nano-TiO_2 was replaced by micron-sized TiO_2, the particles were poorly dispersed. The resultant composites have a performance similar to that of the composites prepared in terms of approaches (1) and (2). This implies, on one hand, that TiO_2 nanoparticles can effectively improve the wear resistance of epoxy, while conventional micron-TiO_2 particles cannot. On the other hand, when the nanoparticles are unevenly dispersed in the matrix, the nanocomposites perform like composites filled with micrometer-sized particulate fillers, in which crack-initiation and coalescence occur more easily in the particulate-rich phases.

Friction and wear properties of polymer nanocomposites are a function of the dispersion state of the inorganic type, nanosized reinforcing nanoparticles. A uniform distribution of the nanoparticles helps to enhance the wear resistance. TiO_2 nanoparticles are able to significantly reduce the wear rate and the frictional coefficient of epoxy. Depending on the dispersion state and content of the nanoparticles, these positive effects do not appear simultaneously. That means, an optimum design of the nanocomposites is required. Mixing of epoxy and the nanoparticles with the aid of chloroform can tentatively lead to a relatively homogeneous phase dispersion, and a continuous stirring associated with heating maintains this microstructure until the composite is fully cured. Further efforts should be made to explore more viable techniques to break up the nanoparticle agglomerates and to improve the interfacial adhesion between the particles and the matrix resin. This is quite important for making full use of nanoparticles in tribological applications.

Clusters and Nanocrystalline Materials

Clusters are groups of atoms or molecules that display properties different from both the smaller atoms or molecules and the larger bulk materials. Many techniques have been developed to produce clusters, beams of clusters, and clusters in a bottle for use in many different applications including thin film manufacture for advanced electronic or optical devices, production of nanoporous structures as thermal barrier coatings, and fabrication of thin membranes of nanoporous materials for filtration and separation. An apparatus developed at the University of Göteborg to measure cluster reactivity and sticking probability as a function of the number of metal atoms in the cluster. The unique properties of nanoparticles make them of interest. For example, nanocrystalline materials composed of crystallites in the 1-10 nm size range possess very high surface to volume ratios because of the fine grain size. These materials are characterised by a very high number of low coordination number atoms at edge and corner sites which can provide a large number of catalytically active sites. Such materials exhibit chemical and physical properties characteristic of neither the isolated atoms nor of the bulk material.

One of the key issues in applying such materials to industrial problems involves discovery of techniques to stabilize these small crystallites in the shape and size desired. This is an area of active fundamental research, and if successful on industrially interesting scales, is expected to lead to materials with novel properties, specific to the size or number of atoms in the crystallite. Schematic drawing of the experimental setup used in Göteborg for studies of chemical reactivity and/or sticking probability of various molecules with the clusters. The production of clusters is via laser vaporisation of metal substrates and detection via photoionisation time-of-flight mass spectrometry (A. Rosen, University of Göteborg, Sweden). A typical objective of nanoscale catalyst research is to produce a material with exceedingly high selectivity at high yield in the reaction product or product state, i.e., chemicals by design, with the option of altering the product or product state simply by changing the surface functionality, elemental composition, or number of atoms in the catalyst particle. For instance, new catalysts with increasing specificity are now being fabricated in which the stoichiometry may be altered due to size restrictions or in which only one or two spatial dimensions are of nanometer size. Five recent examples where nanocrystalline metallic and ceramic materials have been successfully investigated for catalytic applications are discussed below (Trudeau and Ying 1996).

Catalytic Properties of Nanostructured Gold Catalysts

In the study of transition metal catalytic reactions the group at Osaka National Research Institute has discovered that nanoscale gold particles display novel catalytic properties (Haruta 1997). Highly selective low temperature catalytic activity is observed to switch on for gold particles smaller than about 3-5 nanometers in diameter. Accompanying this turn on in catalytic activity is the discovery that these nanoscale gold particles (crystals) also have icosahedral structure and not the bulk fcc structure, again a nanoscale phenomena not available with bulk samples. In fabricating these novel catalytic materials several issues appear to be crucial. For instance, the Osaka group has shown that

(a) the preparation method plays a crucial role for generating materials with high catalytic activity and selectivity; that

(b) the catalytic activity, selectivity and temperature of operation is critically dependent on the choice of catalyst support, and that

(c) water (moisture) even in parts per million (ppm) levels dramatically alters the catalytic properties.

The effect of moisture on the conversion profiles for CO oxidation for nanoscale gold catalysts supported on cobalt oxide. Examples of novel catalytic behaviour of nanoscale gold particles include the following:

- CO oxidation at temperatures as low as -70°C.
- very high selectivity in partial oxidation reactions, such as 100% selectivity at 50°C for oxidation of propylene to propylene oxide as well as near room temperature reduction of nitric oxide with H2 using alumina-supported gold nanoparticles. Effect of moisture on conversion profiles for CO oxidation over Co3O4 and Au/Co3O4.

The fundamental work on gold catalysts has led to "odor eaters" for the bathroom, based on nanoscale gold catalysts supported on a-Fe2O3, a recent commercialisation from Osaka National Research Institute in Japan.

For catalysts based on the layered compound MoS2, maximum hydrodesulfurisation (HDS) activity is obtained only on well-crystallised nanosized materials, while HDS selectivity is determined by the number of layers or "stack height" of the nanocrystalline MoS2. In the hydrodesulfurisation reaction, cyclohexylbenzene occurs only on the MoS2 "rim sites, " or those around the edges of the stack, whereas the pathway to biphenyl requires both "rim" and "edge" sites. Thus, the reaction selectivity can be controlled by controlling the aspect ratio of the nanoparticles of MoS_2. Such control of one- and two-dimensional nanostructures for selective chemical advantage is an exciting new area of research. Of course, a major industrial challenge will be to fabricate such nanocrystals in a commercializable form (Chianelli 1998).

The cerium oxide (CeO_{2-x}) materials have been found to possess a significant concentration of Ce^{3+} and oxygen vacancies, even after high temperature (500°C) calcination. Such nanoclusters give rise to a substantial reduction in the temperature of selective SO_2 reduction by CO and exhibit excellent poisoning resistance against H_2O and CO_2 in the feed stream compared to that for conventional high surface area cerium oxide.

Electrochemical reduction of metal salts is yet another option available to control the size of nanoscale catalyst particles.

A combination of scanning tunneling microscopy (STM) and high-resolution transmission electron microscopy (TEM) has allowed surfactant molecules to be visualised on nanostructured palladium clusters.

Materials with higher hydrogen storage per unit volume and weight are considered by many to be an enabling technology for vehicular fuel cell applications. Scientists at Los Alamos National Laboratories have developed an approach that enables materials such as Mg to be used for hydrogen storage (Schwartz 1998). Magnesium is of interest because it can store about 7.7 wt % hydrogen, but its adsorption/desorption kinetics are slow, i.e., the rate of charge (hydrogen dissociation and hydride formation) is much slower than in metal hydrides. At Los Alamos, high surface area mixtures of nanoscale Mg and Mg_2Ni particles are produced by mechanical means, ball milling. The addition of Mg_2Ni catalyzes the H_2 dissociation such that the rate of hydrogen adsorption increases to a rate comparable to that of $LaNi_5$. Once a hydride is formed, the hydrogen "spillover" leads to magnesium hydride formation. The hydrogen adsorption/desorption characteristics of the mixture of Mg with 23 atomic % Ni. As can be seen, a low pressure plateau at about 1500 torr is obtained for this particular sample. Experiments show that the pressure plateau can be tailored through such alloying. Studies with other nanoscale materials, including other catalysts such as FeTi and $LaNi_5$, are presently ongoing to further improve both the capacity and the rate of hydrogen storage.

A Case Study I: United States of America

Significant work on coatings of dispersions is underway in the United States, as described in the proceedings of the 1997 WTEC workshop report, *R&D status and Trends in Nanoparticles, Nanostructured Materials, and Nanodevices in the United States* (Siegel et al. 1998). Some of the highlights of that volume are summarised below. Similar work is ongoing in other countries.

Controlling the pore structure and synthesising the building blocks are two technical challenges facing future work in this area. However, Aksay has shown that he can grow silicate films onto a wide variety of substrates. The layer structure of the bound surfactant molecules is key. Atomic force microscopy can reveal some details on the morphology of these films. The nanostructural patterns obtained in processing ceramics that contain organic/inorganic composites allow self-assembly to take place at lower temperatures. Another example of thin films or coating is the work of Gell (1998, 124- 130), focused on improving both the physical and the mechanical properties of materials. Nanostructured coating can lead to high diffusivity, improved toughness and strength, and better thermal expansion coefficients, with lower density, elastic modulus, and thermal conductivity.

In comparisons of nanostructures and conventional materials, use of nanostructured WC/cobalt composites have resulted in as much as a two-fold increase in abrasion resistance and hardness. Deposition is often accomplished by utilising a sputtering chamber where the nanomaterial is coated on substrates. This can lead to such benefits as resistance to oxidation and cracks in addition to resistance to wear and erosion. An area that offers exciting possibilities in the area of dispersions is the sol-gel process described by P. Wiltzius (1998, 119-121). Here, a concentrated dispersion of colloids is chemically converted into a gel body. Drying followed by sintering produces a ceramic or glass product.

This process can create nanoparticles, fibers, film, plates, or tubes. All processing is at low temperatures. Lucent Technologies has developed a silica casting process that is reproducible for making tubes of pure silica of one meter in length for use in manufacturing optical fibers. Technical challenges include obtaining pure starting materials, removing refractory particles that lead to breakage in the fiber drawing process, and achieving very tight dimensional tolerances. Colloidal dispersions of this type play a critical role in chemical mechanical polishing. To obtain good yield and high quality, it is necessary to achieve very tight process control. R. Brotzman at Nanophase Technologies Corporation describes the gas phase condensation process for synthesising inorganic and metallic powders (Brotzman 1998, 122-123). Such a process was invented by R. Siegel and his colleagues at Rensselaer Polytechnic Institute in New York (Siegel et al. 1994). The process involves production of physical vapor from elemental or reacted material followed by sudden condensation and reaction of the vapor into small nanometer particles.

The condensation process is rapid and involves dilution to prevent the formation of hard agglomerates and coalescence. Nanophase Technologies has developed a production system where forced convection flow controls the particle/gas stream and enhances particle transport from the particle growth region so as to generate more metal vapor (Parker et al. 1995). The convection flow helps in forming oxides and nitrides from metal crystallites. This process produces commercial quantities of nanosized inorganic powders that have a spherical shape with narrow size distributions. Nanophase Technologies Corporation has developed a coating process that encapsulates nanoparticles with a surface treatment that ensures individual integrity of the particles in subsequent coating steps.

Applications span their use in low dielectric media all the way to water and, if needed, steric stabilizers. Nanophase also has directed efforts in the electronics and industrial catalyst areas. Friedlander (1998, 83-8) describes in good detail aerosol reaction engineering where attention is paid to design of the process and the importance of material properties and process conditions. Key process parameters include time, temperature, and volume concentration. Most commercially produced particles have polydispersity. Important full-scale processes are flame reactors for preparing pigments and powdered materials for optical fibers.

Pyrolysis reactors have long been used to prepare carbon blacks. Kear and Skandan (1998, 102-4) have discussed two divergent approaches toward fabrication of bulk materials. The first is a powder processing route involving preparations by physical or chemical means followed by condensation and sintering. Materials for cutting tools such as tungsten carbide/cobalt powders were prepared in a controllable way to about 50 nm. Liquid phase sintering completes the formation of the solid dispersion phase. However, a gap in fabrication technology remains that of controlling grain growth during liquid phase sintering. A second approach to fabrication of bulk materials is spray forming. This procedure avoids contamination and coarsening of the dispersion and its particles when the process has a controlled atmosphere of inert gas at low pressure. In this approach there is need to establish process/product codesign where we understand cause-effect relationships between processing parameters and properties of nanophase spray/deposited materials.

The Case Studies II: Europe and Japan

Due to time constraints, the WTEC panel was able to visit only a few labs in Europe and Japan that deal with issues associated with dispersions and coatings. Cited below are a visit to the Institute for New Materials at Saarbrücken in Germany and a visit to Japan's Industrial Research Institute of Nagoya, both of which are exploring direct applications of nanostructured dispersions and coatings. Drs. Rudiger Nass and Rolf Clasen at the Institute for New Materials at Saarbrücken make specific use of the sol-gel process in which liquid starting materials are utilised at low temperatures for nanoscale metal, ceramic, glass, and semiconductor nanoparticles (Clasen 1990). The advantages, besides temperature, include the isolation of high purity powders. The institute has focused on four basic areas for spin-off and adaptation towards commercialisation:

1. New functional surfaces with nanomers. Properties such as corrosion protection, wettability, coloration, micropatterned surfaces, porosity, and the ability for selective absorption of molecules produce multifunctionality.
2. New materials for optical applications. This area combines properties of lasers and ceramics with polymers. Such features as optical filters, transparent conducting layers, materials for optical telecommunications, photochromic layers, and holographic image storage are under investigation.
3. Ceramic technologies. In this area, a simple precipitation process such as sol-gel provides for pilot-scale production of agglomerate-free powder.
4. Glass technologies. Chemical incorporation of metal colloids with intelligent properties into glasslike structures is clearly possible.

During the WTEC team's site visits to the Industrial Research Institute of Nagoya, S. Kanzaki and M. Sando described methods for preparing synergy ceramics using nanoporous silica particles that are used to fabricate thin films with one-dimensional throughput channels (Kanzaki et al. 1994). The channels had 10-20 nm sizes. High temperature Fe2SiO4 oxides were prepared as both molecular sieves and particulate fibers. Japan plans to pursue the preparation of nanoporous materials for absorbing oil and identified particulates. The quest for low temperature nanoparticle preparation methods has spanned a wide range of systems. One that has been in existence for decades but has not been put into use in other industries is the method of preparing silver halide particles. Eastman Kodak in France, England, and the United States has utilised solution precipitation technology with well-controlled mixing and nucleation control to produce a wide range of grain sizes. "Lippmann"-type grains have a size of about 50 nm.

Some of the properties of these fine-grain systems are discussed by Trivelli and Smith (1939). There are other methods of creating nanoparticles of organic materials such as filter dye applications in photographic films and spectral sensitising dyes for use in silver halide grains. Ultrafine grinding media are used to almost sandpaper organic crystals to nanoparticle ranges of 20-80 nm (Czekai et al. 1994). Similar technology by Bishop has been utilised in both pharmaceutical preparations and ink jet applications with good success (Bishop et al. 1990). One other exciting area is in polymer science, where dendrimer molecules, often 10 nanometers in diameter, are prepared synthetically. These molecules have been studied in gene therapy, as aids in helping to detect chemical or biological agents in the air, and as a means to deliver therapeutic genes in cancer cells (Henderson 1996). The preparation of dendritic starburst molecules is described by Salamone (1996, 1814). One can imagine applications for coatings in which these molecules with their highly reactive surfaces participate in either classical (sub-nanoscopic chemistry) or novel nanoscopic conversions.

Nanoparticles Pass Muster as Vectors for Gene Therapy

Gene therapy, in which a viral vector is used to modify defective genes or replace missing ones, has shown significant potential as a way of treating disease in animal models. But its use in humans has been hampered by safety concerns, including some fatalities in clinical trials. Researchers have thus been looking into the possibility of using nonviral vectors, which should carry fewer inherent risks, to deliver therapeutic genes. In a paper published online this week by the *Proceedings of the National Academy of Sciences,* scientists report that silicon nanoparticles can perform this task successfully in mice. Paras N. Prasad of the State University of New York (SUNY) at Buffalo and his colleagues manufactured nanoparticles using organically modified silicon. The surface of these particles can be tailored to target specific cells. The team used the tiny units to transport a fluorescent marker gene to dopamine neurons in the brains of mice. After injecting the nanoparticles, the researchers observed brain cells fluorescing using a new imaging technique that works on live animals. According to the report, the study is the first in which a nonviral vector has shown efficacy comparable to that of a viral delivery system in an animal model. What is more, a month later none of the animals had experienced adverse effects from the procedure. The researchers also investigated the possibility of manipulating the behaviour of specific brain cells, instead of solely tagging their presence. In so doing, they discovered that the nanoparticles can be used to reactivate adult stem cells by altering a nuclear growth factor receptor. The team will next test the approach on larger animals. "In the future, " says study co-author Earl J. Bergey, also at SUNY Buffalo, "this technology may make it possible to repair neurological damage caused by disease, trauma or stroke."

The first broad category of enabling technology for molecular manufacturing in the mechanical tradition is biotechnology. Molecular biologists study and modify systems of molecular machines, and genetic engineers reprogramme these systems, sometimes to build novel molecular objects having complex functions. By 1998, biotechnologists could make almost any DNA sequence or poly-peptide chain desired (though not extremely long aperiodic ones), assemble "devices" such as artificial chromosomes and viruses, and were actively pursuing gene therapy, all of which are examples of molecular engineering. While earlier writers had suggested that biomolecules could be used as mechanical components, Drexler evidently was the first to point out, in 1981, that complex devices resembling biomolecular motors, actuators, bearings, and structural components could be combined to build versatile molecular machine systems analogous to machine systems in the macroscopic world. "Development of the ability to design protein molecules, " Drexler wrote, "will open a path to the fabrication of devices to complex atomic specifications, thus sidestepping obstacles facing conventional microtechnology. This path will involve construction of molecular machinery able to position reactive groups to atomic precision."

Perhaps the best-known biological example of such molecular machinery is the ribosome, the only freely programmable nanoscale assembler already in existence. In nature, ~8000 nm^3 ribosomes act as general-purpose factories building diverse varieties of proteins by bonding amino acids together in precise sequences under instructions provided by a strand of messenger RNA (mRNA) copied from the host DNA, powered by the decomposition of adenosine triphosphate (ATP) to adenosine diphosphate (ADP). Each ribosome is a compact ribonucleoprotein particle consisting of two subunits, with each subunit consisting of several proteins associated with a long RNA molecule (rRNA). Biotechnology can employ the ribosomal machinery of bacteria to produce novel proteins, which proteins then might serve as components of larger molecular structures.

Of course, such biological mechanisms have been disparaged as "slipshod devices working in spite of haphazard design." Dan Heidel, a University of Washington biomimetics engineer, agrees that biological systems are sloppy but observes that they are designed to tolerate it, and offers high praise for the lowly ribosome:

"The average ribosome sits in an aqueous environment surrounded by thousands of cosolutes at concentrations just a hair's breadth from precipitating out of solution. The ribosome itself consists of three RNA molecules that spontaneously fold into floppy spaghetti noodle piles of chemically active groups surrounded by >50 proteins of similarly dubious structure. The other solutes are bombarding this motley assemblage at tens to hundreds of miles an hour and the solvating water molecules are a constantly shifting and unpredictable network of hydrogen bonds and polar ionic charges constantly impinging upon the ribosome. The ribosome must recognize one particular cognate transfer RNA (tRNA) for the codon in its A-site out of ~60 other nearly identical tRNAs. This particular tRNA must be discriminated from all its tRNA brethren along with whatever other solutes can fit into the A-site pocket using only three base pairs for discrimination. In fact, most codons use only two of the three base pairs for recognition so some tRNAs are recognised by as few as 4 hydrogen bonds. All of this requires sub-Angstrom precision in the active sites but since the ribosome is a massive complex of subunits held together by only ionic, hydrogen, hydrophobic and Van der Waals bonds, it's like getting submillimeter accuracy from points on a bunch of taped-together water balloons being rolled down a staircase."

"This ridiculous machine can assemble proteins at a 20 Hz subunit-incorporation frequency with a 10^3 error rate. Ribosomes are even more impressive when one considers that the number of amino acids in each ribosome's own protein subunits combined with the 10^3 protein assembly error rate ensures that each ribosome has several errors in it and that each cell almost certainly has no two ribosomes exactly alike. I dare any engineer on this planet to rationally design an assembler at any scale that is able to operate at a comparable level of performance under similarly adverse operating conditions."

Thus ribosomes can manufacture, with >99.9% fidelity to arbitrary specifications, linear strings of amino acids of great length which may then fold up to produce three-dimensional protein structures performing a broad range of highly differentiated functions, providing good analogs for working nanomachines capable of forming self-assembling systems. Much of the great diversity of protein activity flows from the selectivity with which they bind to specific receptor sites on other molecules. Some of these binding mechanisms might be copied to construct molecular robotic arms, while other devices could be made by adapting naturally-occurring proteins, or by synthesising new, custom-built ones. Protein engineering might be used to create proteins with desired qualities as building blocks for constructing the first primitive assemblers.

The field of protein engineering has its own journal of the same name, and has already achieved remarkable results in the synthesis of novel structures. Drexler cites the creation of the first de novo structure by DeGrado, the development of synthetic branched protein-like structures that depart substantially from protein models, and the engineering of a branched, nonbiological protein with enzymatic activity. Since then, numerous alpha-helical peptides have been designed de novo, a 20-residue artificial peptide exhibiting designed beta-sheet secondary structure has been created, and various peptide analogs have also been synthesised. By 1998, advances in computational techniques allowed the design of precise sidechain packing in proteins with naturally occurring backbone structures, and the study of de novo backbone structures had begun. The catalytic task space was being laboriously investigated at the molecular level — for example, one study of comparative enzyme structures revealed that changing as few as four amino acids converted an oleate 12-desaturase to an oleate hydroxylase, and as few as six substitutions could convert an oleate hydroxylase into an oleate desaturase. To further expand the toolkit, Mills and Fahy proposed exploiting the degeneracy of the genetic code (wherein amino acids are specified by up to 6 different codons; by assigning some codons to unnatural amino acids, allowing the creation of artificial proteins with unprecedented structural or catalytic properties. With 61 usable codons available, there is room for up to 41 novel amino acids in addition to the natural 20 amino acids; unnatural amino acids have been introduced into beta-lactamase and T4 lysozyme enzymes in a site-directed fashion, using artificial tRNAs and mRNAs in an in vitro translation system, and into an engineered peptide as a reporter group. Additionally, artificial novel base pairs can be incorporated enzymatically into DNA and other schemes are possible. By incorporating one additional base pair, the number of possible codons is expanded from $4^3 = 64$ codons to $6^3 = 216$ codons, potentially allowing up to 193 novel amino acids to be incorporated into artificial protein structures — or at least 48 new amino acids assuming that present levels of redundancy are retained and the existing 4-letter code is kept as a subset for "upward compatibility". If the genetic alphabet is increased to eight bases ($8^3 = 512$ codons), the number of available codons for novel amino acid-like molecules increases to as many as 489.

One of the greatest challenges in protein engineering (where one objective is to create functional, atomically-precise 3-D aperiodic structures) has been the difficulty of protein fold design, because individual amino acids have no strong, natural complementarity. In 1998, Duan and Kollman successfully folded a solvated 36-residue (~12, 000-atom) protein fragment (i.e., a peptide) by molecular dynamics simulation into a structure that resembles an intermediate native state. During a 1-microsec simulation, the chain folded during 150 nanosec into a compact structure resembling the intermediate native state, as known by NMR, then unfolded and refolded again for a shorter period. In 1998, whole-protein folding to the final stable native state had not yet been computed deterministically using molecular dynamics — the real protein will fold and refold hundreds of times before it stumbles into the stable conformation with the lowest free energy — but research pursuing this objective was active and ongoing. Structural changes that occur during protein function (e.g., enzymatic action) were also being avidly studied.

A different protein-based approach to near-atomically-precise 3-D structures avoids the folding problem by making use of semi-rigid protein nanoshells of deterministic size and shape to guide the ordered assembly of inorganic particles. For example, the capsid of the cowpea chlorotic mottle virus, a protein container with a 28-nm external diameter and an 18-nm internal cavity diameter, is composed of 180 identical coat protein subunits arranged on an icosahedral lattice. Each subunit presents at least nine basic residues (arginine and lysine) to the interior of the cavity, creating a positively charged interior surface which provides an interface for inorganic crystal nucleation and growth, while the outer capsid surface is not highly charged. In one experiment, the empty capsid shell was found to act as a spatially selective nucleation catalyst in paratungstate mineralisation, in addition to its role as a size- and shape-constrained reaction vessel. Noted the authors: "The range of viral morphology and size allow great flexibility for adapting this methodology to control the size and shape of the entrapped material, which is limited only by access to the protein interior and could include inorganic and organic species. The electrostatic environment within the cavity could be altered by site-directed mutagenesis to induce additional specific interactions."

It may also be possible to borrow various existing protein devicesand apply them to new uses. The bacteriophage DNA injection system is self-assembling and could be engineered for other applications. Vogel has attached kinesin motors to flat surfaces in straight grooves; when ATP was added, the motors were activated, passing 25-nm wide microtubules hand over hand down the line in the manner of a ciliary array. Montemagno genetically engineered the 12-nm wide ATPase molecular rotary motor so that one end would adhere to a metal surface and the other end would provide an attachment site for 1-micron fluorescent streptavidin-coated bead payloads. When ATP was added, each bead individually attached to a biomotor in the motor array began twirling at ~10 Hz, generating a >100 pN force. Bead rotation was maintained continuously for more than 2 hours, indicating the high reliability of these ~100% efficient self-assembling biomotors. Montemagno explains that his laboratory's "long-term goal is the integration of the F_1ATPase biological motor with nanoelectromechanical systems (NEMS) to create a new class of hybrid nanomechanical devices." Other natural nanomotors such as the self-assembling bacterial flagellar motor and the hexagonal "packaging RNA" or pRNA pumping mechanism that packs DNA into the capsid shell of the bacteriophage Phi 29 are also being intensively studied.

Yet another material which allows moderately well-defined three-dimensional nanoassemblies is DNA. The ideas behind DNA nanotechnology have been around since 1980, but activity in this field accelerated in the 1990s after numerous experimental difficulties were surmounted. Horn and Urdea reported branched and forked DNA polymers. Niemeyer and Smith exploited DNA specificity to generate regular protein arrays, and both suggested using self-assembled DNA as an early material for molecular nanotechnology. Shi and Bergstrom attached DNA single strands to rigid organic linkers, showing that cyclical forms of various sizes could be formed with these molecules. Mirkin's group attached DNA molecules to 13-nm colloidal gold particles with ~6 nm particle spacings, with the goal of assembling nanoparticles into macroscopic materials; Alivisatos, Schultz and colleagues used DNA to organise 1.4-nm gold nanocrystals in arrays with 2-10 nm spacings. In either case, the specificity of DNA pairing should allow the construction of complex geometries since the use of colloidal particles can potentially add structural elements (which are more rigid than any polymer strand) to the set of building blocks for nanometer-scale structures — although to take full advantage of this capability would require particles which are atomically precise and which possess several chemically distinct anchoring points on each particle. Damha has synthesised V-shaped and Y-shaped branched RNA molecules, branching RNA dendrimers with "forked" and "lariat" shaped RNA intermediates, and even trihelical DNA. Henderson and coworkers have designed a simple DNA decamer that can form an extended linear staggered quadruplex array reaching lengths of >1000 nm, and have made branched oligonucleotides that template the synthesis of their branched "G-wires"; depending on the ratio of linear to branched building blocks, extensive DNA arrays with differing connectivities but irregular interstices can be created.

In 1998 the most intensive work on three-dimensional engineered DNA structures was taking place in Nadrian Seeman's laboratory in the New York University

Department of Chemistry. Seeman originally conceived the idea of rigid 3-D DNA structures in the early 1980s, while examining DNA strands that had arranged themselves into unusual four-armed Holliday junctions. Seeman recognised that DNA had many advantages as a construction material for nanomechanical structures. First, each double-strand DNA with a single-strand overhang has a "sticky end, " so the intermolecular interaction between two strands with sticky ends is readily programmed (due to base-pair specificity) and reliably predicted, and the local structure at the interface is known (sticky ends associate to form B-DNA). Second, arbitrary sequences are readily manufactured using conventional biotechnological techniques. Third, DNA can be manipulated and modified by a large variety of enzymes, including DNA ligase, restriction endonucleases, kinases and exonucleases. Fourth, DNA is a stiff polymer in 1-3 turn lengths and has an external code that can be read by proteins and nucleic acids.

During the 1980s, Seeman worked to develop strands of DNA that would zip themselves up into more and more complex shapes. Seeman made junctions with five and six arms, then squares, stick-figure cubes comprised of 480 nucleotides, and a truncated octahedron containing 2550 nucleotides and a molecular weight of ~790, 000 daltons. The cubes were synthesised in solution, but Seeman switched to a solid-support-based methodology in 1992, greatly improving control by allowing construction of one edge at a time and isolating the growing objects from one another, allowing massively parallel construction of objects with far greater control of the synthesis sequence. By the mid-1990s, most Platonic (tetrahedron, cube, octahedron, dodecahedron, and icosahedron), Archimedean (e.g., truncated Platonics, semiregular prisms and prismoids, cuboctahedron, etc.), Catalan (linked rings and complex knots), and irregular polyhedra could be constructed as nanoscale DNA stick figures.

Seeman's DNA strands that formed the frame figures were strong enough to serve as girders in a molecular framework, but the junctions were too floppy. In 1993 Seeman discovered the more rigid anti-parallel DNA "double crossover" motif, which in 1996 he used to design and build a stiff double junction to keep his structures from sagging. The next goal was to bring together a large number of stick figures to form large arrays or cage-shaped DNA crystals that could then be used as frameworks for the assembly of other molecules into pre-established patterns. These DNA molecules would serve as the scaffolding upon which new materials having precise molecular structure could be assembled.

In 1998, Erik Winfree and colleagues in Seeman's laboratory reported the design and construction of two-dimensional DNA crystals using self-assembling double crossover molecules. Repeating array units were approximately 2 nm x 4 nm x 16 nm in size, and examination of the array with an Atomic Force Microscope (AFM) revealed domains of up to 500, 000 interconnected units, showing that the self-assembly process could be very reliable under ideal conditions. While the work described the construction of two- and four-unit lattices, the number of component tiles that could be used in the repeat unit did not appear to be limited to such small numbers, suggesting that complex patterns could be assembled into periodic arrays, which might also serve as templates for nanomechanical assembly. Noted the authors: "Because oligonucleotide synthesis can readily incorporate modified bases at arbitrary positions, it should be possible to control the structure within the periodic group by decoration with chemical groups, catalysts, enzymes and other proteins, metallic nanoclusters, conducting silver clusters, DNA enzymes, or other DNA nanostructures such as polyhedra." In these experiments, DNA hairpin turns incorporated within selected units inside their structures were visible as topographic features under AFM imaging, proving the ability to build predictable, atomically precise 2-D crystals with design control over every lattice point. Since the density of lattice points was much sparser than atomic spacings, this DNA self-assembly technology could be paired with atomically precise synthesis of diverse nanoscale subassemblies so that each unique double crossover unit in a pattern could be decorated with a unique subassembly of comparable size, possibly allowing the construction of mechanical nanocomputers and other nanomachinery.

In early 1999, Seeman reported yet another breakthrough — the construction of a mechanical DNA-based device that is a possible prototype for a nanoscale robotic arm. The mechanism has two rigid arms a few nanometers long that can be rotated between fixed positions by introducing a positively charged cobalt compound into the solution surrounding the molecules, causing the bridge region to be converted from the normal B-DNA structure to the unusual Z-DNA structure. The free ends of the arms shift position by ~2.0 nm during the structural conversion. Explained Seeman: "Using synthetic DNA as a building material, we have constructed a controllable molecular mechanical system. In the long-term, the work will have implications for the development of nanoscale robots and for molecular manufacturing."

DNA can also serve as an assembly jig in solution phase. Bruce Smith and colleagues are devising a method for the assembly and covalent linkage of proteins into specific orientations and arrangements as determined by the hybridisation of DNA attached to the proteins, called DNA-Guided Assembly of Proteins (DGAP). In this method, multiple DNA sequences would be attached to specific positions on the surface of each protein, and complementary sequences would bind, forcing protein building blocks (possibly including biomolecular motors, structural protein fibers, antibodies, enzymes, or other existing functional proteins) together in specific desired combinations and configurations, which would then be stabilised by covalent interprotein linkages. This technique could also be applied to nonprotein components that can be functionalised at multiple sites with site-specific DNA sequences, although proteins, at least initially, may be more convenient building blocks due to their size, their surface chemistry, the wide variety of functions and mechanical properties they can confer on the resulting assemblies, and the many existing techniques for introducing designed or artificially evolved modifications into natural proteins of known structure. (In 1998, custom DNA and peptide sequences could be ordered online.) Methods for covalently attaching functional proteins to a DNA backbone in a specified manner at ~8.5 nm (25 base-pair) intervals, addressable protein targeting in macro-molecular assembly, and "protein stitchery" were being explored by others. Drexler notes that evolution has not maximised the stability of natural proteins, and that substantially greater stability may be engineered by various means (e.g., increasing folding stability by >100 Kcal/mole.

Building Blocks

Although it seems at first that Nature has provided a limited number of basic building blocks—amino acids, lipids, and nucleic acids—the chemical diversity of these molecules and the different ways they can be polymerised or assembled provide an enormous range of possible structures. Furthermore, advances in chemical synthesis and biotechnology enable one to combine these building blocks, almost at will, to produce new materials and structures that have not yet been made in Nature. These self-assembled materials often have enhanced properties as well as unique applications. The selected examples below show ways in which clever synthetic methodologies are being harnessed to provide novel biological building blocks for nanotechnology. The protein polymers produced by Tirrell and coworkers (1994) are examples of this new methodology. In one set of experiments, proteins were

Top: a 36-mer protein polymer with the repeat sequence (ulanine-glycine)3 – glutamic acid – glycine. Bottom: idealised folding of this protein polymer, where the glutamic acid sidechains (+) are on the surface of the folds. designed from first principles to have folds in specific locations and surfacereactive groups in other places (Krejchi et al. 1994; 1997). One of the target sequences was -((AG)3EG)- 36. The hypothesis was that the AG regions would form hydrogen-bonded networks of beta sheets and that the glutamic acid would provide a functional group for surface modification. Synthetic DNAs coding for these proteins were produced, inserted into an *E. coli* expression system, and the desired proteins were produced and harvested. These biopolymers formed chain-folded lamellar crystals with the anticipated folds. In addition to serving as a source of totally new materials, this type of research also enables us to test our understanding of amino acid interactions and our ability to predict chain folding. Biopolymers produced via biotechnology are monodisperse; that is, they have precisely defined and controlled chain lengths; on the other hand, it is virtually impossible to produce a monodisperse synthetic polymer. It has recently been shown that polymers with well-defined chain lengths can have unusual liquid crystalline properties.

For example, Yu et al. (1997) have shown that bacterial methods for polymer synthesis can be used to produce poly(gamma-benzyl alpha L-glutamate) that exhibits smectic ordering in solution and in films. The distribution in chain length normally found for synthetic polymers makes it unusual to find them in smectic phases. This work is important in that it suggests that we now have a route to new smectic phases whose layer spacings can be controlled on the scale of tens of nanometers. The biotechnology-based synthetic approaches described above generally require that the final product be made from the natural, or L-amino acids. Progress is now being made so that biological machinery (e.g., *E. coli*), can be co-opted to incorporate non-natural amino acids such as b-alanine or dehydroproline or fluorotyrosine, or ones with alkene or alkyne functionality (Deming et al. 1997). Research along these lines opens new avenues for producing controlled-length polymers with controllable surface properties, as well as biosynthetic polymers that demonstrate electrical phenomena like conductivity. Such molecules could be used in nanotechnology applications. Novel chemical synthesis methods are also being developed to produce "chimeric" molecules that contain organic turn units and hydrogen-bonding networks of amino acids (Winningham and Sogah 1997). Another approach includes incorporating all tools of chemistry into the synthesis of proteins, making it possible to produce, for example, mirror-image proteins. These proteins, by virtue of their D-amino acid composition, resist biodegradation and could have important pharmaceutical applications (Muir et al. 1997). Arnold and coworkers are using a totally different approach to produce proteins with enhanced properties such as catalytic activity or binding affinity. Called "directed evolution, " this method uses random mutagenesis and multiple generations to produce new proteins with enhanced properties. Directed evolution, which involves DNA shuffling, has been used to obtain esterases with five- to six-fold enhanced activity against *p*-nitrobenzyl esters (Moore et al. 1997).

The ability of biological molecules to undergo highly controlled and hierarchical assembly makes them ideal for applications in nanotechnology. The self-assembly hierarchy of biological materials begins with monomer molecules (e.g., nucleotides and nucleosides, amino acids, lipids), which form polymers (e.g., DNA, RNA, proteins, polysaccharides), then assemblies (e.g., membranes, organelles), and finally cells, organs, organisms, and even populations (Rousseau and Jelinski 1991, 571-608). Consequently, biological materials assembly on a very broad range of organisational length scales, and in both hierarchical and nested manners (Aksay et al. 1996; Aksay 1998). Research frontiers that exploit the capacity of biomolecules and cellular systems to undergo self-assembly have been identified in two recent National Research Council reports (NRC 1994 and 1996). Examples of self-assembled systems include monolayers and multilayers, biocompatible layers, decorated membranes, organised structures such as microtubules and biomineralisation, and the intracellular assembly of CdSe semiconductors and chains of magnetite. A number of researchers have been exploiting the predictable basepairing of DNA to build molecular-sized, complex, three-dimensional objects. For example, Seeman and coworkers (Seeman 1998) have been investigating these properties of DNA molecules with the goal of forming complex 2-D and 3-D periodic structures with defined topologies.

DNA is ideal for building molecular nanotechnology objects, as it offers synthetic control, predictability of interactions, and well-controlled "sticky ends" that assemble in highly specific fashion. Furthermore, the existence of stable branched DNA molecules permits complex and interlocking shapes to be formed. Using such technology, a number of topologies have been prepared, including cubes (Chen and Seeman 1991), truncated octahedra (Zhang and Seeman 1994), and Borromean rings (Mao et al. 1997). Other researchers are using the capacity of DNA to self-organise to develop photonic array devices and other molecular photonic components (Sosnowski et al. 1997).

This approach uses DNA-derived structures and a microelectronic template device that produces controlled electric fields. The electric fields regulate transport, hybridisation, and denaturation of oligonucleotides. Because these electric fields direct the assembly and transport of the devices on the template surface, this method offers a versatile way to control assembly. There is a large body of literature on the self-assembly on monolayers of lipid and lipid-like molecules (Allara 1996, 97-102; Bishop and Nuzzo 1996). Devices using self-assembled monolayers are now available for analysing the binding of biological molecules, as well as for spatially tailoring. Idealised truncated octahedron assembled from DNA. This view is down the four-fold axis of the squares. Each edge of the octahedron contains two double-helical turns of DNA surface activity. The technology to make self-assembled monolayers (SAMs) is now so well developed that it should be possible to use them for complex electronic structures and molecular-scale devices.

Research stemming from the study of SAMs (e.g., alkylthiols and other biomembrane mimics on gold) led to the discovery of "stamping" (Kumar and Whitesides 1993). This method, in which an elastomeric stamp is used for rapid pattern transfer, has now been driven to < 50 nanometer scales and extended to nonflat surfaces. It is also called "soft lithography" and offers exciting possibilities for producing devices with unusual shapes or geometries. Self-assembled organic materials such as proteins and/or lipids can be used to form the scaffolding for the deposition of inorganic material to form ceramics such as hydroxyapatite, calcium carbonate, silicon dioxide, and iron oxide. Although the formation of special ceramics is bio-inspired, the organic material need not be of biological origin. An example is production of template-assisted nanostructured ceramic thin films (Aksay et al. 1996). A particularly interesting example of bio-inspired self-assembly has been described in a recent article by Stupp and coworkers (Stupp et al. 1997). This work, in which organic "rod-coil" molecules were induced to selfassemble, is significant in that the molecules orient themselves and selfassemble over a wide range of length scales, including mushroom-shaped clusters; sheets of the clusters packed side-by-side; and thick films, where the sheets pack in a head-to-tail fashion. The interplay between hydrophobic and hydrophilic forces is thought to be partially responsible for the controlled assembly.

Dispersion and Coatings

Nanoparticles and nanospheres have considerable utility as controlled drug delivery systems (Hanes et al. 1997). When suitably encapsulated, a pharmaceutical can be delivered to the appropriate site, its concentration can prevented from undergoing premature degradation. Nanoparticles (as opposed to micron-sized particles) have the advantage that they are small enough that they can be injected into the circulatory system. Highly porous materials are also ideal candidates for controlled drug delivery (Schnur et al. 1994) and for tissue engineering (Hubbell and Langer 1995). An example of a controlled drug delivery system comes from the area of microtubules. Phospholipid bilayers can self-assemble into long cylindrical tubes with diameters usually below a micron and lengths up to hundreds of microns (Schnur 1993). During synthesis, drugs can be entrained original patterned silicon *elastomeric stamp elastomeric stamp* substrate. An elastomeric stamp (top left) is made from an original master (bottom left). The stamp is dipped into the biological material (top right) and the pattern is transferred to the substrate (bottom right). Mushroom-shaped clusters formed from self-assembly of rod-coil molecules; these clusters can undergo further packing to form sheets. in these nanotubes, and the final product can be used as a controlled delivery system. Tubules prepared from phospholipid bilayers are ideal for such applications because they are biocompatible. Dendrimers can be prepared so they are of discrete size and contain specific functional groups (Karak and Maiti 1997). They can be functionalised and used in biomedicine. Examples include gene transfer agents for gene therapy, made to carry and control the relaxivity of paramagnetic MRI (magnetic resonance imaging) contrast agents (Toth et al. 1996), and to deliver drugs on a controlled release basis.

Role of Nanoparticles in Health and Pollution

Although beyond the scope of this review, it is important to keep in mind the potential role of atmospheric nanoparticles in photocatalytic and thermal production of atmospheric pollutants. Atmospheric aerosols in heavily polluted areas have the potential to accelerate ozone formation reactions. Furthermore, because they are respirable, they could represent a health hazard. Two controversial studies (the Harvard University six-city study and the American Cancer Society study) have linked the presence of fine particular matter to premature mortality (*Chemical and*

Engineering News 1997). Atmospheric aerosols generally contain two major components: one is composed of amorphous carbon that has fullerene-like particles dispersed in it; the second is inorganic and consists of oxides and sulfides supported on clay minerals. In particular, the iron oxide, manganese oxide, and iron sulfide nanoparticles have band-gaps that could enhance the photocatalytic adsorption of solar radiation. In addition, these materials are acidic and may be coated with water, which would enhance their catalytic ability to crack hydrocarbons and create free radicals (Chianelli 1998). At present this is an underexplored area of research that bears scrutiny. It is interesting to note that some microorganisms produce and sequester CdSe and CdS nanoparticles in response to high levels of toxic Cd++ in their environment (Brus 1996). A large number of organisms also have the ability to precipitate ferrimagnetic minerals Fe_3O_4 and Fe_3S_4.

High Surface Area Materisl

Supported polymeric membranes can be used to remove low molecular weight organics from aqueous solution (Ho et al. 1996). They work by filling the pores of microfiltration or ultrafiltration membranes with functional, polymeric, or oligomeric liquids that have an affinity for the compound of interest. These ideas can be extended to microemulsions and used for separations of large biological molecules. For example, non-ionic microemulsions (i.e., oil-water systems) have been used for hemoglobin extraction. These results augur well for the separation of other biological materials (Qutubuddin et al. 1994). Molecular imprinting can also be used to produce high surface area adsorbents that are enantioselective for amino acids and other biological molecules (Vidyasankar et al. 1997).

Crystalline bacterial cell surface layers (S-layers) are composed of repeating protein units. These layers self-assemble and have a high binding capacity. They have been explored as patterning elements for molecular nanotechnology. For example, they have been used to pattern cadmium sulfide superlattices (Shenton et al. 1997).

Functional Nanostructures

The smallest possible computer would ideally be able to perform computations on a molecular scale. Even though the computation may be carried out on that scale, the issue becomes one of having enough molecules to obtain sufficient signal-to-noise ratio to read out the answer. Consequently, at least at present, these computations must be carried out in bulk or with extremes in temperature. There has been a recent development along this line. Adleman (1994) has shown how the rules of DNA selfassembly, coupled with polymerase chain reaction (PCR) amplification of DNA, can lead to a molecular computer of sorts. The system was used to solve the Hamiltonian path problem, a classic and difficult (NP complete) mathematical problem that involves finding a path between vertices of a graph. The starting and ending vertices of each edge of the graph were encoded as the first and second halves of a strand of DNA. A solution to the problem (what is the path between two specific vertices?) was obtained by using PCR primers for the two vertices. Others have extended these ideas and shown that it is possible to make DNA add (Guarnieri et al. 1996). Although these are exciting demonstrations, at present this technology has a number of drawbacks, including the labor and time it takes to analyze the results of the computation, and the uncertainties associated with wet chemistry.

Optoelectronic Devices

Biological molecules and assemblies, such as the photochemical reaction center, are capable of capturing light with good quantum efficiency and transforming it into chemical energy. If properly exploited, such assemblies have potential applications as biomolecule information processing units. Bacteriorhodopsin, from the purple membrane bacterium *Halobacterium halobium*, is one such system that has been studied extensively and has been commercialised into optical holographic memories (Birge 1995). In the bacterium, the protein bacteriorhodopsin self-assembles into ordered lipid patches. The protein absorbs light and undergoes a cycle involving a complex series of intermediates, resulting in a proton being pumped across the membrane. It was information developed from understanding the basic science behind the way that bacteriorhodopsin works that led to the use of bacteriorhodopsin as a biomolecule information processing unit. To be used for information storage, the protein is placed under nonbiological conditions. It is dispersed and immobilised into a matrix (e.g., collagen or another polymeric substance) and held at liquid nitrogen temperatures. At this low temperature, the protein acts as an optically-driven bistable switch. One form, the light-adapted form of bacteriorhodopsin, absorbs light at 570 nm. When irradiated with green light at 77 K, it switches to a different stable form that absorbs at 630 nm. Using light of different colors enables one to read and write images onto these memories. By subtracting one memory from the other, these memories are especially useful for realtime, rapid comparison of images.

Molecular Motors

The molecular motors found in biology provide for bacterial locomotion, as well as for the active transport and delivery of molecules. For example, the bacterial flagellar motor is about 20 nm in diameter, and is comprised of a complex assembly of more than 10 different proteins (Imae and Atsumi 1989). The role of the motor is to rotate the helical flagella of the bacterium so that it is able to swim. Chemical energy (in this case protons or sodium) is transduced into mechanical energy. Other examples of molecular motors include RNA polymerase (Yin et al. 1995), F1-ATPase (Noji et al. 1997), myosin, and kinesin (Seventh Biophysical Discussions 1995). The fuel that powers these motors is ATP (adenosine triphosphate). A number of researchers have proposed schemes by which such motors could be used to deliver molecules, one at a time, for the purpose of the ground-up assembly of nanoscale devices (NRC 1996). It is envisioned that the highways could be actin or tubulin, which would need to be immobilised onto a surface. Myosin or kinesin, which naturally travel along these highways, could be used to deliver molecular "packages" to a specific assembly site.

Other Forms of Biological Transport Using Nanoparticles

In the human body, the function of the high density lipoproteins is to transport cholesterol. The ~ 7.5 nm discoidal lipoprotein assemblies are sandwiched between discs of phospholipids and stacked, poker-chip style.

The lipoproteins stabilize the cholesterol particles and assist in their transport. Current research involves manipulating and fusing the assemblies and particles with an atomic force microscope (ATM) tip (Sligar 1998).

Bioelectronics

There has been considerable research activity in molecular electronics and bioelectronics, particularly in Japan (Aizawa 1994). Although this area of nanotechnology is still not as well developed as others in this report, it bears watching. For example, Shionoya and coworkers at the Institute for Molecular Science of the Okazaki National Research Institutes in Japan have proposed that novel combinations of DNA, metal ligands, DNA templating, and proteins could produce molecular wires; molecular hoops through which DNA could be threaded; and doublestranded peptides whose helix pitch could be controlled by an entrained metal. The active site containing the metal could be induced to go from Cu(I)tetrahedral to CU(II)square planar, perhaps by electrons delivered by a scanning tunneling microscope (STM) tip.

Consolidated Materials

Ceramics

Biomolecule/inorganic interactions can be used to produce ceramics with increased toughness. Fundamental studies of biomineralisation, in which an organic substance (usually protein or peptide or lipid) interacts with an inorganic phase (e.g., calcium carbonate or hydroxyapatite) have led to the bioinspired synthesis of composite materials. Novel combinations of DNA, metal ligands, DNA templating, and proteins are being investigated for molecular wires, inductors, and switches (photo courtesy of Shionoya and coworkers, Inst. for Molecular Science).

The structure and porosity of the inorganic phase can be controlled by templating with an organic surfactant, vesicular arrays, or liquid crystalline materials. Micelle-templated synthesis can produce ceramics with 20-100 Å pore dimensions (Ying 1998). These tailored pores can be used as catalysts and absorbents, and for gas/liquid separations and thermal and acoustic insulation. Their selectivity makes them very useful for biochemical and pharmaceutical separations. Bioceramics can also be made that are more compatible with teeth and bone. An interesting example of an organic/inorganic composite is the new packaging material that has been developed to replace the polystyrene "clam-shell" for fast food products. Composed of potato starch and calcium carbonate, this foam combines the advantages of good thermal insulation properties and light weight with biodegradability (Stucky 1978).

Biological Bar Magnets

It has been well documented that a very large number of organisms have the ability to precipitate ferrimagnetic minerals such as Fe3O4 and Fe3S4. In addition, linear chains of membrane-bound crystals of magnetite, called magnetosomes, have been found in microorganisms and fish (Kirschvink et al. 1992). For example, the Fe3O4 domain size in the organism *A. magnetotactum* is about 500 Å, and a chain of 22 of those domains has a magnetic moment of 1.3 x 10-15. It is not immediately clear how these particles could be exploited for nonbiological applications.

Opportunities and Challenges

This section of the review describes current limits on the biological aspects of nanotechnology. It outlines selected broad areas of research opportunity, and it sets forth challenges for the intermediate and long-term future.

Enabling Technologies

We owe much of the recent progress in the biological aspects of nanotechnology and nanoparticles to important enabling technologies. Many of these enabling technologies are described in other parts of this report. Those that have had a significant impact on biological measurements include techniques and instrumentation such as optical traps, laser tweezers, and "nano-pokers" to measure femtoNewton forces (Svoboda and Block 1994), and AFM and STM (Wildoer et al. 1998).

There is also a need to move detection away from the ensemble, to the single molecule scale. This includes driving instrumentation toward singlemolecule spectroscopy; single-spin detection (Rugar et al. 1994); chemical analysis of nanoliter volumes (Hietpas et al. 1996, 139-158); nuclear magnetic resonance microcoils; nanoscale electrode arrays and chemical sensing and detecting technology (McConnell 1996, 97-102); separations technology; chemical analysis of single cells (Hietpas et al. 1996); new biological transformations; and chemical probes of nanostructures. Enhanced computational infrastructure, both hardware and algorithm development, will also be required to investigate supramolecular assemblies and to understand interactions that occur over a wide variety of time and length scales. At present, we are fairly well equipped to handle quantum mechanical calculations. At the next level of computational complexity, the molecular dynamics force fields and atomic charges can presently handle up to about a million particles and still retain their accuracy. The next scale of complexity, mesoscale modeling, currently requires the use of pseudo atoms. Much more research is required to improve the accuracy of finite element calculations and the modeling of materials applications (Goddard 1998).

Surface Interactions and the Interface Between Biomolecules and Substrates

Much of the progress in the biological aspects of nanotechnology has come from research on surface interactions. It will be important to continue to develop a better understanding of the interface between biomolecules and surfaces. Of particular interest are questions about whether surfaces cause biomolecules to denature, the optimum length of linker groups, and ways to communicate from the biomolecule, through the linker, to the substrate. Another challenge is to produce nanometer ultrathin films that have stable order (Jaworek et al. 1998).

Robustness of Biomolecules and Their Interactions in Aqueous Solution

Although much progress has been made on ways to confer additional robustness on biomolecules, it will be necessary to continue to improve the stability and reliability of biomolecular assemblies. Such research may be along the lines of understanding the thermal stability of biomolecules, perhaps by examining extremophiles and molecules produced by directed evolution. It will also be necessary to produce molecules that can work in the absence of water, or to devise ways in which aqueous solutions can be used reliably.

Assemblers or Templating

The hierarchical self-assembly of biomolecules is often touted as one of their key attributes for nanotechnology applications. However, there remains much to be learned about how to make periodic arrays of biomolecular assemblies, how to use biological templating in an efficient manner, how to mimic biological self-assembly with nonbiological molecules, and how to exploit differences between biological and nonbiological self-assembly.

Combinatorial Approaches to Nanotechnology

Combinatorial chemistry has had an enormous impact on drug discovery and development in the past several years. Ability to develop molecularly sensitive screening techniques is one of the limiting factors of combinatorial chemistry. It would be intriguing to discover screening methods that are particularly sensitive for picking up new phenomena exhibited by biological nanoparticles.

Biomimetics and Polymer-Biopolymer Chimeras

The largest impact of biological sciences on nanotechnology may prove to be through the lessons one can learn from biology, rather than from directly employing the biological molecules themselves. To this end, it is important to support research on various aspects of biomimetics, including activities in which synthetic and biologically-based molecules are combined in chimeric fashion.

Nanotechnology for Bio Materials and Tissues

Technology Focus

Historically, synthetic materials have not served as sufficient implants.For example, the current average lifetime of an orthopedic implant, such as hip, knee, ankle, etc., is only 15 years. Similarly, for the vascular community, small diameter vascular grafts are only functional 25% of time past 5 years of use. The same limited lifetimes for a number of failing organs could be stated for cartilage, bladder, and the peripheral and central nervous systems. Clearly, conventional materials, or those materials with constituent dimensions greater than 1 micron have not invoked proper cellular responses to regenerate tissue that allows for these devices to be successful for long periods of times.

In contrast, due to their ability to mimic the dimensions of constituent components of natural tissues, like proteins, nanophase materials may be an exciting successful alternative. Nanophase materials are defined as materials with constituent dimension less than 100 nm in at least one direction. Materials investigated to date include nanophase ceramics, metals, polymers, and composites. Nanophase materials have revolutionised many traditional engineering fields such as those involving catalysis, electronics, optics, magnetics, etc. For example, in the catalytic field, higher surface reactivity of nanophase materials allows for faster reaction times. To date, there has also been the emergence of data that nanophase materials may be optimal materials for tissue engineering applications. This is not only due to their ability to simulate dimensions of proteins that comprise tissues, but also because of their higher reactivity for interactions of proteins that control cell adhesion and, thus, the ability to regenerate tissues. There are numerous reasons to consider using nanophase materials as better tissue engineering materials.While most of the promise is conjecture, limited scientific evidence does exist and is generating wide support.

Course Objectives

The first objective of this course is to define current problems with implants. You will be familiar with issues related to implant failure particular for orthopedic, dental, cartilage, vascular, bladder and the nervous systems. The second objective is to understand how cells interact with implanted materials to generate a biological response.We will describe how proteins interact with materials to control cell function leading to either a desirable or undesirable biological response to an implanted material. The third objective is to understand how nanophase materials could be used as better implants. What are their promising properties and what experimental evidence exists? We will sort through the experimental data and separate promise from "hype". The discussion will make you be able to indicate what are the promising properties of nanophase materials and how to make them to control initial protein interactions leading to successful regeneration of tissue.

The last objective is to understand what further information will be needed to elucidate if nanophase materials will make better biomaterials for these organ systems. While there is no golden standard chemistry for various implant applications, this course will discuss the possibility of a standard topography that may be useful to regenerate tissue for a wide range of materials including ceramics, metals, and polymers. You will be able to identify future experiments needed for the further consideration of nanophase materials to regenerate tissue

Course Content

This course will begin with a general overview of current problems with orthopedic, dental, cartilage, vascular, bladder, and central as well as peripheral nervous system devices. These organs will be chosen since they are the ones in which the most implantable devices have been used to date. One day will be spent describing situations in which these devices are needed and why we have not been able to develop sufficient materials to successfully regenerate these tissues. To do this, we will cover the basic anatomy, physiology, and cellular structures of these organs. This will help us understand the complexity of our organs and consequently to design materials that match the most important properties of tissue to promote tissue regeneration.

Next, we will cover how the body responds to implanted materials by first adsorbing a thin layer of proteins and then through select cellular responses. Topics such as how proteins adsorb to surfaces, how cells recognize proteins, how cells adhere and function on materials, and how tissues may or may not regrow will be the focus of this day.The biological response section will be complemented with results from experiments that have delineated fundamental material properties that control such responses. By understanding how the body originally responds to implanted materials, we can begin to develop materials to obtain a desirable response, such a tissue regrowth. We will then interject a new classification of materials that are beginning to be considered to serve as better biomaterials: nanophase materials. One day will be spent defining these materials and highlighting their special properties pertinent for implants. This includes their mechanical, optical, electrical, and surface properties. We will also discuss various routes to synthesize nanophase materials. Although, to date, the special properties of nanophase materials have been used in non-biological applications, we will emphasize their role in biological applications. We will constantly compare theory will experimental evidence in this section. The day will end with the most promising results obtained thus far in terms of increased tissue regeneration on nanophase materials.

Lastly, we will discuss the most promising future directions for the use of nanophase materials in biological applications. This day will focus on key unanswered questions if nanophase materials are to be used in the body. Central to this question is the immunological response to nanoparticles, which will comprise a majority of the day. To enhance learning as well as stimulate discussion, the course will be interactive with participation from the learners throughout.

Current Implant Failures

Orthopedic

- Current Materials Used
- Current Modes of Failure: Poor Bonding to Surrounding Bone – Generation of Wear Debris – Stress and Strain Imbalances at the Tissue Implant Interface
- Current Statistics on Orthopedic Implant Failure

Dental

- Current Materials Used
- Current Modes of Failure: Poor Bonding to Surrounding Bone – Generation of Wear Debris – Stress and Strain Imbalances at the Tissue Implant Interface
- Current Statistics on Dental Implant Failure

Cartilage

- Current Treatment Methods and Materials Used
- Current Modes of Failure: Poor Regeneration of Cartilage Tissue – Generation of Wear Debris – Stress and Strain Imbalances at the Tissue Implant Interface
- Current Statistics on Cartilage Implant Failure

Vascular

- Current Materials Used
- Current Modes of Failure: Poor Bonding to Surrounding Vascular – Endothelium Removal Due to Shear Stress – Generation of Stenosis - Stress and Strain Imbalances at the Tissue Implant Interface
- Current Statistics on Vascular Implant Failure

Bladder

- Current Materials Used
- Current Modes of Failure: Poor Bonding to Surrounding Tissue – Stress and Strain Imbalances at the Tissue Implant Interface
- Current Statistics on Bladder Implant Failure

Central and Peripheral Nervous System

- Current Materials Used
- Current Modes of Failure: Poor Bonding to Surrounding Tissue – Build-up of Glial Scar Tissue – Poor Maintenance of Electrical Properties – Stress and Strain Imbalances at the Tissue Implant Interface
- Current Statistics on Central and Peripheral Nervous System Implant Failure

Biological Response of Implanted Materials

Protein Interactions with Implanted Materials

- Properties of Proteins
- Adsorption
- Conformation/Bioactivity

Cellular Recognition of Proteins Adsorbed on Materials Surfaces

- Adhesion
- Migration
- Differentiation

Cellular Extracellular Matrix Deposition Leading to Tissue Regeneration

- Bone
- Dental Tissue
- Cartilage
- Vascular
- Bladder
- Central and Peripheral Nervous System.

7

Nano Chemistry, Nanoworld Particles and Molecules: Functional Relationship Analysis

Nanoworld Particles and Molecules

Nanotechnology will be a bottom-up technology, building upward from the molecular scale. It will bring a revolution in human abilities like that brought by agriculture or power machinery. It can even be used to reverse many of the changes brought by agriculture or power machinery. But we humans are huge creations with no direct experience of the molecular world, and this can make nanotechnology hard to visualize, hence hard to understand. Scientists working with moleculas face this problem today. They can often calculate how molecules will behave, but to understand this behaviour, they need more than heaps of numbers: they need pictures, movies, and interactive simulations, and so they are producing them at an ever-increasing pace. Molecules are objects that exert forces on one another. If your hands were small enough, you could grab them, squeeze them, and bash them together. Understanding the molecular world is much like understanding any other physical world: it is a matter of understanding size, shape, strength, force, motion, and the like—a matter of understanding the differences between sand, water, and rock, or between steel and soap bubbles. Today's visualisation tools give a taste of what will become possible with tomorrow's faster computers and better "virtual realities, " simulated environments that let you tour a world that "exists" only as a model inside the computer. Before discussing nanotechnology and how it relates to the technologies of today, let's try to get a more concrete understanding of the molecular world by describing a simulation embedded in a scenario. In this scenario, events and technologies described as dating from 1990 or before are historically accurate; those with later dates are either projections or mere scenario elements. The descriptive details in the simulation are written to fit designs and calculations based on standard scientific data, so the science isn't fiction.

We saw Joel Gregory manipulating molecules in the virtual reality of a simulated world using video goggles, tactile gloves, and a supercomputer. The early twenty-first century should be able to do even better. Imagine, then, that today you were to take a really long nap, oversleep, and wake up decades later in a nanotechnological world. In the twenty-first century, even more than in the twentieth, it's easy to make things work without understanding them, but to a newcomer much of the technology seems like magic, which is dissatisfying. After a few days, you want to understand what nanotechnology is, on a gut level. Back in the late twentieth century, most teaching used dry words and simple pictures, but now—for a topic like this—it's easier to explore a simulated world. And so you decide to explore a simulation of the molecular world. Looking through the brochure, you read many tedious facts about the simulation: how accurate it is in describing sizes, forces, motions, and the like; how similar it is to working tools used by both engineering students and professionals; how you can buy one for your very own home, and so forth. It explains how you can tour the human body, see state-of-the-art nanotechnology in action, climb a bacterium, etc. For starters, you decide to take an introductory tour: simulations of real twentieth-century objects alongside quaint twentieth-century concepts of nanotechnology. After paying a small fee and memorising a few key phrases (any variation of "Get me out of here!" will do the most important job), you pull on a powersuit, pocket a Talking Tourguide, step into the simulation chamber, and strap the video goggles over your eyes. Looking through the goggles, you seem to be in a room with a table you know isn't really there and walls that seem too far away to fit in the simulation chamber. But trickery with a treadmill floor makes the walk to the walls seem far enough, and when you walk back and thump the table, it feels solid because the powersuit stops your hand sharply at just the right place. You can even feel the texture of the carvings on the table leg, because the suit's gloves press against your fingertips in the right patterns as you move. The simulation isn't perfect, but it's easy to ignore the defects. On the table is (or seems to be) an old 1990s silicon computer chip. When you pick it up, as the beginners' instructions suggest, it looks like

You feel as though you're falling toward the chip's surface, shrinking rapidly. In a moment, it looks roughly like With your thumb still there holding it. The world grows blurrier, then everything seems to go wrong as you approach the molecular level. First, your vision blurs to uselessness—there is light, but it becomes a featureless fog. Your skin is tickled by small impacts, then battered by what feel like hard-thrown marbles. Your arms and legs feel as though they are caught in turbulence, pulling to and fro, harder and harder. The ground hits your feet, you stumble and stick to the ground like a fly on flypaper, battered so hard that it almost hurts. You asked for realism, and only the built-in safety limits in the suit keep the simulated thermal motions of air molecules and of your own arms from beating you senseless. "Stop!" gives you a rest from the suit's yanking and thumping, and "Standard settings!" makes the world around you become more reasonable. The simulation changes, introducing the standard cheats. Your simulated eyes are now smaller than a light wave, making focus impossible, but the goggles snap your vision into sharpness and show the atoms around you as small spheres. (Real nanomachines are as blind as you were a moment ago, and can't cheat.) You are on the surface of the computer chip, between a cell and two blocky nanocomputers like the ones. The simulation keeps bombarding you with air molecules, but the standard settings leave out the sensation of being pelted with marbles. A moment ago you were stuck tight to the ground by molecular stickiness, but the standard settings give your muscles the effective strength of steel—at least in simulation—by making everything around you much softer and weaker. The tourguide says that the only unreal features of the simulation have to do with you—not just your ability to see and to ignore thermal shaking and bombardment, but also your sheer existence at a size too small for anything so complex as a human being. It also explains why you can see things move, something about slowing down everything around you by a factor of 10 for every factor of 10 enlargement, and by another factor to allow for your being made stronger and hence faster. And so, with your greater strength and some adjustments to make your arms, legs, and torso less sticky, you can stand, see, feel, and take stock of the situation.

The ground underfoot, like everything around you, is pebbly with atom-sized bumps the size of your fingertips. Objects look like bunches of transparent grapes or fused marbles in a variety of pretty but imaginary colors. The simulation displays a view of atoms and molecules much like those used by chemists in the 1980s, but with a sharper 3-D image and a better way to move them and to feel the forces they exert. Actually, the whole simulation setup is nothing but an improved version of systems built in the late 1980s—the computer is faster, but it is calculating the same things. The video goggles are better and the whole-body powersuit is a major change, but even in the 1980s there were 3-D displays for molecules and crude devices that gave a sense of touching them. The gloves on this suit give the sensation of touching whatever the computer simulates. When you run a fingertip over the side of the smaller nanocomputer, it feels odd, hard to describe. It is as if the surface were magnetic–it pulls on your fingertip if you move close enough. But the result isn't a sharp click of contact, because the surface isn't hard like a magnet, but strangely soft. Touching the surface is like touching a film of fog that grades smoothly into foam rubber, then hard rubber, then steel, all within the thickness of a sheet of corrugated cardboard. Moving sideways, your fingertip feels no texture, no friction, just smooth bumps more slippery than oil, and a tendency to get pulled into hollows. Pulling free of the surface takes a firm tug. The simulation makes your atom-sized fingertips feel the same forces that an atom would. It is strange how slippery the surface is—and it can't have been lubricated, since even a single oil molecule would be a lump the size of your thumb. This slipperiness makes it obvious how nano-scale bearings can work, how the parts of molecular machines can slide smoothly. But on top of this, there is a tingling feeling in your fingers, like the sensation of touching a working loudspeaker. When you put your ear against the wall of the nanocomputer, you flinch back: for a moment, you heard a sound like the hiss of a twentieth century television tuned to a channel with no broadcast, with nothing but snow and static—but loud, painfully loud. All the atoms in the surface are vibrating at high frequencies, too fast to see. This is thermal vibration, and it's obvious why it's also called thermal noise.

Individual molecules still move too quickly to see. So, to add one more cheat to the simulation, you issue the command "Whoa!", and everything around seems to slow down by a factor of ten. On the surface, you now can see thermal vibrations that had been too quick to follow. All around, air molecules become easier to watch. They whiz about as thick as raindrops in a storm, but they are the size of marbles and bounce in all directions. They're also sticky in a magnetlike way, and some are skidding around on the wall of the nanocomputer. When you grab one, it slips away. Most are like two fused spheres, but you spot one that is perfectly round—it is an argon atom, and these are fairly rare. With a firm grip on all sides to keep it from shooting away like a watermelon seed, you pinch it between your steel-strong fingers. It compresses by about 10 percent before the resistance is more than you can overcome. It springs back perfectly and instantly when you relax, then bounces free of your grip. Atoms have an unfamiliar perfection about them, resilient and unchanging, and they surround you in thick swarms. At the base of the wall is a churning blob that can only be a droplet of water. Scooping up a handful for a closer look yields a swarm of molecules, hundreds, all tumbling and bumbling over one another, but clinging in a coherent mass. As you watch, though, one breaks free of the liquid and flies off into the freer chaos of the surrounding air: the water is evaporating. Some slide up your arm and lodge in the armpit, but eventually skitter away. Getting rid of all the water molecules takes too much scraping, so you command "Clean me!" to dry off.

Beside you, the smaller nanocomputer is a block twice your height, but it's easy to climb up onto it as the tourguide suggests. Gravity is less important on a small-scale: even a fly can defy gravity to walk on a ceiling, and an ant can lift what would be a truck to us. At a simulated size of fifty nanometers, gravity counts for nothing. Materials keep their strength, and are just as hard to bend or break, but the weight of an object becomes negligible. Even without the strength-enhancement that lets you overcome molecular stickiness, you could lift an object with 40 million times your mass–like a person of normal size lifting a box containing a half-dozen fully loaded oil tankers. To simulate this weak gravity, the powersuit cradles your body's weight, making you feel as if you were floating. This is almost like a vacation in an orbital theme park, walking with stickyboots on walls, ceilings, and whatnot, but with no need for antinausea medication. This looks like a cluster of grapes and is about the same size. It even feels a bit like a bunch of grapes, soft and loose. The parts don't fly free like a gas or tumble and wander like a liquid, but they do quiver like gelatin and sometimes flop or twist. It is solid enough, but the folded structure is not as strong as your steel fingers. In the 1990s, people began to build molecular machinery out of proteins, copying biology. It worked, but it's easy to see why they moved on to better materials. From a simulated pocket, you pull out a simulated magnifying glass and look at the simulated protein. This shows a pair of bonded atoms on the surface at 10 times magnification. The atoms are almost transparent, but even a close look doesn't reveal a nucleus inside, because it's too small to see. It would take 1, 000 times magnification to be able to see it, even with the head start of being able to see atoms with your naked eye. How could people ever confuse big, plump atoms with tiny specks like nuclei? Remembering how your steel-strong fingers couldn't press more than a fraction of the way toward the nucleus of an argon atom from the air, it's clear why nuclear fusion is so difficult. In fact, the tourguide said that it would take a real-world projectile over a hundred times faster than a high-powered rifle bullet to penetrate into the atomic core and let two nuclei fuse. Try as you might, there just isn't anything you could find in the molecular world that could reach into the middle of an atom to meddle with its nucleus. You can't touch it and you can't see it, so you stop squinting though the magnifying glass. Nuclei just aren't of much interest in nanotechnology.

Taking the advice of the tourguide, you grab two molecular knobs on the protein and pull. It resists for a moment, but then a loop comes free, letting other loops flop around more, and the whole structure seems to melt into a writhing coil. After a bit of pulling and wrestling, the protein's structure becomes obvious: It is a long chain–longer than you are tall, if you could get it straight—and each segment of the chain has one of several kinds of knobs sticking off to the side. With the multicolored, glassy-bead portrayal of atoms, the protein chain resembles a flamboyant necklace. This may be decorative, but how does it all go back together? The chain flops and twists and thrashes, and you pull and push and twist, but the original tight, solid packing is lost. There are more ways to go wrong in folding up the chain than there are in solving Rubik's Cube, and now that the folded structure is gone, it isn't even clear what the result should look like. How did those twentieth-century researchers ever solve the notorious "protein folding problem"? It's a matter of record that they started building protein objects in the late 1980s. This protein molecule won't go back together, so you try to break it. A firm grip and a powerful yank straightens a section a bit, but the chain holds together and snaps back. Though unfolding it was easy, even muscles with the strength of steel—the strength of Superman—can't break the chain itself. Chemical bonds are amazingly strong, so it's time to cheat again. When you say, "Flimsy world–one second!" while pulling, your hands easily move apart, splitting the chain in two before its strength returns to normal. You've forced a chemical change, but there must be easier ways since chemists do their work without tiny superhands. While you compare the broken ends, they thrash around and bump together. The third time this happens, the chain rejoins, as strong as before. This is like having snap-together parts, but the snaps are far stronger than welded steel. Modern assembler chemistry usually uses other approaches, but seeing this happen makes the idea of molecular assembly more understandable: Put the right pieces together in the right positions, and they snap together to make a bigger structure. Remembering the "Whoa!" command, you decide to go back to the properly scaled speed for your size and strength. Saying "Standard settings!, " you see the thrashing of the protein chain speed up to hard-to-follow blur.

At your feet is a ribbed, ringed cylindrical object about the size of a soup can—not a messy, loosely folded strand like the protein (before it fell apart), but a solid piece of modern nanotechnology. It's a gear like the one. Picking it up, you can immediately feel how different it is from a protein. In the gear, everything is held in place by bonds as strong as those that strung together the beads of the protein chain. It can't unfold, and you'd have to cheat again to break its perfect symmetry. Like those in the wall of the nanocomputer, its solidly attached atoms vibrate only slightly. There's another gear nearby, so you fit them together and make the atomic teeth mesh, with bumps on one fitting into hollows on the other. They stick together, and the soft, slick atomic surfaces let them roll smoothly. Underfoot is the nanocomputer itself, a huge mechanism built in the same rigid style. Climbing down from it, you can see through the transparent layers of the wall to watch the inner works. An electric motor an arm-span wide spins inside, turning a crank that drives a set of oscillating rods, which in turn drive smaller rods. This doesn't look like a computer; it looks more like an engineer's fantasy from the nineteenth century. But then, it is an antique design–the tourguide said that the original proposal was a piece of exploratory engineering dating from the mid-1980s, a mechanical design that was superseded by improved electronic designs before anyone had the tools to build even a prototype. This simulation is based on a version built by a hobbyist many years later.

The mechanical nanocomputer may be crude, but it does work, and it's a lot smaller and more efficient than the electronic computers of the early 1990s. It's even somewhat faster. The rods slide back and forth in a blur of motion, blocking and unblocking each other in changing patterns, weaving patterns of logic. This nanocomputer is a stripped-down model with almost no memory, useless by itself. Looking beyond it, you see the other block–which contains a machine powerful enough to compete with most computers built in 1990. This computer is a millionth of a meter on a side, but from where you stand, it looks like a blocky building looming over ten stories tall. The tourguide says that it contains over 100 billion atoms and stores as much data as a room full of books. You can see some of the storage system inside: row upon row cf racks containing spools of molecular tape somewhat like the protein chain, but with simple bumps representing the 1s and 0s of computer data. These nanocomputers seem big and crude, but the ground you're now standing on is also a computer–a single chip roughly as powerful as the smaller, stripped-down nanocomputer at your side. As you gaze out over the chip, you get a better sense for just how crude things were a few decades ago. At your feet, on the smallest scale, the chip is an irregular mess. Although the wall of the nanocomputer is pebbly with atomic-scale bumps, the bumps are as regular as tile.

The chip's surface, though, is a jumble of lumps and mounds. This pattern spreads for dozens of paces in all directions, ending in an irregular cliff marking the edge of a single transistor. Beyond, you can see other ridges and plateaus stretching off to the horizon. These form grand, regular patterns, the circuits of the computer. The horizon–the edge of the chip–is so distant that walking there from the center would (as the tourguide warns) take days. And these vast pieces of landscaping were considered twentieth-century miracles of miniaturisation?

Even back then, research in molecular biology had revealed the existence of smaller, more perfect machines such as the protein molecules in cells. A simulated human cell–put here because earlier visitors wanted to see the size comparisons–its on the chip next to the smaller nanocomputer. The tourguide points out that the simulation cheats a bit at this point, making the cell act as though it were in a watery environment instead of air. The cell dwarfs the nanocomputer, sprawling across the chip surface and rearing into the sky like a small mountain. Walking the nature trail around its edge would lead across many transistor-plateaus and take about an hour. A glance is enough to show how different it is from a nanocomputer or a gear: it looks organic, it bulges and curves like a blob of liver, but its surface is shaggy with waving molecular chains. Walking up to its edge, you can see that the membrane wrapping the cell is fluid (cell walls are for stiff things like plants), and the membrane molecules are in constant motion. On an impulse, you thrust your arm through the membrane and poke around inside. You can feel many proteins bumping and tumbling around in the cell's interior fluid, and a crisscrossing network of protein cables and beams. Somewhere inside are the molecular machines that made all these proteins, but such bits of machinery are embedded in a roiling, organic mass. When you pull your arm out, the membrane flows closed behind. The fluid, dynamic structure of the cell is largely self healing. That's what let scientists perform experimental surgery on cells with the old, crude tools of the 20th century: they didn't need to stitch up the holes they made when they poked around inside. Even a single human cell is huge and complex. No real thinking being could be as small as you are in the simulation: A simple computer without any memory is twice your height, and the larger nanocomputer, the size of an apartment complex, is no smarter than one of the submoronic computers of 1990. Not even a bendable finger could be as small as your simulated fingers: in the simulation, your fingers are only one atom wide, leaving no room for the slimmest possible tendon, to say nothing of nerves. For a last look at the organic world, you gaze out past the horizon and see the image of your own, full-sized thumb holding the chip on which you stand. The bulge of your thumb rises ten times higher than Mount Everest. Above, filling the sky, is a face looming like the Earth seen from orbit, gazing down. It is your own face, with cheeks the size of continents. The eyes are motionless. Thinking of the tourguide's data, you remember: The simulation uses the standard mechanical scaling rules, so being 40 million times smaller has made you 40 million times faster. To let you pull free of surfaces, it increased your strength by more than a factor of 100, which increased your speed by more than a factor of 10. So one second in the ordinary world corresponds to over 400 million here in the simulation. It would take years to see that huge face in the sky complete a single eyeblink. Enough. At the command "Get me out!", the molecular world vanishes, and your feeling of weight returns as the suit goes slack. You strip off the video goggles—and hugely, slowly, blink.

Nucleic Acids and Proteins

Perhaps the most specific biomolecular self-assembly occurs when two single strands of Deoxyribonucleic Acid (DNA) assemble into the infamous double helix. Based on Watson and Crick base pairing, two complimentary strands of DNA will find each other within an enormously complex mixture. Double helical DNA strands can be separated from one another through simple heating, resulting in single stranded DNA (ssDNA). If the solution is allowed to cool, the two strands will re-assemble into a stable double helix. While the Watson and Crick base pairing then allows the specific assembly of the double helix, which is stabilised by relatively strong *hydrophobic* interactions between the aromatic bases within the center of the helix also known as *pi-stacking*.

Nanoscientists can take advantage of the double helix's specific stability in numerous ways. One promising approach is to attach single-stranded DNA to small particles that one wishes to assemble. These components could range from diamondoid manifolds to metallic clusters that function as quantum dots. When DNA is used to create functional surfaces, the technique is called *DNA Directed Immobilisation (DDI)*. Such patterned surfaces have applications for both nanoelectronic devices and biosensors. *Nanogen*, a biosensor manufacturer has a versatile method for that takes advantage of DNA's negative charge. By applying a voltage to various parts of a DNA containing solution, the concentrations of DNA over various parts surfaces can be regulated.

As the primary components of the molecular machinery that makes life possible, proteins have a vast array of resources available to them for biomolecular self-assembly. Take for instance the eukaryotic nucleus of a multicellular organism (such as a human). The organisation of the information contained within this tiny compartment is so vast that scientists are only now beginning to attempt to understand it. Containing the compartment is the nuclear membrane, which is filled with proteins that pump nutrients in and messages out. Entwined within what is called the "nuclear matrix" is the entire human genome, containing about 3 meters worth of DNA, packed into an area of only a few cubic micrometers. Keeping it all under control are then *histones* and thousands of other (non-histone) proteins that carry out processes such as *transcription* and *translation*. Each of these proteins has the appropriate information encoded within its peptide sequence for getting it to the location where it can do its job. In order for each of our cells to produce those proteins required for its specific task, only certain genes can be transcribed in certain cells. Thus, a very important amount of information is contained not only in the sequence of the DNA, but in its spatial organisation within the nucleus. For instance, when the nucleus divides, all the contents must be unraveled, copied, and then reassembled back into two daughters of the original organised state. All this is accomplished by means of biomolecular self-assembly.

Structural proteins are generally called "fibrous" because they often have a fiber like nature. With names such as collagen, keratin, elastin, tubulin and fibroin, fibrous proteins self-assemble (often covalently) into large polymers that provide various nano- meso- and even macro-scale structures. Even our most modern synthetic polymers can not compete with fibrous proteins such as fibroin (silk) when it comes to strength, flexibility and overall molecular scale order. Thus, serious efforts are underway to better understand and take advantage of fibrous self-assembly.

Enzymes (biological catalysts) are generally "globular" as opposed to "fibrous" and they often work together in the form of multi-enzyme complexes. This means that both intramolecular self-assembly and intermolecular self-assembly must take place in order for the enzyme to reach the conformation that allows it to do its job. Intramolecular self-assembly means that the newly synthesised protein chain must fold into a functional state. Most often this is accomplished by the aggregation of hydrophobic residues in the center of the protein, thus forcing hydrophilic residues to the surface of the protein. Once a single protein chain is adequately folded, it will often begin intermolecular self-assembly. In order to assemble a molecular machine known as a "multienzyme complex," the numerous subunits (or proteins) contact one another in ways that cause new surfaces to appear, thus allowing further subunit components to bind. Since all the information for this self-assembly is often contained within the protein sequence, it is possible to reproduce this assembly in a test tube. For instance, the molecular machine that builds proteins in E. Coli (the ribosome) can be taken apart into 58 isolated protein chains (as well as a few RNA molecules. If these components are mixed back together with the appropriate timing then functional ribosomes can be reassembled.

Membrane bound proteins *are incorporated into the cellular membranes that surround cells and their compartments. In order for these proteins to perform their numerous functions (signal transduction, ion balance maintenance, molecular transport, electron transport, structure, etc.), they must be appropriately oriented and organised within the membrane. Determining their structure within the membrane is not an easy task for biochemists, since the membranes are often destroyed during purification. However, membrane bound proteins often have a common structural characteristic that suggests a mechanism for their assembly. That is, several hydrophobic residues exposed on a certain area of the protein. Since the center of the membrane is hydrophobic, it can be expected that these hydrophobic residues stick to the inside of the membrane. Again, the hydrophobic effect is most effective for this purpose. Finally, once the molecule is incorporated into the membrane, changes in the membranes constitution as well as interactions with neighboring membrane bound proteins govern the organisation of these proteins.*

Bilayer Lipid Membranes and Existing Protein Machines and Design

Cellular membranes define the boundries between various cellular compartments (such as the nucleus, mitochondria, endoplasmic reticulum, etc.) as well as enclosing the cell from the external environment. Similar to soap bubbles, cellular membranes are made of specific types of *surfactants* that tend to form bilayers. The primary surfactants are phospholipids, steroids (primarily cholesterol), and glycolipids. Two nanoscale bodies of water can be effectively separated from one by the formation of a bilayer of these molecules. The main barrier is the hydrophobic region in the center of a lipid bilayer membrane. Water and polar molecules dissolved in water can not pass through such a nonpolar barrier without the assistance of ion channels or molecular pumps.

These protein hormones and enzymes selectively stick to other molecules. An enzyme changes its target's structure, then moves on; a hormone affects its target's behaviour only so long as both remain stuck together. Enzymes and hormones can be described in mechanical terms, but their behaviour is more often described in chemical terms.

But *other proteins serve basic mechanical functions*. Some push and pull, some act as cords or struts, and parts of some molecules make excellent bearings. The machinery of muscle, for instance, has gangs of proteins that reach, grab a "rope" (also made of protein), pull it, then reach out again for a fresh grip; whenever you move, you use these machines. Amoebas and human cells move and change shape by using fibers and rods that act as molecular muscles and bones. A reversible, variable-speed motor drives bacteria through water by turning a corkscrew-shaped propeller. If a hobbyist could build tiny cars around such motors, several billions of billions would fit in a pocket, and 150-lane freeways could be built through your finest *capillaries*.

Simple molecular devices combine to form systems resembling industrial machines. In the 1950s engineers developed machine tools that cut metal under the control of a punched paper tape. A century and a half earlier, Joseph-Marie Jacquard had built a loom that wove complex patterns under the control of a chain of punched cards. Yet over three billion years before Jacquard, cells had developed the machinery of the *ribosome*. Ribosomes are proof that nanomachines built of protein and RNA can be programmed to build complex molecules.

Then consider viruses. One kind, the *T4 phage*, acts like a spring-loaded syringe and looks like something out of an industrial parts catalog. It can stick to a bacterium, punch a hole, and inject viral DNA (yes, even bacteria suffer infections). Like a conqueror seizing factories to build more tanks, this DNA then directs the cell's machines to build more viral DNA and syringes. Like all organisms, these viruses exist because they are fairly stable and are good at getting copies of themselves made.

Whether in cells or not, nanomachines obey the universal laws of nature. Ordinary chemical bonds hold their atoms together, and ordinary chemical reactions (guided by other nanomachines) assemble them. Protein molecules can even join to form machines without special help, driven only by thermal agitation and chemical forces. By mixing viral proteins (and the DNA they serve) in a test tube, molecular biologists have assembled working T_4 viruses. This ability is surprising: imagine putting automotive parts in a large box, shaking it, and finding an assembled car when you look inside! Yet the T_4 *virus* is but one of many *self-assembling structures*. Molecular biologists have taken the machinery of the ribosome apart into over fifty separate protein and RNA molecules, and then combined them in test tubes to form working ribosomes again.

To see how this happens, imagine different T_4 protein chains floating around in water. Each kind folds up to form a lump with distinctive bumps and hollows, covered by distinctive patterns of oiliness, wetness, and electric charge. Picture them wandering and tumbling, jostled by the thermal vibrations of the surrounding water molecules. From time to time two bounce together, then bounce apart. Sometimes, though, two bounce together and fit, bumps in hollows, with sticky patches matching; they then pull together and stick. In this way protein adds to protein to make sections of the virus, and sections assemble to form the whole.

Protein engineers will not need nanoarms and nanohands to assemble complex nanomachines. Still, tiny manipulators will be useful and they will be built. Just as today's engineers build machinery as complex as player pianos and robot arms from ordinary motors, bearings, and moving parts, so tomorrow's biochemists will be able to use protein molecules as motors, bearings, and moving parts to build robot arms which will themselves be able to handle individual molecules.

Designing With Protein and Disassemblers

How far off is such an ability? Steps have been taken, but much work remains to be done. Biochemists have already mapped the structures of many proteins. With gene machines to help write DNA tapes, they can direct cells to build *any protein they can design*. But they still don't know how to design chains that will fold up to make proteins of the right shape and function. The forces that fold proteins are weak, and the number of plausible ways a protein might fold is astronomical, so designing a large protein from scratch isn't easy.

The forces that stick proteins together to form complex machines are the same ones that fold the protein chains in the first place. The differing shapes and kinds of stickiness of *amino acids*—the lumpy molecular "beads" forming protein chains—make each protein chain fold up in a specific way to form an object of a particular shape. Biochemists have learned rules that suggest how an amino acid chain might fold, but the rules aren't very firm. Trying to predict how a chain will fold is like trying to work a jigsaw puzzle, but a puzzle with no pattern printed on its pieces to show when the fit is correct, and with pieces that seem to fit together about as well (or as badly) in many different ways, all but one of them wrong. False starts could consume many lifetimes, and a correct answer might not even be recognised. Biochemists using the best computer programmes now available still cannot predict how a long, natural protein chain will actually fold, and some of them have despaired of designing protein molecules soon.

Yet most biochemists work as scientists, not as engineers. They work at predicting how natural proteins will fold, not at *designing* proteins that will fold predictably. *These tasks may sound similar*, but they differ greatly: the first is a scientific challenge, the second is an *engineering* challenge. Why should natural proteins fold in a way that scientists will find easy to predict? All that nature requires is that they in fact fold correctly, not that they fold in a way obvious to people.

Proteins *could* be designed from the start with the goal of making their folding more predictable. Carl Pabo, *writing in the journal Nature*, has suggested a design strategy based on this insight, and some biochemical engineers have designed and built *short chains of a few dozen pieces* that fold and nestle onto the surfaces of other molecules as planned. *They have designed from scratch a protein* with properties like those of melittin, a toxin in bee venom. They have modified existing enzymes, *changing their behaviours in predictable ways*. Our understanding of proteins is growing daily.

In 1959, *according to biologist Garrett Hardin*, some geneticists called genetic engineering impossible; today, it is an industry. Biochemistry and computer-aided design are now exploding fields, and as *Frederick Blattner wrote in the journal Science*, "computer chess programmes have already reached the level below the grand master. Perhaps the solution to the protein-folding problem is nearer than we think." *William Rastetter* of *Genentech*, writing *in Applied Biochemistry and Biotechnology* asks, "How far off is *de novo* enzyme design and synthesis? Ten, fifteen years?" He answers, "Perhaps not that long."

Forrest Carter of the *U.S. Naval Research Laboratory*, Ari Aviram and Philip Seiden of *IBM*, Kevin Ulmer of Genex Corporation, and other researchers in university and industrial laboratories around the globe have already begun theoretical work and experiments aimed at developing molecular switches, memory devices, and other structures that could be incorporated into a protein-based computer. The U.S. Naval Research Laboratory has held *two international workshops on molecular electronic devices*, and a meeting sponsored by the *U.S. National Science Foundation* has *recommended support for basic research* aimed at developing molecular computers. Japan has reportedly begun a multimillion-dollar programme aimed at developing self-assembling molecular motors and computers, and *VLSI Research Inc.*, of San Jose, *reports* that "It looks like the race to bio-chips [another term for molecular electronic systems] has already started. *NEC*, *Hitachi*, *Toshiba*, *Matsushita*, *Fujitsu*, *Sanyo-Denki* and *Sharp* have commenced full-scale research efforts on bio-chips for bio-computers."

Biochemists have other reasons to want to learn the art of protein design. New enzymes promise to perform dirty, expensive chemical processes more cheaply and cleanly, and novel proteins will offer a whole new spectrum of tools to biotechnologists. We are already on the road to protein engineering, and as Kevin Ulmer notes in the quote from *Science* that heads this chapter, this road leads "toward a more general capability for molecular engineering which would allow us to structure matter atom by atom."

Molecular computers will control molecular assemblers, providing the swift flow of instructions needed to direct the placement of vast numbers of atoms. Nanocomputers with molecular memory devices will also store data generated by a process that is the opposite of assembly. Assemblers will help engineers synthesize things; their relatives, *disassemblers*, will help scientists and engineers analyze things. The case for assemblers rests on the ability of enzymes and chemical reactions to form bonds, and of machines to control the process. The case for disassemblers rests on the ability of enzymes and chemical reactions to break bonds, and of machines to control the process. Enzymes, acids, oxidizers, alkali metals, *ions*, and reactive groups of atoms called *free radicals*—all can break bonds and remove groups of atoms. Because nothing is absolutely immune to corrosion, it seems that molecular tools will be able to take anything apart, a few atoms at a time. What is more, a nanomachine could (at need or convenience) apply mechanical force as well, in effect prying groups of atoms free. A nanomachine able to do this, while recording what it removes layer by layer, is *a disassembler*. Assemblers, disassemblers, and nanocomputers will work together. For example, a *nanocomputer* system will be able to direct the disassembly of an object, record its structure, and then direct the assembly of perfect copies, And this gives some hint of the power of nanotechnology.

Assemblers will take years to emerge, but their emergence seems almost inevitable: Though the path to assemblers has many steps, each step will bring the next in reach, and each will bring immediate rewards. The first steps have already been taken, under the names of "genetic engineering" and "biotechnology." Other paths to assemblers seem possible. Barring worldwide destruction or worldwide controls, the technology race will continue whether we wish it or not. And as advances in computer-aided design speed the development of molecular tools, the advance toward assemblers will quicken.

To have any hope of understanding our future, we must understand the consequences of assemblers, disassemblers, and nanocomputers. They promise to bring changes as profound as the industrial revolution, antibiotics, and nuclear weapons all rolled up in one massive breakthrough. To understand a future of such profound change, it makes sense to seek principles of change that have survived the greatest upheavals of the past. They will prove a useful guide.

Carbon Nanotubes, Nanoparticles, Nano-material & Nano-structuring Techniques

One third of the tubes are metallic or quasi, the others are semiconductors with moderate band-gaps.

If solid particles become small enough, in the nanometer range, they become electronically comparable to atoms and molecules. Instead of the laws of classical physics, they become quantum systems with discrete energy states. This is the basis for a number of useful applications. Quantum dots are reminiscent of the rhetorical question raised by Greek philosophers in antiquity: consider a sand-heap. If one removes from it a single grain of sand, obviously it is still a sand-heap. One continues doing so. At which stage does the sand-heap become something else?

One of the most exciting applications of quantum dots is the single electron tunneling transistor. An electrometer measures electric charge. A single electron transistor is, at present, the most sensitive electrometer. Small alterations of the charge distribution beneath a small isolated island of aluminum atoms markedly affect the conductivity across that island. When a gate voltage is applied to an electron donor material, such as tellurium, placed underneath the island, changes in conductivity can be monitored.

Nanocrystals are indeed important in single electron devices operating at room temperature, such as supersensitive electrometers and memory storage. Capped nanocrystals of both metals and semiconductors by virtue of their size possess capacitance in the range of attofarads (1 aF = 10^{-18} F). Charging a nanocrystal with an extra electron perturbs it to such an extent that the next electron requires an appreciable change in the charging potential. This is often seen as a 'Coulomb staircase' in the current-voltage tunneling spectra. Indeed, the charging energy varies linearly with the inverse of the diameter of the nanocrystal. Such sensitivity to single electron charging makes a nanocrystal an ideal candidate for use in single electron transistors and memory devices.

There are metallic and there are semiconducting nanoparticles. The former enjoy industrial and medical applications, for their magnetic properties in particular. The latter have uses in electronics and optoelectronics. One can also write with nanoparticles, dip-pen lithography is one of the techniques. More generally, the challenge is to organise nanoparticles in three dimensions.

The natural and the artificial once again meet! Biosystems, such as enzymes, antibodies, and nucleic acids have sizes commensurate with those of man-made nanoparticles. Hence, it becomes tempting to marry and hybridize them into nanocomposites, which is one of the most recent developments.

Two of the main thrusts of nanotechnology have been self-assembled ordered arrays of atoms with specific properties (electrical conductivity, magnetism, *etc.*) and well-defined, molecule-like clusters of atoms endowed with specific properties. The keyword here is specificity.

Founding Moment

The standard account of the birth of nanotechnologies dates it to the lecture by Richard P. Feynman on December 29, 1959 at the annual meeting of the American Physical Society held at Caltech. The books under review rightfully salute Feynman as the Founding Father of the field. As is well-known, the brilliant Caltech physicist ended his talk with a dare, announcing two prizes financed with his own money for the miniaturisation of the pages of a book and for the building of a miniscule electric motor, only 1/64 inch cube.

Feynman's lecture makes interesting reading. Its perspective is identical to that in Jules Verne's books, not visionary in the sense of a prophecy, imagining revolutionary developments and devices. The outlook is much more down to earth. Jules Verne and Richard P. Feynman took heed of recent scientific and technological advances to extrapolate to what the future would almost certainly witness, in a prospective.

In his lecture Feynman took as his starting point the feasibility of scaling down the contents of the world's libraries. He guesstimated their holdings at about 24 million volumes. He figured that their total information could be contained on an area of about 35 pages of the *Encyclopaedia Britannica*.

The discussions of potential economic, medical, and environmental benefits may have given the false impression that nanotechnology will create a wondrous utopia in which all human problems are solved and we all live happily ever after. This is even more mistaken than the idea that new technologies always cause more problems than they solve. Many of the main constraints and difficulties faced by people are based not on technology or its lack, but instead by the very nature of the world we live in and the essence of our humanness. Increasing affluence based on molecular manufacturing won't end economic problems any more than past increases in affluence have. Wilderness can still be destroyed; people can be oppressed; financial markets can be unstable; trade wars can be waged; inflation can soar; individuals, companies, and nations can go into debt; bureaucracy can stifle innovation; tax levels can become crippling; wars and terrorism can rage. None of these will automatically be stopped by advanced technology. What is more, the potential benefits of new technologies aren't automatic. Nanotechnology could be used to restore the environment, to spread wealth, and to cure most illness. But will it? This depends on human action, working within the limits set by the real world. This chapter first describes some of the limits to what nanotechnology can accomplish, and then some of the adverse side effects of its basically good applications. The next will discuss the problem of accidents, which seems manageable, and then the far greater problem of potential abuse of new capabilities.

The world imposes limits on what we can do. Technology in general (and nanotechnology in particular) can provide padding for us as we throw ourselves against these hard, sharp limitations, and can sometimes help us slip past old limits through previously unknown gaps. Eventually, though, we will encounter new limits. In the end, solid constraints will limit human action no matter how much we juggle atoms and molecules, or the bits and bytes of information. Let's look at some of these, starting with the most abstract and long term—the most definite and hardest to avoid—and moving toward the more personal and near term.

Many problems differ fundamentally from the material problems of limited matter and energy: they involve information. Some of the most precious stores of information in the world today are the genetic codes of the biosphere. This information, different for virtually every individual organism, is the product of millions of events that we are incapable of modeling or recreating. When this information is lost, it is lost forever. When the atoms encoding this information are thoroughly scattered, there seems to be no way to retrieve it. With any species, most genetic information is shared in common, found in all members of that species. But the variations in genetic code between individuals are important, both to the individuals themselves and to the health and prospects of the species as a whole. Consider the northern white rhino, whose numbers have dropped to an estimated thirty-two animals, or the California condor, of which only forty remain, all in captivity. Even if biologists succeed in reestablishing these species—eight condors were hatched in 1989—much of the diversity of their genetic information has been lost. Worse yet are extinctions of species for which no tissue samples were saved. The future may see some amazing recoveries: dry skin and bones may yield a complete set of genes when sifted by molecular machinery, and even current techniques have been used to recover genes from an ancient leaf, almost 20 million years old. Our eyes and instruments cannot yet tell us how much information from the past remains, but we do know that genetic information is being lost every day, and once lost, it is irretrievable.

People have often been wrong about physical limits, confusing the limits of their technology with the limits of the possible. As a result, learned men first dismissed the idea of heavier-than-air flight, and then dismissed the idea of flying to the Moon. Yet physical limits are real, and all technology—past, present, and future—will stay within those limits. There is even reason to suspect that some of those limits are where the learned now believe them to be. Nanotechnology will make it possible to push closer to the real limits set by natural law, but it will not change those laws or the limits they set. It will not affect the law of gravity, the gravitational constant, the speed of light, the charge of the electron, the radius of the hydrogen atom, the value of Planck's constant, the effects of the uncertainty principle, the principle of least action, the mass of the proton, the laws of thermodynamics, or the boiling point of water. Nanotechnology won't make energy or matter from nothing. It seems a good bet that no one will build a faster than light spacecraft, or an antigravity machine, or a cable twice as strong as diamond. There are limits. Science today may be wrong about some limits, but scientific knowledge is practically defined to be our best information about how the world works, so it isn't wise to bet against it. There will be claims that nanotechnology will be able to do things that it can't, or that capabilities are around the corner when they aren't. Sometimes these will be innocent errors, sometimes they will be culpably stupid errors, and sometimes they will be what amounts to fraud. Among the problems that nanotechnology cannot solve is that of misguided claims, by people calling themselves "scientists, " "engineers, " or "businesspeople, " that they have a big technical breakthrough worth a fortune. Every interesting new technology, particularly in its early days, is a chaotic mix of competent workers and charlatans. For every Thomas Edison inventing useful products such as light bulbs or the precursor of movie projectors, there were people promoting electric hairbrushes to cure baldness, and electric shoes, electric belts, electric hats—the list goes on—that authoritatively claimed cures for infertility, overweight, underweight, and all the ills and discomforts of mankind. Today, we laugh at the credulity of our forefathers who bought these gadgets; we shouldn't, unless we laugh at our own times as well.

Natural law imposes limits, but so does the nature of human beings. These will continue as long as people do. Reproduction is a deeply ingrained instinct enforced by the march of time, which ruthlessly discards the genetic material of all who neglect it. Many would argue that the Earth is already overpopulated. While nanotechnology could enable the current population, and even a greatly increased one, to live more lightly on the Earth, there will still be limits to Earth's capacity. The norms of human life are shaped by ancient patterns: high rates of infant or childhood mortality have been facts of life for millennia, and having many, many children has been a way to ensure that one or two will survive to work on the farm, and to care for you in your old age. Large families naturally become traditional. When modern medicine and reliable food supplies change those conditions—as they have, in cultural terms, virtually overnight—behaviour does not shift as quickly. The result is the Third World population boom. In Western countries, where there has been time for behaviour to adapt, a huge family is the exception. It might seem that our problem is solved. Molecular manufacturing can make everyone wealthy, and wealthy populations today have stable or shrinking populations. The Earth can support more people with advanced technologies, and these will also open up the vast room and resources of the world beyond Earth. Would that this were true. If 99 per cent of the people in a population respond to wealth by reducing childbearing, the population will indeed stabilize or shrink, for a while. But populations are not uniform. What of the 1 per cent, say, who are members of a minority with different values? If that minority has a growth rate of 5 per cent per year, then in ninety-five years they will be the majority, and in one thousand years their population will have grown by a factor of 1, 500, 000, 000, 000, 000, 000, 000, if resource limits or genocide haven't intervened. Note that the Hutterites of North America, a reasonably wealthy religious group viewing fertility control as a sin and high fertility as a blessing, have managed an average of ten children per woman. Given enough time, exponential growth of even the smallest population can consume all the resources in reach. The right to reproduce is often regarded as basic, as illustrated by the outrage at reports of forced abortion in the People's Republic of China. The Hutterites and many others regard it as part of their freedom of religion. But what happens when parents have more children than they can support—does redistribution solve the problem? If reproduction is not forcibly suppressed, and if resources are forcibly and repeatedly redistributed so that each human being has a roughly equal share, then each person's share will steadily shrink. Even given the most optimistic assumptions regarding available resources, with a policy of resource redistribution and unlimited reproduction, the amount per person would eventually be insufficient to sustain life. This policy must be avoided, because if it is followed, it will kill everyone. As soon as we grant that any entity is entitled to certain rights—whether that entity be a human child, an animal, or some future artificial intelligence—the question arises of who is responsible for providing resources to support it when it can't do so for itself. The above argument indicates that a policy of coercion by some central power to compel the entire population to support an exponentially exploding population of these individuals would lead directly to disaster. Ultimately, this responsibility must rest with the entities' initiator: the designer of the artificial intelligence, the owner of the pet, the parents of the child. No new technology can magically remove the limits imposed by natural law, and thereby lift the burden of human responsibility.

Every time a technology solves a problem, it creates new problems. This doesn't mean that the change is neutral, or for the worse, of course. The Salk and Sabin vaccines for polio virtually destroyed the iron-lung industry, and the pocket calculator virtually destroyed the slide-rule industry, but these advances were worth the price of some economic adjustment. Molecular manufacturing and nanotechnology will bring far greater changes, placing far greater strains on our ability to adapt. We shouldn't be surprised when basically beneficial applications make someone miserable. Our lives are largely centered around problems. If we can solve many of these problems, the centers of our lives will shift, creating fresh problems. This section sketches some of the issues of change and adaptation more to raise questions than to offer solutions.

Molecular manufacturing offers the possibility of drastic change, a change in the means of production more fundamental than the introduction of industry, or of agriculture. Our economic and social structures have evolved around assumptions that will no longer be valid. How will we handle the changes in the way we work and live? Nanotechnology will have wide-ranging impact in many areas, including economic, industrial, and social patterns. What do historical patterns in similar circumstances tell us about the future? Any powerful technology with broad applications revolutionizes lives, and nanotechnology will be no exception. Depending on one's point of view, this may sound exciting or it may sound disturbing, but it most certainly does not sound comfortable. In comparison to many projections of the twenty-first century, though, nanotechnology may lead to comparatively comfortable change. The changes most often projected—for a future not including nanotechnology—have been ecological disaster, resource shortages, economic collapse, and a slide back into misery. The rise of nanotechnology will offer an alternative—green wealth—but that alternative will bring great changes from the patterns of recent decades. Times of rapid technological change are disconcerting. For most of humanity's existence, people lived in a stable pattern. They learned to live as their parents had lived—by hunting and gathering, later by farming—and changes were small and gradual. A knowledge of the past was a reliable guide to the future. Sudden changes, when they did occur, were apt to be ruinous: invasions or natural disasters. These sudden changes were fought or repaired or survived as best one could. Making major changes by choice was rare, and radical innovations were generally for the worse: the old ways at least ensured the ancestors' survival, the new might not. This made cultures conservative. It is only natural that there be efforts to resist change, but before undertaking such an effort, it makes sense to examine the record of what works and what doesn't. The only examples of successful change fighters have been communities that have created and maintained barricades to isolate themselves from the outside world socially, culturally, and technologically. For the two centuries before 1854, Japan turned its back on the outside world, following a deliberate policy of seclusion. The leaders of Albania restricted contacts for many years; only recently have they started to open up. Isolation attempts have worked better on a smaller scale, when participation is voluntary rather than decreed by government. Today, within the Hawaiian island chain, the tiny, privately owned island of Niihau, sixteen miles long and six miles wide, is deliberately kept as a preserve of the nineteenth-century Hawaiian lifestyle. Over two hundred full-blooded Hawaiians there speak the Hawaiian language and use no telephones, plumbing, television, and no electricity (except in the school). The Amish of Pennsylvania have no surrounding ocean to help maintain their isolation, but rely instead on tight social, religious, and technological rules aimed at keeping external technology and culture out, and themselves grouped in; those who leave the fold are excluded. On a national scale, attempts to take only one part of the package—whether social or technological—haven't done well at all. For decades, the Soviet Union and the Eastern bloc nations welcomed Western technology but attempted tight restrictions on the passage of people, ideas, and goods. Yet illegal music, thoughts, literature, and other knowledge still crept in—as they do into the Islamic countries. Fighting technological change in society at large has had little success, where that change gave some large group what it wanted. The most famous fighters of technological change—the Luddites—were unsuccessful. They smashed "automated" textile machinery that was replacing old hand looms during the early industrial revolution in England, but people wanted affordable clothing, and smashing equipment in one place just moved the business elsewhere. Change has sometimes been postponed, as when a later group, under the banner of "Captain Swing, " smashed hundreds of threshing machines in a wide area of southern England in 1830. They succeeded in keeping the old, labor-intensive ways of harvesting for over a generation.

In previous centuries, when the world was less tightly connected by international trade, communications, and transportation, delays of years and even decades could be enforced through violence or legal maneuvers such as tariffs, trade barriers, regulations, or outright banning. Attempting to stop or postpone change is less successful today, when technology moves internationally almost as easily as people do—and human travel is so easy that 25 million people cross the Atlantic each year. Change fighters find that the problems they create mount with time. Products made using the old, high-cost techniques are uncompetitive. There is no way to bring back the "old jobs": they no longer make sense. But old habits die hard, and these same responses to the prospect of technological change continue today—ignoring it, denying it, and opposing it. Societies that have fought change, as Britain did, have fallen behind in a cloud of coal smoke. Why did the Luddites respond violently? Perhaps their response can be attributed to three factors: First, the change in their lives was sudden and radical; second, it affected a large group of people at one time, in one area; and third, in a world unprepared for rapid technological change, there was no safety net to catch the unemployed. While local economies might have been able to absorb a trickle of hungry laid-off workers, they lacked the size and diversity needed to offer other employment options quickly to large numbers of unemployed. In the twentieth century, however, societies have of necessity become somewhat better adapted to change. This has been a matter of necessity, because sluggish communities soon fall behind. In the ancient days of peasant stability, there was no need for institutions like Consumer Reports to study and rate new products, or regulators like the Environmental Protection Agency to watch over new hazards. We developed the needs, and we developed the institutions. These mechanisms represent important adaptions, not so much to the technologies of the twentieth century, but to the increasing change in technology during the twentieth century. There is great room for improvement, but they can perhaps provide a basis for adapting to the next century as well. Even with the best of institutions to cushion shocks and discourage abuse, there will be problems. The very act of solving problems of production—of increasing wealth—will create problems of economic change.

Over centuries, the trend has seemed to be toward centralisation, beginning with the rise of factories and industrial towns. What drove these developments was the high cost of machinery and plant operations, the need to be near power sources, the impracticality of transportation among many small, dispersed sites, and the need for face-to-face communication. Beginning with the first industrial revolution, factories employed large numbers of people in one place, leading to overcrowding and making local economies dependent on one industry and sometimes on a single company. Costly equipment necessitated central locations for textile production, rather than the cottage industries where a lone woman could earn a livelihood carding wool and creating thread on a spinning wheel (providing the origin of the term *spinster*). By the 1930s, the belief in the virtures of centralisation and central planning—the supposed efficiencies and economies of scale—led to nationwide or continentwide experiments in centralisation. But over the last decade, these large-scale experiments have been dismantled, from Britain's privatisation of nationalised utilities to the beginning of a return to the market system in Eastern European countries. Because the old limits on transportation, energy sources, and communication have fallen, business is now decentralising. Between 1981 and 1986, the Forbes 500 companies cut their employees by 1.8 million. But during those same years, total civilian jobs went up by 9.2 million. Start-up companies created 14 million jobs; small companies created another 4.5 million. Telecommuting is booming, as are new businesses, independent professionals, and cottage industries. We've also seen the resurgence of small, but highly diverse stores: gourmet food shops, specialty ethnic shops, tea and coffee purveyors, organic and health food stores, bakeries, yogurt shops, gourmet ice-cream stores, convenience stores offering twenty-four-hour access, shops selling packaged food plus snacks.

These stores epitomize something fundamental: At some point, what we want is not a standard good at an ever cheaper price, but special things customised to meet our own individual tastes or needs. The trend for advanced technologies seems to be leading away from centralisation. Will nanotechnology counter or accelerate this trend? By reducing the cost of equipment, by reducing the need for large numbers of people to work on one product, and bringing greater ability to produce the customised goods that people want, nanotechnology will probably continue the twentieth-century trend toward decentralisation. The results, though, will be disruptive to existing businesses. The computer industry perhaps provides a clue to what might happen as costs are lowered by nanotechnology. The computer-software industry is characterised by the garage-shop start-up. When your equipment is cheap—inexpensive PCs built around low-cost chips—and you can make a product by throwing in some ingenuity and human labor, it's possible to start a new industry on a shoestring. In 1900, when cars were simple, there were many car manufacturers. By the 1980s, if you weren't an industrial giant like General Motors or Ford, Honda or Nissan, you had to be John De Lorean to even get a shot at acquiring the capital to play in the business. If molecular manufacturing can slash the capital costs for producing cars or other plant-intensive equipment, we will see the equivalent of garage-shop businesses springing up to offer new products, and hiring workers away from the industrial giants of today just as the personal computer has destroyed the dominance of the mainframe.

The U.S.A. dream is to be an entrepreneur, and the technological trends of the twentieth century point in that direction. Nanotechnology probably continues it. In one area, however, the late twentieth-century trend has been toward uniformity. The nations of Western Europe are in the process of uniting under one set of economic rules, and parts of Eastern Europe are anxious to join them. More and more supranational and transnational organisations knit the world together. The growth of trade has motivated economic integration. Molecular manufacturing will work against this trend as well, permitting radical decentralisation in economic terms.

This will help groups that wish to step aside from the stream of change, enabling them to be more independent of the turbulent outside world, picking and choosing what technologies they use. But it will also help groups that wish to free themselves from the constraints of the international community. Economic sanctions will have little force against countries that need no imports or exports to maintain a high standard of living. And export restrictions will likewise do little to hamper a military buildup. By weakening the ties of trade, molecular manufacturing threatens to weaken the glue that holds nations together. We need that glue, though, to deal with the arms control issues raised by molecular manufacturing itself. This problem, caused by the potential for decentralisation, may loom large in the coming years.

Lester Milbrath, professor of sociology and political science, observes, "Nanotechnologies will create the problem of how to meaningfully and sustainably occupy the time of people who need not perform much work in order to have a sufficiency of life's goods. Our society has never faced this problem before, and it is not clear what social restructuring will be required to have a good society in those circumstances. We face much deep social learning." The world has had little experience with what anthropologists call "abundance economies." The native American tribes of the Pacific Northwest were one of those rarities. Ruth Benedict, in her classic book *Patterns of Culture*, wrote, "Their civilisation was built upon an ample supply of goods, inexhaustible, and obtained without excessive expenditure of labor." The Kwakiutls became famous for their "potlatches": contests in which they sought to shame their rivals by heaping more gifts upon them than they could ever return. The potlatches would often be a year in preparation, last for days, and occasionally involve destruction of entire buildings. It was certainly a colorful form of keeping up with the Joneses.

What will motivate us, once we have achieved an abundance economy? What will we regard as worthwhile goals to pursue? Increased knowledge, new art, improved philosophy, eliminating human and planetary ills? Will we find ourselves creating a better, wiser world, or sunk in boredom and jaded now that we have all and want nothing? If boredom gets out of hand, the lively spectacle of wealthy donors seeking to outdo each other to endow the arts, aid the poor, and do other good deeds for the sake of prestige would be welcome. What will happen as life spans continue to lengthen and the time needed to make a living decreases? Even today, there are people who, when confronted with the prospect of a significantly longer life span, exclaim that they couldn't imagine what they would do with all that time. This response can be hard to understand, when it would take a thousand years to walk all the world's roads, more thousands of years to read all the world's books, and another ten thousand years to have a dinner conversation with each of the world's people—but tastes differ, and even a few decades of bad television might make anyone long for the peace of the grave.

A major concern, and certainly the single area of greatest upheaval, is employment (which may become hard to distinguish from leisure). Once, people had little choice of employment. To keep a full belly, most had to work at the only job available: peasant farming. Eventually, people will have a complete choice of employment: they will be able to keep a full belly and a wealthy lifestyle while doing whatever they please. Today, we are about halfway between those extremes. In advanced economies, many different jobs are deemed useful enough that other people will offer an adequate income in exchange for the result. Some people can make a living doing something they enjoy—is this work, or leisure?

The impact of nanotechnology on patterns of employment will depend on when it arrives. Current demographics show a shrinking supply of young people entering the work force. Agriculture, the assembly line, and entry level service jobs are experiencing a labor shortage, and no relief is in sight. If these trends continue, nanotechnology may show up in the midst of a shortage of labor. If it arrives late enough, it may compete with industries that are already nearing full automation; "job displacement" may mean replacing an industrial robot with a nanomachine. Employment patterns have shifted radically in the past. One hundred and fifty years ago, the United States was an agricultural nation—69 per cent of all people worked the land and a growing percentage worked in industry doing things like building steam locomotives for Baldwin Locomotives Works or tanning leather for the giant Central Leather monopoly.

By the early twentieth century, agriculture was waning in numbers but increasing in productivity; most people worked in industry, and the tiny information and service sector was beginning to grow. Today, the picture has reversed: 69 per cent of employed Americans work in information or service jobs, only 28 per cent work in industrial production, and 3 per cent in agriculture. This tiny fraction feeds the other 97 per cent of Americans, exports hugely to other countries, and receives subsidies and price support payments to stop them from growing even more food. Manufacturing, even without nanotechnclogy, seems to be heading toward a similar condition. With an ever-declining percentage of our population working in manufacturing, we have as everyday products things that were once available only to kings and the high nobility.

Yet owning multiple suits of clothes, having personal portraits of ourselves and family members, having music upon our command, having a personal bedroom, and having a coach awaiting our need—these are now regarded as being among the bare necessities of life. It may be possible to adjust to even greater wealth with even less required labor, but the adjustment will surely cause problems. In a world in which nanotechnology reduces the need for workers in agriculture and manufacturing still further, the question will be asked, "What jobs are left for people to do once food, clothing, and shelter are very inexpensive?" Again, the twentieth century provides some guidelines. As technology has reduced costs by efficiently producing many units of an identical item, people have begun to demand customisation to meet individual needs or preferences. As a result, there are ever more jobs in producing custom goods. Today, semi-custom goods that try to help us meet our needs or express our taste abound: designer linens, ready-to-wear fashions, cosmetics, cars, trucks, recreational vehicles, furniture, carpeting, shoes, televisions, toys, sports equipment, washing machines, microwave ovens, food processors, bread bakers, pasta makers, home computers, telephones, answering machines—are all available in large and ever-changing variety.

Just as varied is the fabulous wealth and diversity of information produced in the twentieth century. Information products are a large factor in the economy: Americans buy 2.5 billion books, 6 billion magazines, and 20 billion newspapers each year. In recent years, new magazines have been invented and launched at the rate of one every business day of the year. A visit to a well-stocked magazine rack shows only a hint of the wealth of highly specialised publications, each one focused on a specialised interest or attitude: hotdog skiing, low-fat gourmet cooking, travel in Arizona, a magazine for people with a home office and a computer, and finely tuned magazines on health, leisure, psychology, science, politics, movie stars and rock stars, music, hunting, fishing, games, art, fashion, beauty, antiques, computers, cars, guns, wrestling. Motion pictures, which started as a flock of independent production companies and then consolidated into the great studios of the 1930s, have since followed the decentralisation and diversification trends of recent years. Now an expanding range of film entertainment comes via network TV, cable channels, private networks, videotapes, music videos. Independent producers are aided by the technology innovations of cable, direct broadcast satellites, videotape technology, laser disks, videocameras.

The arts have burgeoned, with the general public as the new patron of the arts. Any artist or art form that could find and satisfy a market boomed in the twentieth century. Not just the traditional arts of actors, writers, musicians, and painters, but all forms of "domestic" artistry have grown to unprecedented levels: landscape and interior design, fashion design, cosmetics, hairstyling, architecture, bridal consulting. Providing for these demands are some of the "service and information" jobs created in the late twentieth century. "Service" jobs include many ways of helping other people: from nursing to computer repairs to sales. In "information" jobs, projected to have the fastest percentage growth over the next decade, people find, evaluate, analyze, and create information. A magazine columnist or TV news producer obviously has an "information" job. But so do programmers, paralegals, lawyers, accountants, financial analysts, credit counselors, psychologists, librarians, managers, engineers, biologists, travel agents, and teachers. "Increasingly, " states *Forbes* magazine, "people are no longer laborers; they are educated professionals who carry their most important work tools in their heads. Dismissing them from their jobs, cutting them off from their places of employment may hurt them emotionally and financially. But it doesn't separate them from their vocation in the same way that pushing a farmer off his freshly seeded land does. For centuries workers were more dependent on a particular physical setting than they are now. Modern occupations generally give their practitioners more independence—and greater mobility—than did those of yesteryear." These human skills that people carry with them will continue to be valued: managing complexity, providing creativity, customising things for other people, helping people deal with problems, providing old services in new contexts, teaching, entertaining, and making decisions. A reasonable guess would be that many of the service and information industries of the twentieth century will continue to evolve and exist in a world with nanotechnology. What is harder to imagine would be what new industries will come into being once we have new capabilities and lower costs. Along with the old economic law of supply and demand is another governing factor: price elasticity effects. People's desire for something is "elastic": it expands or contracts when the cost of something valuable goes down or up. If the price of a flight to Europe is five hundred dollars, more people will take a European vacation than if the price is five thousand dollars. When you had to hire a highly trained mathematician to do equations, calculation was slow and expensive. People didn't do much of it unless they absolutely had to. Today, computers make calculation cheap and automatic. So now businesses do sophisticated financial modeling, chemists design protein molecules, students calculate orbital trajectories for spaceships, children play video games, moviemakers do ever more amazing special effects, and the cartoon—virtually extinct because of high labor costs—has returned to movie theaters, all because computers permit cheap calculation. Nanotechnology will offer new, affordable capabilities to these and other people. Today, it's as hard to predict what new industries will be invented as it would have been for the creators of the ENIAC computer to have predicted cheap, handheld game computers for children.

So rather than producing drastic unemployment, nanotechnology seems likely to continue the trend already seen today, away from jobs that can be automated and into jobs where the human perspective is vital. But the true possibilities are, as always in the modern world, beyond predicting.

Major shifts in demographics always cause disruptions. Even when we know they are coming, we never prepare for them. Our plans are based on expectations of what will happen. If things don't go as expected, we find that we have "malinvested." Houston real estate was valuable and looked to become even more so when times were good for the oil business there; when the fortunes of the oil business changed, Houston real estate was found to have been overbuilt, overpriced, and many millions of dollars were lost. Lengthening life spans push people toward taking a longer-term perspective, but rapid rates of change force a shorter-term perspective in investments. Turbulence in technology and in governmental monetary policy have already shortened time horizons. Businesspeople once routinely built plants with a thirty-year useful life. Today, the rate of change is too fast, and uncertainty regarding inflation and potential changes in tax laws is too great for such investments to make sense. Faster change will shrink time horizons further. Governments have taken on themselves the burden of looking a lifetime ahead, and the Social Security Administration is in for some rough times. When Otto von Bismarck, Germany's Iron Chancellor, came up with the notion of a guaranteed old age pension, it was a cynically clever and low-cost way to gain popular goodwill. So few people lived to age sixty-five that the amounts paid out in pensions were a pittance. After watching the German experiment for a handful of years, other governments began following suit. None of them expected a world like ours where a baby girl born in the United States today has an average life expectancy of 78.4 years—double that of Bismarck's time—and even this estimate is based on the faulty assumption that her medical care will be no better than her great-grandmother's was.

At present, the Social Security Administration has two models: one they call "positive" and one they call "negative." In the "positive" model, people work like dogs until old age, retire, and promptly die—presumably before they've had a chance to collect substantial social security or medical benefits. In the "negative" model, people retire early, develop illnesses that require medical intervention, and then live a long time making doctor visits and hospital stays during those years. Plans based on these models deserve to be disrupted. A better, more realistic scenario would have people living and able to support themselves for a long time, with illnesses that can be handled easily and inexpensively. Present social security benefits are enough to provide a certain standard of living—food, housing, transportation, and so forth. In a future of great material wealth, these benefits will be easy to provide, and present projections of economic woe resulting from an aging population will seem quaint.

Back in the seventies, author Alvin Toffler brought out a book called *Future Shock*, describing how disturbing rapid change is for people. The book was a best-seller, but how much actual future shock has been seen in the past decade? Most people seem to have come through the last two decades pretty much all right, not in a state of shock at all. Rather than being shocked by technology, they are instead annoyed about pollution and traffic. Does this mean Toffler was wrong in predicting future shock? It's true that technology has been advancing rapidly in many areas over the past twenty years. But consider the average person's home life: How much of this rapid technological advance has shown up there? A great deal, yet most of it is hidden, unlike the earlier part of the century, where obvious change was the norm. Electric lights and appliances, automobiles, telephones, airplanes, radio, and television affected almost everyone's private life. One person's life could span the time from horse-and-buggy travel to watching the Moon landings on television. In contrast, the past twenty years have seen new technologies move more quietly into the home. The

VCR and microwave oven don't seem nearly as revolutionary as earlier inventions. Telephone answering machines are useful but haven't caused major changes in lifestyles. Fax machines are handy, but they're much like having very fast mail, and as this is written, fax machines aren't yet in most homes. So it's not surprising that the average person has felt little future shock lately. New medicines taken as pills—which may be radically improved—look just like the earlier pills. The computerised bills that come in the mail aren't any more exciting to pay than the old human-prepared bills. This situation is unlikely to last. How much longer can technology advance so rapidly in so many fields without major effects on our lifestyles? There's been a respite from future shock in the last three decades; people have had a chance to catch their breath. When nanotechnology arrives, will future shock arrive with it? Some segments of society today are already getting practice in dealing with rapid technological advance. Those getting the most vigorous workout are in the computer field, where a machine two years old is regarded as obsolete, and software must be updated every few months to keep abreast of the new developments. But has this terrific rate of progress been dizzying or overwhelming?

Not for the consumer—on the contrary, computers have become easier to use. In the 1960s, the New Math that was introduced into American grade schools and junior high schools included extensive study of arithmetic using numbers written in something other than the familiar base 10. This was to prepare the "Adults of Tomorrow" for "The Computer Age" in which we would all be writing assembly language computer programmes in binary (base 2) code. But customers now purchase software rather than write it themselves—they need never deal with computer languages at all, much less a primitive assembly language. The rapid increase of computer speed has helped make computers easier to use.

This progression has occurred many times before: Cars started off with external hand cranks, then advanced to starters you could yank from the comfort of the driver's seat; now starters perform invisibly when you turn the key in the ignition. This pattern will surely continue. First, some people will adapt to the technology, but in the long run the technology will adapt to us. The more flexible and powerful the technology, the more easily it will adapt. Seen from a distance, seemingly trivial patterns of adaptation form part of a larger process that has marked the last century: the Western world has begun to invent mechanisms to handle a world of persistent change. Our mechanisms are by no means perfect or painless, as any unemployed person can testify. Employment agencies and headhunters for job seekers; unemployment and severance packages to ease job transitions; on-the-job training, continuing education, retraining, specialised seminars to update professional skills, professional associations, networking, community resources centers, government training programmes, and volunteer agencies are just a few of the inventions dealing with change and transition. Consumer information services, regulatory agencies, and environmental organisations are others. The most effective will endure. More options will continue to be invented. intellectual exercise of Feynman's had its obvious seed in the then recent (1953) discovery of the DNA double helix by Watson and Crick. If the genome has such a small size, Feynman thought, given the awesome amount of information it encodes, is there anything preventing scientists and engineers from emulating this natural feat?

Let us keep in mind such inheritance. Biology thus gave birth to the whole field of nanotechnologies. To go by the empirical criterion, Feynman's call was that of a pioneer, it went unheeded for a generation.

After Heinrich Rohrer and Gerd Binnig devised in 1981 a scanning tunneling microscope, a collective leap forward ensued. About that time, George B. Whitesides initiated a series of studies which showed the kind of contributions chemists might make to nanosystems. The nanotechnology bandwagon started rolling in the 1990s. It was pulled by two horses. There was Moore's Law. Engineers and scientists had for decades looked at the horizon of computers foregoing silicon chips in favour of molecular devices. And there was the demographic factor. The steady increase in the number of scientists, which continued well into the 1980s, was unmatched by a corresponding increase in public money. Accordingly, young scientists as a group partitioned themselves between industrial and pharmaceutical work.

An Offshoot from Chemistry or from Biology?

That biology gave rise to nanotechnology raises a counterclaim. Whenever a group of chemists discuss nanotechnologies, they remind one another "this is nothing but chemistry, after all". The declaration can be seen as narcissistic, self-centered, and narrow-minded. Nevertheless, it has a degree of truth. Moreover, like any such stereotypic assertion on the part of a group of scientists. it is worthy of in-depth scrutiny. Let us then assume for the new sub-discipline to be yet another spin-off from chemistry, in like manner to biochemistry, solid-state physics, molecular biology, or materials science. Because of some unfortunate connotations of the word 'chemistry', its ancient roots on the one hand, its age-old associations with explosives, poisons, and pollutants on the other, people have refrained from using it as a label on a field holding both novelty and promise of lucrative applications. Hence, the renaming as nanotechnologies.

Chemists are prompt to recognize in nanotechnologies lines of forces to them familiar and maybe even *passé*. These include colloids and Langmuir-Blodgett films, which have regained a new hold on both scientists' imagination and grantsmanship. At a more fundamental level, though, chemists have injected into nanotechnologies a key notion, that of specificity. Contrary to what is routinely the case in physics, with fractals an example, chemical phenomena fail to obey scale invariance. They are liable to show quite different features depending on the scale of the observation. Indeed, this serves as a dividing line between chemistry as the science and chemistry as the industry. The latter is interested only in bulk matter and in bulk phenomena, whether heat transfer or equations for overall transformations, with more concern for the bottom line in financial reports than for the specific and microscopic (or nanoscopic) mechanisms involved, in catalysis for instance. However, what makes nanotechnologies a unique and viable hybrid is that their other parent has been, unmistakably, biology. This is an important legacy, all the more so in view of many of the present applications being in areas distant from biology such as electronics, optoelectronics, and other highly applied applications, from cosmetics to self-cleaning windows.

Biological phenomena are a continued inspiration and biological applications are around the corner. The former? While it was known that many organisms orient themselves from the earth's magnetic field, Blakemore's 1975 finding of bacteria from marine marsh muds heading rapidly north was seminal. Blakemore then used transmission electron microscopy to reveal iron-rich crystals, above 100 nm in diameter, in these bacteria. Several crystals would align into chains along their major axis. Frankel, Blakemore, and Wolfe proceeded to report that an unclassified magnetotactic spirillum (MS-1), when cultured in solutions containing ferric salts, precipitated uniform single crystals that consisted of mostly spinel Fe_3O_4. Microbial magnetite and the ferrimagnetic greigite g-Fe_3S_4, present in magnetotactic bacteria grown in marine, sulfidic environments, are inorganic crystallisations impressive for their quality. The magnetic particles have the optimal single domain size for the specific mineral. Each magnetic particle contains only a single magnetic domain. Moreover, within the magnetosome, the crystals are aligned so that their assembly bears a permanent magnetic dipole, making an effective compass. Biogenic magnetite provides great inspiration to laboratories active in nanotechnology, as for instance the book by Günther Schmid testifies to. The latter? Innocuous quantum dots (QD) bioconjugates have been injected into *Xenopus* embryos to monitor the development to the tadpole stages. QD label *Xenopus* embryos at different stages and with specific QD intracellular localisations.

New Technology on Old Way

But let us return to the attitude of chemists who discard nanotechnologies as a combination of old, well-known chemistry and a public relations ploy to raise money from industry and from granting agencies. Could it be true that the field of nanotechnologies amounts only to renaming a subset of well-established chemical knowledge? Does it lack originality and impetus? Are we dealing with old wine, which time may have started to sour and to make unpalatable and thus needed a new label? Conversely, is this new wine whose novel taste suggests brand-new flasks?

Consider colloidal gold as a test case. Gold nanoparticles have been the subject of considerable attention over the ages. A back-of-the-envelope calculation shows that a small nanocrystal of one nanometer diameter will have as many as 30% of its atoms on the surface while a larger nanocrystal of 10 nm (ca. 1000 atoms) will have about 15% of its atoms on the surface. Metal particles with dimensions in the nanometer scale thus display unusual properties. The mean free path of an electron in a metal at room temperature is in the range 10-100 nm. As a metallic particle shrinks to that dimension, unusual effects might be observed. Indeed, gold nanoparticles of diameters about 100 nm or less appear red, not gold, when suspended in transparent media.

They enjoy an interesting history dating back to the pioneering work of Faraday on the synthesis of gold hydrosols, gold nanoparticles dissolved in water. As one may recall, early in 1856 Michael Faraday (1791-1867) became interested in the interaction of gold with light. Gold leafs show different colors in transmitted light, green, blue, and purple, than they do in reflected light. This observation intrigued Faraday. Together with his friend, Warren de la Rue, he prepared gold films. He started preparation of gold precipitates from reducing with ferrous sulfate a solution of auric salts, made presumably by attack of the metal with *aqua regia*. In so doing, Faraday discovered colloidal gold. This was the subject of intense experimenting and thought. Between the onset of this work, February 2, and its completion at the end of the same year 1856, Faraday wrote in his diary no fewer than 300 manuscript pages (1, 160 numbered entries) on this topic. He submitted an article to the Royal Society on November 15, it appeared on February 15, 1857. In general and since Faraday, metal sols possess fascinating colors and have long been used as dyes.

Smaller gold nanoparticles of diameters about 3 nm are no longer 'noble' and unreactive, but can catalyze chemical reactions. Consider nanocatalysts comprised of gold atom clusters, Au_8, Au_4, Au_3Sr, adsorbed at an F center of a MgO (100) plane. The optimal adsorption geometries of the O_2 molecule to these nanocatalysts differ markedly. Another example is from the temperature-programmed reaction spectra of the different products of the polymerisation of acetylene on small supported, monodispersed palladium clusters. Striking atom-by-atom size-dependent reactivities and selectivities are observed. Only the three reaction products C_6H_6 (benzene), C_4H_8 (butene), and C_4H_6 (butadiene) form. Remarkably no C_3H_n, C_5H_n, and C_8H_n are detected, indicating the absence of C-C bond scission, as already observed on palladium single crystals and on palladium particles. Up to Pd_3, only benzene formation is catalysed, reflecting a high selectivity for cyclotrimerisation of acetylene. Pd_n clusters ($4 < n < 6$) reveal a second reaction channel by catalysing in addition the formation of butadiene. The third reaction product, butene, desorbing at a rather low temperature of 200 K, is clearly observed for Pd_8. For this cluster size the abundance of the three reaction products is similar. For even larger clusters ($13 < n < 30$) benzene formation increases with cluster size, whereas the conversion of acetylene to butene reaches a maximum for Pd_{20}. Note that Pd_{30} selectively suppresses the formation of butadiene (see for instance the Nanocatalysis chapter in the book by Rao, Müller, and Cheetham).

What this example of metallic nanoparticles shows, which many of us have always suspected to be true but which conventional wisdom and standard practice – why bother reading any scientific publication older than say five years? – have both condemned summarily, is that casks of old wine connect with containers of the precious liquid from the most recent harvest. New science can be at its most innovative when it grafts itself onto its historical roots. Prediction starts with retrodiction, a point to which I will return.

Beyond the Hype

Who is most responsible for inflating the nanobubble? Are scientists, journalists, or venture capitalists the guiltiest? Whatever the answer, a host of start-up companies have seen the light of day, and have issued shares. The Yellow Pages will have to make room for quite a few names with the nano- prefix: Nanometrics, Nanogen, Nanologic, Nanopierce, Nanoproprietary, Nanosys, Nanotechnology Development Corporation are just a few. Some universities have invested heavily into setting-up industrial parks specialised in nanotechnology. SUNY Albany and the University of South Carolina are two examples. The promise is there, because some of the devices already made actually work. IBM was able to use a nanotechnique to increase by a factor 20 the amount of memory on a hard drive.

John Wolfe, who started the first nanotech venture-capital firm, compares it to the Internet circa 1993, before Netscape went public. Is it to say that this new bubble is doomed from the start and that it will burst, sooner or later? Assuredly. Is it to say that it will sink without leaving a trace? Assuredly not. Some companies will survive and be lasting. Already, one can differentiate between fly-by-night operations buoyed only by conning the gullible, and reputable firms such as Nanosys doing solid work, and worth investing into. Nanosys collaborates with well-established corporate giants such as Intel, Matsushita, and DuPont. Peugeot-Citroën and Ford inject sub-10 nm CeO_2 nanoparticles into diesel fuel before it is burnt in the engine. This results in an *in situ* catalyst that not only improves fuel efficiency but also greatly reduces the emission of particulates. More than a million automobiles will incorporate that system by 2005 already. The American Federal government is also among the investors in the Nano Rush. Washington is spending $ 3.7 billion on nanotech research over the next four years.

To this writer, the key issue is representation. Social groups create self-images, they project somewhat different images to the public.

Nanotechnology, as mentioned at the beginning of this essay, arose from one of the icons of the 20th century, the DNA double helix. It was successful, early on, in giving itself a similar and logo-like icon. IBM scientists created it and the advertising arm of the multinational corporation was ready to pounce and was quick to hammer it irreversibly into public consciousness. This is the STM image of a corral of 48 iron atoms placed in a circle of 7.3 nm radius. This particular image has become emblematic of nanotechnologies, at least for the time being. With it, scientists exhibited to the public, not without intense pride, an atom-made fence. What a feat! But this construction can also be seen as an unselfconscious picture of the proverbial ivory tower. In yet another image, nanotechnology proudly depicted in 1990 on the cover of *Nature* its archetype with an STM image of DNA.

Another graphic depiction of nanotechnologies brings up my next point, representation as re-presentation. The cover page in issue number 24 of *Angewandte Chemie International Edition in English,* volume 37 for 1998, was adorned with a schematic representation of the Mo_{132} cluster. Its similarity to Kepler's early model of the universe, featuring the ubiquity of the icosahedron, was emphasised in the illustration.

Thus, nanotechnology in the time of the Millennium rediscovered the microcosm-macrocosm correspondence prevailing in medieval metaphysics. With some modification, needless to say. Humanism is out: a metallic cluster as the nanocosm replaces man as the microcosm. Just as for medieval theologians, the stress is on the unity of nature, on the uniformity of physical law throughout the universe. But physical science has replaced God in Western thought.

One can interpret also the bow to Kepler as one of the numerous reincarnations of the Platonic archetype of the five regular solids, which has marked the history of chemistry in the braudelian *longue durée*: Kepler and the snowflake, the constituent particles of crystalline solids as proposed by René-Just Haüy, molecular shape according to André-Marie Ampère, Achille Le Bel's and Jacobus Henricus van't Hoff's tetrahedral carbon, Alfred Werner's complexes, Aaron Klug's icosahedral viruses and, closer yet to us, Richard Smalley's and Harold Kroto's buckminsterfullerene. Thus, nanochemists gave themselves a self-image of continuity within the geometric, Pythagorean tradition.

Let me sum up with representation as re-presentation. I have been at pains in this review to point to all the aspects of nanotechnology that predated it. Nanoscientists have moved existing pieces to create new patterns. There is nothing wrong with this efficient heuristics. Discovery is not the only path to knowledge, fame, and riches. But this raises an interesting question in philosophy and historiography, which will be my final point.

Science rewrites its own history, not only as history but as science fodder too. Which should not be construed in the weak sense of Whig history-telling as indulged in by scientists. While it is very much true that scientists, in a Whig perspective, tend to see the past as heralding the present, their purpose is not the mundane one of setting up an artificial history, approximating to a linear genealogy. The nanotechnologies paradigm shows a considerably more interesting attitude. Science reconfigures past events and developments in order to refocus itself into new directions. Such a reshuffling of the past, which of course may be deeply upsetting to professional historians who see their work voided and having to be restarted anew, the undersigned included, is an essential part of the dynamics of science. And it shows historians of chemistry that there is life after Partington.

Positional Control and the Synthesis of Diamond

We can in principle make a great many very useful structures from diamond, including very powerful computers and molecular robotic devices able to position molecular components to within a fraction of an atomic diameter. But how can we synthesize diamond? One answer to this question comes from looking at how we grow diamond today using *Chemical Vapor Deposition (CVD)*. In a process somewhat reminiscent of spray painting, we build up layer after layer of diamond on a surface by holding that surface in a cloud of very reactive molecules and atoms — like H; CH_3, C_2H, etc. When these reactive molecules bump into the surface they change it, either by adding, removing, or re-arranging atoms. By carefully controlling the pressure, temperature, and the exact composition of the gas, we can create conditions that favour the growth of diamond on the surface.

While these are the right chemical reactions to make diamond, randomly bombarding the growing surface with reactive molecules doesn't offer the finest control over the growth process. It would be like trying to build a wristwatch using a sandblaster. We want the chemical reactions to take place at precisely the places on the surface that we specify — not wherever random gusts of turbulent gas might dictate.

A second problem is that the hydrogenated diamond surface is chemically inert: it's difficult to add carbon (or anything else) to it. We could overcome this problem by removing a hydrogen atom from the surface, leaving behind a very reactive dangling bond. This is a critical first step during CVD diamond growth but it happens at random, anywhere on the growing diamond surface, when the right kind of reactive gas molecule happens to strike the surface in the right way. We want to do something similar but more controlled: remove a *specific* hydrogen atom from a *specific* spot on the diamond surface. To do this, we'll need a "hydrogen abstraction tool".

What might such a tool look like? If we base it on our understanding of diamond CVD, it should be a highly reactive radical with a high affinity for hydrogen. At the same time it must be possible to position it, so it must have a stable region which can serve as a "handle." The tool would be held by (for example) the molecular robotic arm discussed earlier, and would be positioned directly over the hydrogen we wish to abstract.

A simple strategy for finding such radicals is to look through the table of bond strengths in the *Handbook of Chemistry and Physics* and pick the molecule with the strongest bond to hydrogen. The radical created by removing the hydrogen should then have a very high affinity for hydrogen, as desired. Unfortunately, this strategy first produces fluorine. While atomic fluorine does indeed have a very high affinity for hydrogen, there is no obvious way to attach a "handle" to it. Interest therefore turns to the molecule with the second strongest bond to hydrogen: acetylene. Here, we are in luck. Not only does the acetylene radical have a high affinity for hydrogen, it also has a chemically stable region which can be modified into a "handle" (see illustration above).

Drexler realised this sometime in the early 1980's. The author subsequently and independently followed the same line of logic and reached the same conclusions. An initial check of the validity of this idea was provided by semi-empirical calculations at Xerox PARC (easily done on a workstation using freely available programmes). Discussions with *Bill Goddard and members of his molecular modeling group at Caltech* resulted in *more accurate calculations using high level ab initio quantum chemistry methods* on a more powerful computer. (These methods have the interesting property that the calculations will converge on the correct answer if enough computing power is used). *Drexler published this observation in Nanosystems* in 1992. A *group at NRL interested in diamond growth* found the proposal intriguing and modeled the abstraction of a hydrogen at room temperature using a qualitatively different computational model of molecular behaviour (which permitted them to model the behaviour of several hundreds of atoms in the vicinty of the abstraction). As the results of three different computational approaches have all produced the same qualitative answer, and as the results are also in accord with our intuition about what *should* happen, it's safe to say that this hydrogen abstraction tool will, when finally implemented, be able to abstract hydrogen.

The hydrogen abstraction tool illustrates some important ideas common to other proposals for molecular tools for the synthesis of diamond. First, the tool has a reactive end which is brought into contact with the molecular workpiece and an inert end which is held by a molecular positional device. Second, the point of application of the tool is controlled by the molecular positional device. Third, the environment around the tool is itself inert (typical proposals involve the use of either vacuum or a noble gas). The inert environment prevents the tool (which is quite reactive) from reacting with anything undesired.

While the hydrogen abstraction tool creates a very reactive spot on the surface we also need a tool that will deposit one or more carbon atoms on the growing surface. One proposal is the dimer deposition tool. (a "dimer" is just two of something stuck together. In this case, two carbon atoms are stuck together by a triple bond, hence the two carbon atoms form a "dimer"). A dimer deposition tool is simply a dimer with two weak bonds to a supporting structure. A proposal by Drexler is shown to the left. A second proposed dimer deposition tool, which has particularly weak bonds to the dimer, is illustrated at the right. [Subsequent note: further investigation of the proposal at right suggests it might not be stable except at very low temperatures. A variety of alternative dimer deposition tools are possible, so even if this particular structure should prove unsastisfactory for room temperature operation, other tools able to carry out the same function should be feasible.]

To illustrate how we might combine the use of the hydrogen abstraction tool with the dimer deposition tool, we proceed in three steps. First, we start with the hydrogenated diamond (111) surface. (The notation (111) refers to a particular surface of the diamond crystal. When diamond shatters, it shatters along certain planes and not along others. As these different planes have surfaces with different properties, it's important to specify which plane you're talking about). Second, we use the hydrogen abstraction tool to remove two adjacent hydrogen atoms. Third, we use the dimer deposition tool to deposit two carbon atoms on the surface — the two ends of the carbon dimer are originally single-bonded to the rest of the tool. The two dangling bonds on the surface are very reactive, and react with the ends of the carbon dimer. When this happens, the bonds holding the dimer to the dimer deposition tool break, and the bonds in the deposition tool rearrange to eliminate what would otherwise be two radicals. Result: the carbon dimer is transferred from the tool to the surface. The following illustration shows the dimer deposition tool adding a dimer to a small cluster which represents the diamond (111) surface. (Because accurate ab initio calculations are computationally less expensive for a small cluster of atoms, the following illustrations show only the atoms directly involved in the reaction).

This sequence of steps starts with a flat diamond surface and adds two carbon atoms to it. It could be repeated at other sites. A computational investigation of the final step was reported by *Stephen Walch at the recent workshop on computational nanotechnology* sponsored by the NAS programme at NASA. Because the energy released during the reaction is much larger than thermal noise (e.g., the reaction is quite exothermic), the reverse reaction is very unlikely to occur. The barrier to the reaction is also small. Taken together, these mean the reaction can take place easily and irreversibly. The dimer will "snap" onto the surface and stay there. (Calculations at the 6-31G* MP2 level show all positive vibrational frequencies for this tool. Further investigation of its stability is in progress).

A third proposed tool is the carbene insertion tool. "It is no exaggeration to claim a major role for carbenes in the modern chemists's attitude that he can very probably make anything he wants." (W. J. Baron). A positionally controlled carbene should be correspondingly more useful, as it could be made to react at any position on a growing molecular workpiece. Carbenes are very reactive and can insert readily into double and triple bonds. A plausible reaction would be to insert a carbene into the dimer described above.

A fourth proposal is for a hydrogen deposition tool. Where the hydrogen abstraction tool is intended to make an inert structure reactive by creating a dangling bond, the hydrogen deposition tool would do exactly the opposite: make a reactive structure inert by terminating dangling bonds. This tool could be used during the synthetic process to stabilize a structure that might otherwise undergo a spontaneous and undesired rearrangement or reconstruction, or it could be used at the end of the synthetic process to stabilize the finished product. There are many possible candidates for such a tool: any molecular structure that has a weak bond to a hydrogen. One particularly attractive candidate is the use of tin. The hydrogen-tin bond is quite weak, and so a tin-based hydrogen deposition tool should be quite effective.

These four tools let us (a) make an inert surface reactive by removing a hydrogen from one or more specific sites, (b) add one or two carbon atoms to a surface at selected sites and (c) make a reactive surface inert by adding hydrogen to it — which lets us prevent undesired surface rearrangements.

The Stiff Hydrocarbons as a Model System for Study

These four molecular tools — by themselves — should be sufficient to make a remarkably wide range of stiff hydrocarbons. This suggests that we define a simplified model of nanotechnology that we might call "hydrocarbon-based nanotechnology." This simplified version would only be able to make things from hydrogen and carbon — a much less ambitious goal than making things from the approximately 100 elements found in the full periodic table. But in exchange for confining ourselves to this more limited class of structures, we make it much easier to analyze the structures that can be fabricated and the synthetic reactions needed to make them. This narrower proposal can be more readily and more thoroughly investigated than full nanotechnology.

At the same time, hydrocarbon-based nanotechnology retains several of the key features of general nanotechnology. Diamond and shatterproof variants of diamond still fall within its purview; and it can still make all the parts needed for the operation of basically mechanical devices, including: struts, bearings, gears, robotic arms, and the like.

It's often useful to define a somewhat broader class than the stiff hydrocarbons, but a class which is still much more restrictive than the class of "most structures consistent with physical law." We'd like to be able to build molecular structures that are similar to diamond but which incorporate some other elements. We might want to add impurities (as in diamond electronic devices) or terminate a bearing surface with fluorine. Perhaps we're bending the diamond structure, and want to add some silicon internally to relieve the strain, (the silicon-carbon bond is longer than the carbon-carbon bond, so adding silicon to a region will "puff it up"). Or we find that adding some nitrogen to the internal surface of a bearing will relieve some strain (the nitogen-carbon bond is slightly shorter than the carbon-carbon bond). The term "diamondoid" is used to cover these (and related) possiblities. Diamondoid materials are made primarily of the elements hydrogen; first row elements like carbon, nitrogen, oxygen or fluorine; and second row elements like silicon, phosphorous, sulfur or chlorine; all held together by many strong covalent bonds. While including only about a tenth of the elements in the periodic table, the diamondoid materials include many structures of striking utility.

Self replication: making things inexpensively

Positional control combined with appropriate molecular tools should let us build a truly staggering range of molecular structures — but a few molecular devices built at great expense would hardly seem to qualify as a revolution in manufacturing. How can we keep the costs down?

Potatoes are a miracle of biology with tens of thousands of genes and proteins and intricate molecular machinery; yet we think nothing of eating this miracle, mashed with a little butter. Potatoes, along with many other agricultural products, cost less than a dollar a pound. The key reason: provide them with a little dirt, water and sunlight and a potato can make more potatoes.

If we could make a general purpose programmable manufacturing device which was able to make copies of itself (the *author does* work at *Xerox*, after all....), then the manufacturing costs for both the devices and anything they made could be kept quite low — likely no more than the costs for growing potatoes.

Drexler called such devices "assemblers."

The first serious analysis of self replicating systems was by *von Neumann* in the 1940's. *He carried out a detailed analysis of one such system in a theoretical cellular automata model* (The best known example of a cellular automata model is *Conway's Game of Life*, which is played out on a giant checkerboard and uses a simple set of rules to decide what happens at each square of the checkerboard at each time step). In von Neumann's cellular automata model he used a universal computer for control and a "universal constructor" to build more automata. The "universal constructor" was a robotic arm that, under computer control, could move in two dimensions and alter the state of the cell at the tip of its arm. By sweeping systematically back and forth, the arm could "build" any structure that the computer instructed it to. In his three-dimensional "kinematic" model, von Neumann retained the idea of a positional device (now able to position in three dimensions rather than two) and a computer to control it.

A *NASA study in 1980* extended the general conclusions of von Neumann and concluded that a fully automated lunar mining and manufacturing operation, able to extend itself using its own mining and manufacturing capabilities, would be feasible and could be built given a multi-billion dollar budget and a few decades of work. (Note that this kind of budget and time frame was in keeping with other proposed aerospace projects of the time).

The architecture for Drexler's assembler is a specialisation of the more general architecture proposed by von Neumann. As before, there is a computer and constructor, but now the computer has shrunk to a "molecular computer" while the constructor combines two features: a robotic positional device (such as the robotic arm discussed earlier) and a well defined set of chemical operations that take place at the tip of the positional device (such as the hydrogen abstraction reaction and the other reactions involved in the synthesis of diamond).

The complexity of a *self replicating system* need not be excessive. In this context the complexity is just the size, in bytes, of a "recipe" that fully describes how to make the system. The complexity of an assembler needn't be beyond the complexity that can be dealt with by today's engineering capabilities. As shown in the following table, there are several self replicating systems whose complexity is well within current capabilities. What is the "complexity" of a living system? We'll take this to mean the number of bytes in the DNA "blueprints." As each base pair in DNA can be one of four possibilities, it encodes two bits. One byte (8 bits) can be encoded in four base pairs. This means we count the number of base pairs in the DNA and divide by four to get the number of bytes in the "blueprints." For Mycoplasma genitalia, which has 580, 070 base pairs, this results in 145, 017.5 bytes. For humans, with roughly 3.2 billion base pairs, this results in 800 million bytes. (We're using the haploid base pair count: each cell in your body has DNA from your mother and DNA from your father — but much of this is similar. We're counting only the DNA from one parent). The complexity and sophistication of most living systems goes well beyond anything we might need just to achieve low cost manufacturing. Much of the complexity of the human genome is unrelated to issues of self-replication: people do more than just self-replicate!

The complexity of the internet worm is just an estimate of the number of bytes in its C programme. Because the environment in which it operates is highly structured and provides relatively easy access to complex and sophisticated software, it could be argued that its complexity might be much less than the complexity of a system that operated in a "simple" environment. We'll leave it in our table anyway, as it's an interesting data point. This argument is less applicable to von Neumann's universal constructor which operates in a simple environment: just a large two-dimensional checkerboard with a finite number of states at each square. Its complexity is an estimate of the number of bytes needed to describe the constructor. The complexity of the NASA Lunar Manufacturing Facility was estimated in the NASA study.

What will we be able to make?

Today, most airplanes are made from metal despite the fact that diamond has a strength-to-weight ratio over 50 times that of aerospace aluminum. Diamond is expensive, we can't make it in the shapes we want, and it shatters. Nanotechnology will let us inexpensively make shatterproof diamond (with a structure that might resemble diamond fibers) in exactly the shapes we want. This would let us make a Boeing 747 whose unloaded weight was 50 times lighter but just as strong.

Today, travel in space is *very* expensive and reserved for an elite few. *Nanotechnology will dramatically reduce the costs and increase the capabilities of space ships and space flight*. The strength-to-weight ratio and the cost of components are absolutely critical to the performance and economy of space ships: with nanotechnology, both of these parameters will be improved by one to two orders of magnitude. Improvements in these two parameters alone (ignoring other advantages provided by nanotechnology) should improve the overall cost/performance ratio by over three orders of magnitude. This has led the *National Space Society* (NSS) to adopt *a position paper supporting nanotechnology*; Dan Goldin, *NASA*'s chief administrator, to support nanotechnology; and *NAS at NASA Ames Research Center to start a project (now with perhaps half a dozen researchers) to examine molecular manufacturing systems and molecular machines using computational models*.

Beyond inexpensively providing remarkably light and strong materials for space ships, nanotechnology will also provide extremely powerful computers with which to guide both those ships and a wide range of other activities in space.

Today, computer chips are made using lithography — literally "stone writing." It's our finest manufacturing technology, but there seem to be fundamental limits in how much further we can improve it. In lithography, we draw fine lines on a silicon wafer by using methods borrowed from photography. A light-sensitive film — called a "resist" — is spread over the silicon wafer. The resist is exposed to a complex pattern of light and dark, like a negative in a camera, and developed. The exposed resist is then washed away, and a particular chemical is sprayed over the surface. Where the resist has been washed away, the chemical reaches the surface of the silicon and can diffuse a short distance into it. Where the resist has not been washed away, the spray is blocked. Finally, the remaining resist is washed away, along with any chemicals that it prevented from reaching the silicon surface. The result: a fine pattern of some desired chemical is laid out on the silicon surface. By repeating this process, an intricate set of interlocking patterns can be made that defines the complex logic elements of a modern computer chip.

If the computer hardware revolution is to continue at its current pace, in a decade or so we'll have to move beyond lithography to some new post lithographic manufacturing technology. Making patterns on a resist and spraying chemicals around simply can't arrange atoms with the ultimate precision that should be feasible. More precise methods will be needed. Ultimately, each logic element will be made from just a few atoms. Designs for computer gates with less than 1, 000 atoms have already been proposed — but each atom in such a small device has to be in exactly the right place. To economically build and interconnect trillions upon trillions of such small and precise devices in a complex three dimensional pattern we'll need a manufacturing technology well beyond today's lithography: we'll need nanotechnology. With it, we should be able to build mass storage devices that can store more than a hundred billion billion bytes in a volume the size of a sugar cube; RAM that can store a mere billion billion bytes in such a volume; and massively parallel computers of the same size that can deliver a billion billion instructions per second.

Today, "smart" weapons are fairly big — we have the "smart bomb" but not the "smart bullet." In the future, even weapons as small as a single bullet could pack more computer power than the largest supercomputer in existence today, allowing them to perform realtime image analysis of their surroundings and communicate with weapons tracking systems to acquire and navigate to targets with greater precision and control. We'll also be able to build weapons both inexpensively *and much more rapidly*, at the same time taking full advantage of the remarkable materials properties of diamond. Rapid and inexpensive manufacture of great quantities of stronger more precise weapons guided by massively increased computational power will alter the way we fight wars. Changes of this magnitude could destabilize existing power structures in unpredictable ways.

Military applications of nanotechnology raise a number of concerns that prudence suggests we begin to investigate before, rather than after, we develop this new technology. While molecular manufacturing will not arrive for many years, its obvious military potential will increasingly attract the interest of strategic planners. For example, in a talk titled *Nanotechnology and global security* at the *Fourth Foresight Conference on Molecular Nanotechnology, Admiral David E. Jeremiah,*

USN (Ret), former Vice Chairman of the Joint Chiefs of Staff, said "Military applications of molecular manufacturing have even greater potential than nuclear weapons to radically change the balance of power." As it seems implausible that military applications of this technology will never be developed and deployed, it would seem safer to encourage the relatively early interest of those organisations less prone to the abuse of power and more likely to curb its abuse by others. To make power today we dig coal and oil from the ground, we dam rivers, and we burn nuclear fuel in nuclear power plants. Yet the sunshine all around us could provide orders of magnitude more power than we use — and do so more cleanly and less expensively — if only we could make low cost solar cells and batteries. We already know how to make solar cells that are efficient enough for this application: it just costs too much to make them. Nanotechnology will cut costs both of the solar cells and the equipment needed to deploy them, making solar power economical. In this application we need not make new or technically superior solar cells: making inexpensively what we already know how to make expensively would move solar power into the mainstream.

Today, our surgical tools are large and crude at the molecular scale — yet the cellular and molecular machinery in our tissue is small and precise. There is a fundamental mismatch between the capabilities of our tools and what's needed to cure the injuries in our tissue. Today's scalpels are, as seen by a cell, large crude scythes that are more suited to cut and tear than to heal and mend. Ripping through tissue, they leave dead and maimed cells in their wake. The only reason that modern surgery works is the remarkable ability of cells to regroup, bury their dead, and heal over the wound.

It is not modern medicine that does the healing, but the cells themselves: we are but onlookers. If we had surgical tools that were molecular both in their size and precision, we could develop a medical technology that for the first time would let us directly heal the injuries at the molecular and cellular level that are the root causes of disease and ill health. With the precision of drugs combined with the intelligent guidance of the surgeon's scalpel, *we can expect a quantum leap in our medical capabilities*.

Nanotechnology should let us make almost every manufactured product faster, lighter, stronger, smarter, safer and cleaner. We can already see many of the possibilities as these few examples illustrate. New products that solve new problems in new ways are more difficult to foresee, yet their impact is likely to be even greater. Could Edison have foreseen the computer, or Newton the communications satellite?

Summary of the core technology

We can now see the fundamental shape of a molecular manufacturing technology. Self replicating assemblers, operating under computer control, let us inexpensively build more assemblers. The assemblers can be reprogrammed to build other products. The assemblers use programmable positional control to position molecular tools and molecular components, permitting the inexpensive fabrication of most structures consistent with physical law. Diamondoid materials in particular become inexpensive and commonplace, and their remarkable properties usher in what has been called the Diamond Age.

Developmental pathways

In the preceding discussion of nanotechnology we gave scant attention to possible developmental pathways that would let us migrate from our current technology to what we might call "mature" systems able to inexpensively manufacture most diamondoid structures. The proposals advanced so far have been driven largely by the desired goal: a system able to inexpensively synthesize most diamondoid structures. The proposal that emerged was a general purpose programmable manufacturing system which uses positionally controlled highly reactive tools in vacuum and is able to self replicate.

No constraint was imposed requiring that the proposed system be easy to fabricate given our current technology — and indeed, this problem appears non trivial. But we must eventually build such systems if they are to be of any use — how can we do this? To give the reader some feeling for the magnitude of the effort involved, we can compare the requirements for a mature system with present capabilities.

A mature system must be able to build an assembler with hundreds of millions or billions of atoms with no atom out of place. If we are to do this, and if each unit operation typically handles one or a few atoms (a hydrogen abstraction, a carbene insertion, etc), then either the error rate per unit operation must be low (less than one in a billion) or we must adopt error detection and correction methods.

A general rule of thumb is that making something right the first time is easier than making it wrong and fixing it. Given the choice between a manufacturing process that has a low enough error rate that most finished products will simply work correctly the first time versus a process which requires elaborate error detection and correction to make any working products at all, the former is greatly to be preferred. Because analysis of the fundamental causes of errors during the manufacture of diamondoid products in general and assemblers in particular supports the idea that error rates substantially lower than one in a billion should be achievable, and hence that an assembler can make another assembler with a high probability of success while using no error detection or correction, existing design proposals adopt the simpler and higher performance approach. Had the opposite conclusion been reached, i.e., that error rates could *not* be driven below one in one billion, then design proposals would have adopted a smaller module size (consistent with the error rate) and incorporated appropriate mechanisms for module testing, the rejection of bad modules, and the assembly of working modules into larger systems. This line of logic results in proposals for assemblers in which error rates are very low. These low error rates, however, are predicated on the prior existence of diamondoid assemblers. Unfortunately, we do not have assemblers today. Worse, current technology would be hard pressed to achieve such error rates. Chemists view a synthesis that provides 99% yield as very good. The synthesis of proteins from amino acids by ribosomes has an error rate of perhaps one in 10, 000. Replication of DNA, by using extensive error detection and correction, along with built in redundancy (DNA has two complementary strands), achieves an error rate of roughly one base in a billion (which varies depending on particular circumstances).

The size of structure that can be built by today's SPMs without error is smaller than has been achieved by chemists. The first such work *arranged 35 xenon atoms on a nickel surface at 4 kelvins in vacuum to spell out "IBM."* Later work in Japan spelled out "NANO SPACE" at room temperature in vacuum by removing individual sulfur atoms from the surface. More recently, *6 molecules were arranged in a hexagonal pattern at room temperature in vacuum*. As each molecule had 173 atoms, this modular approach was able to position over 1, 000 atoms. While these are great successes we have not yet seen a long sentence spelled out in this fashion, let alone a paragraph or a book. Further, SPMs can "see" the structure that's being built. Current successes use this ability to provide constant feedback to the human operator, who can detect and correct errors.

In short, today's SPMs can build structures that are only a small fraction of the size of proposed assemblers and have error rates high enough that they must use relatively sophisticated error detection and correction methods.

The situation is made more difficult by the additional requirement that unit operations be fast. If an assembler is to manufacture a copy of itself in about a day, and if this takes one hundred million to a billion operations, then each unit operation must take place in a fraction of a millisecond. Ribosomes take tens of milliseconds to add a single amino acid to a growing protein. Today's SPMs can take hours to arrange a few atoms or molecules.

At the same time, the use of vacuum prevents us from using self assembly or any of the other solution-based techniques developed by chemistry. Abandoning these powerful tools does not seem like a good first step.

While speed, reliability and operation in vacuum do not pose fundamental problems, the conclusion is obvious: directly building a diamondoid assembler using existing technology is a daunting task. Perhaps, rather than attempting to solve all the problems in a single giant leap, we should instead approach them in a more incremental fashion. To this end, we need to define one or more intermediate systems. Almost by definition, an intermediate system is easier to build from current technology, but is still able to make systems that are more powerful than itself. A series of intermediate systems, each able to build the next system in the chain, would let us adopt more incremental methods.

One approach is to eliminate the requirement that the assembler be made from diamondoid structures. We wanted to make diamondoid structures because of their remarkable strength, stiffness, electrical properties, etc. But an intermediate system need only be able to make a more advanced system, and perhaps products that are impressive in comparison with today's products. It doesn't have to be diamondoid itself.

This suggests what might be called "building block based nanotechnology." Rather than building diamond, we'll build some other material from molecular building blocks that are relatively large: tens, hundreds or even thousands of atoms in size. Larger building blocks reduce the number of assembly steps, so fewer unit operations are needed and they need not be as reliable. Soluble building blocks, with "linkage" groups that are selective (the building blocks only stick to other building blocks, not to the solvent or low concentrations of contaminants) eliminate the need for vacuum. We certainly have many choices: any of the wide range of molecules that chemists have synthesised or could reasonably synthesize which has the desired properties. *Krummenacker* concluded that each molecular building block should have at least three sites where it can link to other building blocks (two sites leads to the familiar polymers so ubiquitous in biological systems: proteins, DNA, RNA, etc. Addition of a third site makes the design of three dimensional stiff structures much easier). While such building blocks could be linked to each other using any one of a variety of reactions, a particularly attractive possibility is the Diels-Alder reaction. This reaction, well known to chemists, works in most solvents (or even in vacuum). It involves a reaction between a diene and a dieneophile. The reaction is specific: the diene and the dieneophile react with each other but seldom with other groups (see illustration above, courtesy of Krummenacker). Because it works in vacuum and doesn't produce any small molecules (which would destroy the vacuum), it could be used both in intermediate systems in solution, and later in systems that use vacuum to make diamondoid materials. Solution based systems could use positional control to assemble the building blocks, but can also use the methods of self assembly. In particular, the self assembly of a positional device should be feasible. If you ask someone familiar with the field if it would be feasible to self assemble a robotic arm, the usual response is some variation of "no." If, instead, you ask whether it would be feasible to self assemble an octahedron, the usual response is to point out that *Nadrian Seeman* of New York University has already self assembled a *truncated octahedron from DNA* and plans to do the same for a regular octahedron (both for reasons unrelated to positional control). (The computer generated graphic of the truncated octahedron at right is courtesy of Nadrian Seeman). For this work he received *The 1995 Feynman Prize in nanotechnology*.

As we have seen, an octahedron is the basic structure required for the Stewart platform. The truncated octahedron, because it has sides that aren't triangular, is floppier than a regular octahedron. Even the regular octahedron is likely to be too floppy for this application because of the limited stiffness of DNA. Further, the length of the edges cannot be changed (a requirement if we wish to control the position of the platform with respect to the base). However, self assembling a Stewart platform able to assemble molecular building blocks seems much less difficult than directly building a diamondoid assembler able to synthesize diamond in vacuum. Several ways to deal with the problem of inadequate stiffness and adjustable length edges are possible. One way would be to attach selectively sticky strands of DNA to a stiff molecular structure. This would let us use DNA (which has been intensively studied and whose selective stickiness is well known) to guide the process of self assembly, while the stiff molecular structure would would let us deal with concerns about inadequate stiffness. To change the length of an edge, we would need to use molecular structures that change shape in response to light, pressure, temperature, chemicals, or some other external signal. Many such molecular structures are known. Self assembling a Stewart platform whose edges are stiff enough to make a useful positional device, and the length of whose edges can be changed by a suitable signaling mechanism, is no longer a challenge that seems beyond our reach.

A second approach for positioning molecular building blocks in solution is to use an Atomic Force Microscope (AFM). This type of SPM relies on "touch" to create an image. By pushing on the structure being scanned, and feeling how hard it pushes back, the AFM can build up an image of stiff structures (it doesn't work very well when the applied force is strong enough to deform the structure). Because the AFM touches the surface of the structure being probed, it can also change that surface. To do this, it's very useful if the precise molecular structure of the tip can be controlled, so that the precise nature of the tip-surface interaction can be well defined.

This line of reasoning leads to the "molecular manipulator: " an AFM with a reactive fragment of an *antibody* bound to its tip. By changing the *antibody*, we can change what type of molecule will stick to the AFM tip. If the antibody both sticks selectively to a particular molecular building block, and also is bound to the AFM tip, we can now position the molecular building block by positioning the AFM tip. We can use any one of a wide range of possible molecular building blocks (*antibodies* can today be created that will bind to most small molecules). Again, this no longer seems beyond our reach.

Another developmental pathway is defined by the desire of the semiconductor industry to make ever smaller transistors despite the fact that optical lithography — the current workhorse manufacturing method for making computer chips — will reach a limit in a few more years. Transistors are made today by "drawing" very fine lines on silicon — but the optical methods in use today are limited by the wavelength of light: a few hundred nanometers. SPMs could let us make much smaller circuits by drawing much finer lines. Already demonstrated for making the finest and most critical lines in a transistor, if this approach can be made low cost and reliable it would let the entire semiconductor industry be retooled to use these finer lines to make more and smaller transistors. This, by itself, is not molecular nanotechnology; but once we learn to draw the smallest and finest lines of all, lines where every atom is in the right place (and there will be strong economic incentives to move in this direction), we could use the same techniques to make molecular machines. At first, these molecular machines would be very expensive and useful only for applications where their remarkable precision would justify the cost — but they should eventually lead to machines sophisticated enough to make simple assemblers.

Whether through self-assembly, by improvements in SPMs, some hybrid approach, or perhaps by some other path; we are moving from an era of expensive and imprecise products to an era of inexpensive products of molecular precision. We are going to replace most of the manufacturing base of the world with a fundamentally new and *much* better manufacturing technology.

The critics

Arguments that nanotechnology is infeasible are relatively rare and uniformly poor in technical quality. The best known are the arguments of David Jones, a *Nature* columnist and chemist. Quoted extensively in a *Scientific American* article on nanotechnology (see *http: / / www.foresight.org / SciAmDebate / SciAmOverview.html* for commentary) he advanced arguments like the following: "Single atoms ... are amazingly mobile and reactive. They will combine instantly with ambient air, water, each other, the fluid supporting the assemblers, or the assemblers themselves." However, as the proposals involving reactive molecular tools specify that the environment should be inert (i.e., vacuum) there is no "ambient air" to react with. As the molecular tools are positionally controlled, they will not react with each other or the assembler itself for the same reason that a hot soldering iron does not react with the skin of the person doing the soldering. Jones' criticisms raise greater questions about his understanding of the field than about the feasibility of nanotechnology. Other arguments against the feasibility of nanotechnology have likewise had obvious flaws, reminiscent of the 1920 New York Times editorial which said that rockets to the moon were impossible because there was no air to push against in space.

As should be clear from this article there are major technical challenges that must be overcome if we are to develop molecular manufacturing systems able to synthesize diamondoid structures. The serious questions, however, are about which development pathways should be pursued and how long it will take.

How long?

The single most frequently asked question about nanotechnology is: *how long?*. How long before it will let us make molecular computers? How long before inexpensive solar cells let us use clean solar power instead of oil, coal, and nuclear fuel? How long before we can explore space at a reasonable cost?

The scientifically correct answer is: *I don't know.*

Having said that, it is worth pointing out that the trends in the development of computer hardware have been remarkably steady for the last 50 years. Such parameters as

- the number of atoms required to store one bit
- the size of a transistor
- the energy dissipated by a single logic operation
- the resolution of the finest machining technology
- the cost of a computer gate

have all declined with great regularity, even as the underlying technology has changed dramatically. *From relays to vacuum tubes to transistors to integrated circuits to Very Large Scale Integrated circuits (VLSI) we have seen steady declines in the size and cost of logic elements and steady increases in their performance.*

Extrapolation of these trends suggests we will have to develop molecular manufacturing in the 2010 to 2020 time frame if we are to keep the computer hardware revolution on schedule.

Of course, extrapolating past trends is a philosophically debatable method of technology forecasting. While no fundamental law of nature prevents us from developing nanotechnology on this schedule (or even faster), there is equally no law that says this schedule will not slip.

Much worse, though, is that such trends imply that there is some ordained schedule — that nanotechnology will appear regardless of what we do or don't do. Nothing could be further from the truth. How long it takes to develop this technology depends very much on what we do. If we pursue it systematically, it will happen sooner. If we ignore it, or simply hope that someone will stumble over it, it will take much longer. And by using theoretical, computational and experimental approaches together, we can reach the goal more quickly and reliably than by using any single approach alone.

While some advances are made through serendipitous accidents or a flash of insight, others require more work. It seems unlikely that a scientist would forget to turn off the bunsen burner in his lab one afternoon and return to find he'd accidentally made a Space Shuttle.

Like the first human landing on the moon, the Manhatten project or the development of the modern computer the development of molecular manufacturing will require the coordinated efforts of many people for many years. How long will it take? A lot depends on when we start.

Engines of Healing: The Coming Era of Nanotechnology

We will use *molecular technology* to bring health because the human body is made of molecules. The ill, the old, and the injured all suffer from mis-arranged patterns of *atoms*, whether mis-arranged by invading viruses, passing time, or swerving cars. Devices able to rearrange atoms will be able to set them right. *Nanotechnology* will bring a fundamental breakthrough in medicine.

Physicians now rely chiefly on surgery and drugs to treat illness. Surgeons have advanced from stitching wounds and amputating limbs to repairing hearts and re-attaching limbs. Using microscopes and fine tools, they join delicate blood vessels and nerves. Yet even the best micro-surgeon cannot cut and stitch finer tissue structures. Modern scalpels and sutures are simply too coarse for repairing *capillaries*, *cells*, and molecules. Consider "delicate" surgery from a cell's perspective: a huge blade sweeps down, chopping blindly past and through the molecular machinery of a crowd of cells, slaughtering thousands. Later, a great obelisk plunges through the divided crowd, dragging a cable as wide as a freight train behind it to rope the crowd together again. From a cell's perspective, even the most delicate surgery, performed with exquisite knives and great skill, is still a butcher job. Only the ability of cells to abandon their dead, regroup, and multiply makes healing possible.

Yet as many paralysed accident victims know too well, not all tissues heal. Drug therapy, unlike surgery, deals with the finest structures in cells. Drug molecules are simple molecular devices. Many affect specific molecules in cells. Morphine molecules, for example, bind to certain receptor molecules in brain cells, affecting the neural impulses that signal pain. Insulin, beta blockers, and other drugs fit other receptors. But drug molecules work without direction. Once dumped into the body, they tumble and bump around in solution haphazardly until they bump a target *molecule*, fit, and stick, affecting its function.

Surgeons can see problems and plan actions, but they wield crude tools; drug molecules affect tissues at the molecular level, but they are too simple to sense, plan, and act. But molecular machines directed by nanocomputers will offer physicians another choice. They will combine sensors, programmes, and molecular tools to form systems able to examine and repair the ultimate components of individual cells. They will bring surgical control to the molecular domain.

These advanced molecular devices will be years in arriving, but researchers motivated by medical needs are already studying molecular machines and molecular engineering. The best drugs affect specific molecular machines in specific ways. Penicillin, for example, kills certain *bacteria* by jamming the nanomachinery they use to build their cell walls, yet it has little effect on human cells.

Biochemists study molecular machines both to learn how to build them and to learn how to wreck them. Around the world (and especially the Third World) a disgusting variety of viruses, bacteria, protozoa, fungi, and worms parasitize human flesh. Like penicillin, safe, effective drugs for these diseases would jam the parasite's molecular machinery while leaving human molecular machinery unharmed. Dr. Seymour Cohen, professor of pharmacological *science* at SUNY (Stony Brook, New York), *argues that biochemists should systematically study the molecular machinery of these parasites*. Once biochemists have determined the shape and function of a vital protein machine, they then could often design a molecule shaped to jam it and ruin it. Such drugs could free humanity from such ancient horrors as schistosomiasis and leprosy, and from new ones such as AIDS.

Drug companies are already redesigning molecules based on knowledge of how they work. Researchers at Upjohn Company have *designed and made modified molecules of vasopressin*, a hormone that consists of a short chain of *amino acids*. Vasopressin increases the work done by the heart and decreases the rate at which the kidneys produce urine; this increases blood pressure. The researchers designed modified vasopressin molecules that affected receptor molecules in the kidney more than those in the heart, giving them more specific and controllable medical effects. More recently, they designed a modified vasopressin molecule that binds to the kidney's receptor molecules without direct effect, thus blocking and *inhibiting* the action of natural vasopressin.

Medical needs will push this work forward, encouraging researchers to take further steps toward protein design and molecular engineering. Medical, military, and economic pressures all push us in the same direction. Even before the *assembler* breakthrough, molecular technology will bring impressive advances in medicine; trends in biotechnology guarantee it. Still, these advances will generally be piecemeal and hard to predict, each exploiting some detail of biochemistry. Later, when we apply assemblers and technical AI systems to medicine, we will gain broader abilities that are easier to foresee.

To understand these abilities, consider cells and their self-repair mechanisms. In the cells of your body, natural radiation and noxious chemicals split molecules, producing reactive molecular fragments. These can misbond to other molecules in a process called *cross-linking*. As bullets and blobs of glue would damage a machine, so radiation and reactive fragments damage cells, both breaking molecular machines and gumming them up.

If your cells could not repair themselves, damage would rapidly kill them or make them run amok by damaging their control systems. But *evolution* has favoured organisms with machinery able to do something about this problem. The self-replicating factory system sketched in this encyclopaedia repaired itself by replacing damaged parts; cells do the same. So long as a cell's *DNA* remains intact, it can make error-free tapes that direct ribosomes to assemble new protein machines.

Unfortunately for us, DNA itself becomes damaged, resulting in mutations. Repair *enzymes* compensate somewhat by detecting and repairing certain kinds of damage to DNA. These repairs help cells survive, but existing repair mechanisms are too simple to correct all problems, either in DNA or elsewhere. Errors mount, contributing to the aging and death of cells—and of people.

8

Soft Nanotechnology, Nano-Vasculoid and Biomimetic Self-Assembly: Industrial Applications and Processes

Understanding Nanotechnology for Life, Mind, and Machines

Does it make sense to describe cells as "machinery, " whether self-repairing or not? Since we are made of cells, this might seem to reduce human beings to "mere machines, " conflicting with a holistic understanding of life.

But *a dictionary definition of holism* is "the theory that reality is made up of organic or unified wholes that are greater than the simple sum of their parts." This certainly applies to people: one simpler sum of our parts would resemble hamburger, lacking both mind and life. The human body includes some ten thousand billion *billion* protein parts, and no machine so complex deserves the label—mere." Any brief description of so complex a system cannot avoid being grossly incomplete, yet at the cellular level a description in terms of machinery makes sense. Molecules have simple moving parts, and many act like familiar types of machinery. Cells considered as a whole may seem less mechanical, yet biologists find it useful to describe them in terms of molecular machinery.

Biochemists have unraveled what were once the central mysteries of life, and have begun to fill in the details. They have traced how molecular machines break food molecules into their building blocks and then reassemble these parts to build and renew tissue. Many details of the structure of human cells remain unknown (single cells have billions of large molecules of thousands of different kinds), but biochemists have mapped every part of some viruses. Biochemical laboratories often sport a large wall chart showing how the chief molecular building blocks flow through bacteria. Biochemists understand much of the process of life in detail, and what they don't understand seems to operate on the same principles. The mystery of heredity has become the industry of genetic engineering. Even embryonic development and memory are being explained in terms of changes in biochemistry and cell structure.

In recent decades, the very quality of our remaining ignorance has changed. Once, biologists looked at the process of life and asked, "How can this be?" But today they understand the general principles of life, and when they study a specific living process they commonly ask, "Of the many ways this could be, which has nature chosen?" In many instances their studies have narrowed the competing explanations to a field of one. Certain biological processes—the coordination of cells to form growing embryos, learning brains, and reacting immune systems—still present a real challenge to the imagination. Yet this is not because of some deep mystery about how their parts work, but because of the immense complexity of how their many parts interact to form a whole.

Cells obey the same natural laws that describe the rest of the world. Protein machines in the right molecular environment will work whether they remain in a functioning cell or whether the rest of the cell was ground up and washed away days before. Molecular machines know nothing of "life" and "death."

Biologists—when they bother—sometimes define life as the ability to grow, replicate, and respond to stimuli. But by this standard, a mindless system of replicating factories might qualify as life, while a conscious *artificial intelligence* modeled on the human brain might not. Are viruses alive, or are they "merely" fancy molecular machines? No experiment can tell, because nature draws no line between living and nonliving. Biologists who work with viruses instead ask about viability: "Will this *virus* function, if given a chance?" The labels of "life" and "death" in medicine depend on medical capabilities: physicians ask, "Will this patient function, if we do our best?" Physicians once declared patients dead when the heart stopped; they now declare patients dead when they despair of restoring brain activity. Advances in cardiac medicine changed the definition once; advances in brain medicine will change it again.

Just as some people feel uncomfortable with the idea of machines thinking, so some feel uncomfortable with the idea that machines underlie our own thinking. The word "machine" again seems to conjure up the wrong image, a picture of gross, clanking metal, rather than signals flickering through a shifting weave of neural fibers, through a living tapestry more intricate than the mind it embodies can fully comprehend. The brain's really machinelike machines are of molecular size, smaller than the finest fibers.

A whole need not resemble its parts. A solid lump scarcely resembles a dancing fountain, yet a collection of solid, lumpy molecules forms fluid water. In a similar way, billions of molecular machines make up neural fibers and synapses, thousands of fibers and synapses make up a neural cell, billions of neural cells make up the brain, and the brain itself embodies the fluidity of thought.

To say that the mind is "just molecular machines" is like saying that the Mona Lisa is "just dabs of paint." Such statements confuse the parts with the whole, and confuse matter with the pattern it embodies. We are no less human for being made of molecules.

Analysing Protein Design as a Pathway to Molecular Manufacturing

Feynman's 1959 talk entitled "There's Plenty of Room at the Bottom" discussed microtechnology as a frontier to be pushed back, like the frontiers of high pressure, low temperature, or high vacuum. He suggested that ordinary machines could build smaller machines that could build still smaller machines, working step by step down toward the molecular level; he also suggested using particle beams to define two-dimensional patterns. Present microtechnology (exemplified by integrated circuits) has realised some of the potential outlined by Feynman by following the same basic approach: working down from the macroscopic level to the microscopic.

Present microtechnology handles statistical populations of atoms. As the devices shrink, the atomic graininess of matter creates irregularities and imperfections, so long as atoms are handled in bulk, rather than individually. Indeed, such miniaturisation of bulk processes seems unable to reach the ultimate level of microtechnology — the structuring of matter to complex atomic specifications. In this paper, I will outline a path to this goal, a general molecular engineering technology. The existence of this path will be shown to have implications for the present.

Although the capabilities described may not prove necessary to the achievement of any particular objective, they will prove sufficient for the achievement of an extraordinary range of objectives in which the structuring and analysis of matter are concerned. The claim that devices can be built to complex atomic specifications should not, however, be construed to deny the inevitability of a finite error rate arising from thermodynamic effects (and radiation damage). Such errors can be minimised through the use of free energy in error-correcting procedures (including rejection of faulty components before device assembly); the effects of errors can be minimised through fault-tolerant design, as in macroscopic engineering.

The emphasis on devices that have general capabilities should be taken in the spirit of early work on the theoretical capabilities of computers, which did not attempt to predict such practical embodiments as specialised or distributed computation systems. The present argument, however, will proceed from step to step by close analogies between the proposed steps and past developments in nature and technology, rather than by mathematical proof. We commonly accept the feasibility of new devices without formal proof, where analogies to existing systems are close enough: consider the feasibility of making a clock from zirconium. The detailed design of many specific devices to render them describable by dynamical equations would be a task of another order (consider designing a clock from scratch) and appears unnecessary to the establishment of the feasibility of certain general capabilities.

Protein design

Biochemical systems exhibit a "microtechnology" quite different from ours: they are not built down from the macroscopic level but up from the atomic. Biochemical microtechnology provides a beachhead at the molecular level from which to develop new molecular systems by providing a variety of "tools" and "devices" to use and to copy. Building with these tools, themselves made to atomic specifications, we can begin on the far side of the barrier facing conventional microtechnology. What can be built with these tools? Gene synthesis and recombinant DNA technology can direct the ribosomal machinery of bacteria to produce novel proteins, which can serve as components of larger molecular structures. One might think assembly of such components into complex systems would require a preexisting technology able to handle molecules and assemble them; fortunately, biochemistry demonstrates that intermolecular attraction between complementary surfaces can assemble complex structures from solution. For example, the complex machinery of the ribosome self-assembles from more than 50 different protein molecules and can do so in vitro.

At present, the design of protein systems as complex as a ribosome seems an awesome task. Indeed, chemists cannot yet predict the three-dimensional conformation of a natural protein from its amino acid sequence, an ability that might seem requisite to the design of new proteins. Two considerations suggest that this obstacle is surmountable: first, the continuing improvement in protein science and, second, the difference between natural science and design engineering.

Regarding the first, computer simulation of protein molecules in solution (5) shows promise. As computer technology and chemical knowledge improve, simulations will increase in accuracy, speed, and size. Improvement promises new insight into protein behaviour and may permit the designer to modify (simulated) molecules quickly and to observe their behaviour directly.

Regarding the second consideration, natural scientists seek a more general understanding than design engineers require. Science seeks the ability to predict the conformations of all natural polypeptides. In attempting this, protein chemists can search for a minimum-energy chain conformation (in hope that the protein assumes not a local but a global minimum-energy conformation) (6) or can attempt to follow the chain-folding mechanism to find the final conformation (7). Prediction will be easier if the natural conformation has outstanding stability or if its folding mechanism proceeds in a sequence of strongly preferred steps. Unfortunately, natural selection accepts polypeptides that have natural conformations of low stability (in energetic terms) so long as they exhibit long lifetimes on the cellular time scale (or renature readily). Similarly, natural selection accepts any folding process so long as the chain reaches its natural conformation with essentially 100% yield. Moreover, random mutations are unlikely to enhance the stability of a particular conformation (or the predictability of its folding mechanism). Thus, natural proteins tend to accumulate disruptive changes until they reach the threshold of poor stability or reduced yield of the natural conformation; only then does natural selection come into play. Thus, it is little wonder that chemists cannot yet predict the conformations of natural proteins; they are not designed to fold predictably.

Engineers (in contrast to scientists) need not seek to understand all proteins but only enough to produce useful systems in a reasonable number of attempts. An engineer designing a protein that has 1000 amino acids may choose among some 10^{1300} different amino acid sequences. It might be that only one in 10^9 (or even 10^{700}) randomly selected sequences would yield a predictable conformation, yet this tiny fraction represents a vast number of proteins. Through use of strategically placed charged groups, polar groups, disulfide bonds, hydrogen bonds, and hydrophobic groups, the engineer should be able to design proteins that not only fold predictably to a stable structure (sometimes) but that serve a planned function as well. Even a low success rate will lead to an accumulation of successful designs. Thus, the difficulties encountered in predicting the conformations of natural proteins do not seem insurmountable obstacles to protein engineering.

Computer modeling and chemical understanding of biological targets have already found use in pharmaceutical design (8), and an artificial 34-residue polypeptide designed to interact with RNA has been synthesised and found active (9). It has been proposed to give microcircuitry special sensitivities by adsorbing engineered proteins onto selected surfaces (10). The promise of enzyme design in chemical engineering is evident. As protein science has great promise and difficulties in understanding natural proteins need not block engineering, the substantial payoffs for improved capabilities should lead to development of protein design technology. It would be foolish to underestimate the time and effort that will be required to develop basic design capabilities and then a broad family of working molecular devices; still, the path seems clear to achieving the capabilities exhibited by existing biochemical systems, by copying their features if need be.

Self-Assembly Processes for Nanotech Applications

Researchers at the University of Massachusetts Amherst have developed a series of novel techniques in nanotechnology that hold promise for applications ranging from highly targeted pharmaceutical therapies, to development of nutrition-enhanced foods known as "nutraceuticals, " to nanoscopic sensors that might one day advance medical imaging and diagnostics. The research, published in the Jan. 10 issue of Science, was funded by the U.S. Department of Energy and the National Science Foundation. The team included faculty members Thomas Russell and Todd Emrick of the department of polymer science and engineering, and Anthony Dinsmore of the department of physics, and graduate students Yao Lin and Habib Skaff, both of polymer science and engineering. "Our findings open new avenues to revolutionize technology by the controlled fabrication of nanoscopic materials having unique optical, magnetic and electronic properties, " said Russell.

The study details three major findings:

- A novel method to create robust capsules from nanometer-sized particles
- A new technique to make nanoscopic particles water-soluble
- Functionalising regions of the capsules with tailored properties, such as luminescence.

Emrick's research explores the behaviour of nanoparticles to which ligands organic molecules and polymers have been attached. Russell is an expert in the surface and interfacial properties of polymers, and polymer-based nanostructures. Dinsmore specializes in colloidal assemblies and interface physics. "This is a productive collaboration in that we really have all the bases covered in terms of synthesis, understanding of interfacial activity and mediation, and the physics issues including surface tension and particle interactions, " Emrick said. The study details a new method for assembling nanoparticles into robust, three-dimensional structures by encapsulating and stabilising water droplets. Nanoparticles suspended in oil will self-assemble around a droplet of water, fully coating it with a shell. Although scientists have long known that particles tend to assemble at fluid interfaces, "the idea of using liquid interfaces as scaffolds is exciting and tremendously useful since researchers can tailor or modify the nanoparticles from both sides of the interface, " explained Dinsmore. "We have much more surface area to work with for adding or removing specific particles."

"Nanoparticles have exciting properties due to their small size, and they can be prepared in various shapes and sizes. What's really key is that you attach ligands that extend from the nanoparticles like hairs, in order to preserve the nanoscopic integrity of the particles and prevent them from clustering, " Emrick said. "Changing the nature of these organic ligands can really modify the behaviour of the particles. You can endow the nanoparticles, and thus the capsules that they form upon interfacial assembly, with a wide range of properties based on which ligands are attached." The effect of the ligands on the interactions of nanoparticles with the surrounding environment is crucial in medical applications. "These organic molecules will dictate the solubility, miscibility, and charge transport properties of the particles, " Emrick said.

UMass researchers also developed a method to take these nanoparticles, which are oil-soluble, and make them water-soluble, simply by shining light on them. "Developing nanoparticles that are water-soluble has significant implications for medicine in the biosensors area, " Emrick said. "Using luminescent material, as we did, could lead to advances in very sophisticated medical-imaging techniques as the fluorescent nature of these particles allows them to be viewed and tracked over time."

Finally, the UMass team discovered that when nanoparticles of different sizes compete for assembly at the interface, the bigger ones win, and segregate or cluster into patches on the droplet surface. "This opens a range of possibilities for developing nanoscopic capsules that have certain properties in specific areas, " said Dinsmore. "You could build in an area with permeability, magnetism, or conductivity, so that one area would be functionally distinct."

In just a few decades physicians could be sending tiny machines into our bodies to diagnose and cure disease. These nanodevices will be able to repair tissues, clean blood vessels and airways, transform our physiological capabilities, and even potentially counteract the aging process. It is the year 2031, and the age of advanced nanomedicine has arrived. A young man arrives at his physician's office with a mild fever, nasal congestion, discomfort and a cough. The physician pulls from her pocket a lightweight, handheld device resembling a pocket calculator. She unsnaps from it a cordless, self-sterilising, pencil-sized probe and inserts it into the patient's mouth as if it were a tongue depressor. On the tip of the probe are billions of molecular assay receptors, mounted on hundreds of self-guiding retractile stalks. Each receptor is sensitive to the chemical signature of a specific kind of bacterium or virus. "Ahhh, " says the patient, and a few seconds later a three-dimensional, color-coded map of his throat appears on the display panel of the device. Beneath the map scroll columns of data, revealing the unique molecular signature of a known and unwelcome bacterial pathogen. With the diagnosis complete, the infectious microorganism can be exterminated. No need for antihistamines, cough drops and a week-long course of antibiotics. The physician keeps several generic classes of nanorobots in her office for just such a circumstance. Using a desktop appliance in her office, she programmes billions of nanorobots to find, recognize and destroy the particular microbial strain. The nanomachines are suspended in a carrier fluid that the patient inhales into his lungs, after which the mobile devices march down the patient's throat, propelled on tiny legs. Following a search pattern, the nanorobots ingest and destroy the harmful bacteria they encounter using mechanical and chemical phagocytosis. The patient feels nothing: nanorobots are the size of bacteria, which constantly crawl on and inside the body without ever being noticed. After several minutes, the physician activates an acoustic homing beacon to guide the nanorobots back into the patient's mouth, where she retrieves them through a collection port on the tip of the homing device. A further survey with the original diagnostic probe reveals no evidence of the pathogen. This is a wonderful vision of medicine in the future, but how do we get there from here? Nanotechnology ("nano" from the Greek word for dwarf) involves engineering and manufacturing at the molecular, or nanometer, level. One nanometer is one billionth of a meter, about the width of six bonded carbon atoms. How does one build at this scale? Scientists at New York University have adopted the self-assembly approach, and have succeeded in producing complementary strands of DNA that can zip themselves into complex structures. They have built cubes, octahedra and other solids made of just a few thousand nucleotides each, by the billions per batch. Meanwhile at Cornell University, researchers have genetically engineered a natural biomotor normally found in the enzyme ATPase to allow it to incorporate nonbiological parts such as a silicon nitride bar 100 nanometers in length, making the first artificial hybrid nanomotor. In a microscopic video presentation, dozens of bars attached to artificial molecular motors in a large precise array could be seen spinning at 200 revolutions per minute, like a field of tiny propellers.

Another way to build at the molecular level is by positional assembly—picking and placing molecular parts exactly where you want them. A device capable of positional assembly would work much like the robot arms that manufacture cars on automobile assembly lines, or like the ribosome that assembles amino acids one by one into proteins in our cells. The US engineering firm Zyvex Corp. is aiming to be the first to create an artificial molecular assembler using positional assembly to manufacture atomically precise structures.

Elsewhere, researchers at Harvard University have constructed the first general-purpose "nanotweezers, " using a pair of electrically controlled carbon nanotubes These have successfully grasped 300-nanometer clusters of polystyrene spheres and extracted a single semiconductor wire 20 nanometers wide from a mass of entangled wires. The scientists hope eventually to produce nanotweezers small enough to grab individual large molecules. Such nanoresearch will eventually lead to applications affecting almost every aspect of modern life, including the home, transport, computers, energy and the environment. Some of the most important and life-changing applications, however, are likely to be seen in medicine. We may see the first nanomedical materials and devices in use within the next few years. Scientists at the University of Michigan's Center for Biologic Nanotechnology are currently pursuing the use of dendrimers as a safer and more effective genetic therapy agent. Dendrimers are branching synthetic molecules that can be grown nanometer by nanometer to reach the desired size. They take on a spherical shape, and have sufficiently large openings and cavities to carry small molecules—dendrimers could therefore be used to sneak DNA into cells without triggering an immune response. At the same center, researchers have reported using dendrimer "nanodecoys" to trap and deactivate influenza virus particles. Relatively simple nanodevices could soon offer cures for major conditions such as diabetes. Researchers at Ohio State University and the University of Illinois have managed to construct silicon-based microcapsules highly perforated with "nanopores, " each as small as 20 nanometers in diameter. These pores are large enough to allow small molecules such as oxygen, glucose and insulin to pass through, but are small enough to impede the passage of much larger immune system molecules such as immunoglobulins. Microcapsules containing islet cells could be implanted beneath the skin of some diabetes patients, temporarily restoring the body's delicate glucose control feedback loop without the need for powerful immuno-suppressants that can leave the patient at serious risk for infection. The same method of encapsulation could be used to treat other enzyme- or hormone-deficiency diseases. Encapsulated neurons could also be implanted in the brain and then electrically stimulated to release neurotransmitters, possibly as part of a future treatment for Alzheimer's or Parkinson's disease.

Artificial "biobots" could be in our bodies within five to ten years. Advances in genetic engineering are likely to allow us to construct an artificial microbe—a basic cellular chassis—to perform certain functions. These biobots could be designed to produce vitamins, hormones, enzymes or cytokines in which the host body was deficient, or they could be programed to selectively absorb and break down poisons and toxins. A new company called EngeneOS, Inc., founded in late 2000, has already announced plans to develop artificial Engineered Genomic Operating Systems using the techniques of molecular biology. These systems will comprise a library of component device modules and proprietary modular components. This will allow the engineering and construction of programmable biobots with novel form and function. The greatest power of nanomedicine will emerge in the longer term, perhaps 10-20 years from now, when we learn how to design and construct complete artificial nanorobots using strong diamond-like materials, nanometer-scale parts, and onboard subsystems including sensors, motors, manipulators, power plants, and molecular computers. One example is an artificial mechanical cell called a respirocyte, which could be used to keep a patient's tissues safely oxygenated for up to about four hours (at maximum dosage) if their heart has stopped beating. These cells (side panel, 4) could also enable a healthy person to sprint at top speed for at least 15 minutes without breathing, or to sit underwater at the bottom of a swimming pool for hours. Still entirely theoretical, the respirocyte is a micron-wide spherical nanorobot made of 18 billion atoms precisely arranged in a diamondoid structure to form a tiny 1, 000-atmosphere pressure tank. Several billion molecules of oxygen and carbon dioxide can be absorbed into, or released from, this tank using computer-controlled molecular pumps powered by serum glucose and oxygen. External gas concentration sensors would allow respirocytes to mimic the action of the natural hemoglobin-filled red blood cells, with oxygen released and carbon dioxide absorbed in the tissues, and vice versa in the lungs. Each respirocyte would be able to hold 200 times more gas per unit volume than a natural red cell, so a few cubic centimeters injected into the human bloodstream would exactly replace the gas carrying capacity of the patient's entire 5.4 liters of blood.

Other proposed medical nanorobots offer equally astonishing performance improvements over nature. For instance, micronsize artificial mechanical platelets could allow complete hemostasis (control of bleeding by clot formation) in just one second, even for moderately large wounds, a response time 100-1, 000 times faster than the natural system. This would be achieved by rapidly unfurling a compactly stowed onboard biodegradable mesh under the control of an onboard computer. These "clottocytes" would be about 10, 000 times more effective as clotting agents than an equal volume of natural platelets. Nanorobotic phagocytes called microbivores could patrol the bloodstream, seeking out and digesting unwanted pathogens including bacteria, viruses or fungi. Each nanorobot could completely destroy one pathogen in just 30 seconds—about 100 times faster than natural leukocytes or macrophages—releasing a harmless effluent of amino acids, mononucleotides, fatty acids and sugars. No matter that a bacterium has acquired multiple drug resistance to antibiotics or to any other traditional treatment. The microbivore will eat it anyway, achieving complete clearance of even the most severe septicemic infections in minutes to hours, as compared to weeks or even months for antibiotic-assisted natural phagocytic defenses, without increasing the risk of sepsis or septic shock. Related nanorobots could be programed to recognize and digest cancer cells, or to clear circulatory obstructions within minutes in order to rescue stroke patients from ischemic damage. More sophisticated medical nanorobots will be able to intervene at the cellular level, performing surgery within cells. Physician-controlled nanorobots could extract existing chromosomes from a diseased cell and insert newly manufactured ones in their place, a process called chromosome replacement therapy. This would allow a permanent cure of any pre-existing genetic disease, and permit cancerous cells to be reprogramed to a healthy state.

In recent years, many gerontologists have begun to think of aging as a genetic disease that might be cured. Evidence from the Human Genome Project suggests that just a few hundred genes at most may be directly involved in aging. If we can completely understand these few genes and how they work, we may be able to alter them to eliminate this unwanted syndrome. Using cytosurgical nanorobots, corrected genes could be installed in every one of the 10 trillion tissue cells in our bodies. We would then no longer naturally age, and our bodies would again repair themselves as well as they did when we were children. Although nanotechnology is in its infancy, researchers are steadily making major breakthroughs. If we can learn to harness and precisely control the ability to manipulate molecules, then many aspects of our lives will change forever. In particular, the ability to carry out medical procedures at the molecular level will revolutionize medical practice. The next few decades will be very interesting indeed.

A new era on medicine are expected to happen in the coming years. Due to the advances in the field of nanotechnology, nanodevice manufacturing has been growing gradually. From such achievements in nanotechnology, and recent results in biotechnology and genetics, the first operating biological nanorobots are expected to appear in the coming five years, and more complex diamondoid based nanorobots will become available in about ten years. In terms of time it means a very near better future with significant improvements in medicine. In this work we present a practical approach taken on developing nanorobots for medicine in the sense of using computational nanomechatronics techniques as ancillary tools for investigating manufacturing design, nanosystems integration, sensing and actuation for medicine applications. Thus, the work describes pathways that could enable design testability, but also help scientists and profit corporations in providing the helpful information needed to test and design integrated devises and solutions towards manufacturing biomedical nanorobots. The use of robots in surgery has provided additional tools for surgeons enabling minimally invasive intervention or even long distance teleoperated surgeries. Indeed we may trust on human creativeness and technical capabilities that can ever be improved in terms of technical achievements. In recent years the medicine has enabled significant wellness for the life quality and longevity of the world population. And for the coming years, we may be prepared to experiment even more benefits, as results from advances that are being pursued step by step in new fields of science, such as nanobiotechnlogy. With the expected miniaturisation of devices provided by several works on nanoelectromechanical systems (NEMS), nanomanufacturing has actually become a reality. Hence, with the NEMS recent advances on building nanodevices, and the development of interdisciplinary works, altogether may be translated in few years through the development of integrated nanomachines, also known as nanorobots. With the use of techniques that are advancing rapidly, such as nano-transducers, and biomolecular computing, nanorobots are expected to be able to operate in a well defined set of behaviours performing pre-programed tasks. Thus, in the coming few years, nanorobots being teleoperated to perform surgery, or even nanorobots continually supervising the human body in order to assist organs that may require some kind of repair, is one of the most expected revolutionary tools for biomedical engineering problems.

The development of nanorobots is an emerging field with many aspects under investigation. Simulation is an essential tool for exploring alternatives in the organisation, configuration, motion planning, and control of nanomachines exploring the human body. Nanorobot applications could be focused mainly on two major areas, as follows: nanorobots for surgical interventions, as well as their utilisation for patients that need constant monitoring. The nanorobots require specific controls, sensors and actuators, basically in accordance with each kind of biomedical problem. Advanced simulations can include various levels of detail, giving a trade-off between physical accuracy and the ability to control large numbers of nanorobots over relevant time scales with reasonable computational effort. Another advantage is that simulation can be done in advance of direct experimentation. It is most efficient to develop the control technology in tandem with the fabrication technologies, so that when we are able to build these devices, we will already have a good background in how to control them.

Computational mechatronic approaches were proposed as a suitable way to enable the fast development of nanorobots operating in a fluid environment relevant for medical applications. Unlike the case of larger robots, the dominant forces in this environment arise from viscosity of low Reynolds number fluid flow and Brownian motion and such parameters are been implemented throughout a set of different investigations. Practical and innovative paradigms have been developed based on the Nanorobot Control Design (NCD) simulator that allows fast design testability comparing various control algorithms for nanorobots and their application for different tasks. Also such information generated by the NCD can be useful as parameters for building nanodevices, such as transducers and actuators.

In future decades the principal focus in medicine will shift from medical science to medical engineering, where the design of medically-active microscopic machines will be the consequent result of techniques provided from human molecular structural knowledge gained in the 20th and early 21st centuries. For the feasibility of such achievements in nanomedicine, two primary capabilities for fabrication must be fulfilled: fabrication and assembly of nanoscale parts. Through the use of different approaches such as biotechnology, supramolecular chemistry, and scanning probes, both capabilities had been demonstrated to a limited degree as early as 1998. Despite quantum effects which impart a relative uncertainty to electron positions, the quantum probability function of electrons in atoms tends to drop off exponentially with distance outside the atom.

Even in most liquids at their boiling points, each molecule is free to move only ~0.07 nm from its average position. Developments in the field of biomolecular computing have demonstrated positively the feasibility of processing logic tasks by biocomputers, a promising first step toward building future nanoprocessors with increasing complexity. There has been progress in building biosensors and nanokinetic devices, which also may be required to enable nanorobotic operations and locomotion. Classical objections related to the feasibility of nanotechnology, such as quantum mechanics, thermal motions and friction, have been considered and resolved and discussions of techniques for manufacturing nanodevices are appearing in the literature with increasing frequency.

One important challenge that has become evident as a vital problem in nanotechnology industrial applications is the automation of atomic-scale manipulation. The starting point of nanotechnology to achieve the main goal of building systems at the nanoscale is the development of control automation for molecular machine systems, which could enable the massively parallel manufacture of nanodevice building blocks. Governments all around the world are directing significant resources toward the fast development of nanotechnology. At least 30 countries have initiated activities in this field, and beyond that, with government investments to nanotechnology of US$ 800 million in Japan and US $ 774 million in USA. The U.S. National Science Foundation has launched a programme in "Scientific Visualisation" in part to harness supercomputers in picturing the nanoworld. A US $ 1 trillion market consisting of devices and systems with some kind of embedded nanotechnology is projected by 2015. More specifically, the firm 'display search' predicts rapid market growth from US $ 84 million today to $ 1.6 billion in 2007. A first series of commercial nanoproducts has been announced as foreseeable by 2007. To reach the goal of building organic electronics, firms are forming collaborations and alliances that bring together new nanoproducts through the joint efforts of companies such as IBM, Motorola, Philips Electronics, Xerox/PARC, Hewlett-Packard, Dow Chemical, Bell Laboratories, and Intel Corp., among others. For such goals, new methodologies and theories to explore the nanoworld are the key technology.

A useful starting point for achieving the main goal of building nanoscale devices is the development of generalised automation control for molecular machine systems which could enable a manufacturing schedule for positional nanoassembly manipulation. In our work we consider more specialised scheduling problems with a focus on nanomedicine: describing in a detailed fashion the nanorobot control designs and the surrounding virtual workspaces modelling that are required for the main kinematics aspects in the physically-based simulations. Here the biomolecular assembly manipulation could be automatically performed by smart agents, which are given a set of possible tasks for biomedical engineering problems, embedded in a complex 3D environment. Virtual Reality could be considered as a suitable technique for nanorobot design and for the use of macro and microrobotics concepts given certain theoretical and practical aspects that focus on its domain of application. The collective nanorobotics approach proposed is one possible method to perform a massively-parallel positional nanoassembly manipulation . We constructed and demonstrated the applicability of multi-robot teams in timely sequenced set of works with practical applications that could enable the establishment of generalised control guidelines for nanorobotics. The use of multirobot teams working cooperatively to achieve a single global task applied to nanotechnology is a field of research that is very new and challenging. Research on collective robotics suggests that we should consider emulating the methods of the social insects to build decentralised and distributed systems that are capable of accomplishing tasks through the interaction of agents with the same structures and pre-programed actions and goals. Thus, a careful decomposition of the main problem task into subtasks with action based on local sensor-based perception could generate multirobot coherent behaviours. Several techniques was applied for such aims, as Neural Networks algorithms, Evolutionary computation, chemical based sensors and actuators, and even temperature time-gradients, just to quote a few. Among other interesting nanorobot applications, we could foresee their use to process specific chemical reactions in the human body as ancillary devices for injured organs. Nanorobots equipped with nanosensors could be used to detect glucose demand in diabetes patients. Moreover, nanorobots could also be applied in chemotherapy to combat cancer through superior chemical dosage administration, and a similar approach could be taken to enable nanorobots to deliver anti-HIV drugs. Successful nanorobotic systems must be able to respond efficiently in real time to changing aspects of microenvironments not previously examined from a control perspective. Unlike some prior simulators for simple robots, in the present work we have developed nanorobots that are not restricted to a fixed grid nor behave as simple cellular automata with very simple environments. Also by contrast, most CAD approaches provide only animation or visualisation tools, while the NCD is a physically-based simulator. The set of experiments we have carried out include the main physical properties existent in the environment where the nanorobots are being currently projected to be operating in the coming years. Hence, the majority of results from our investigation should be useful as well on analysing integrated manufacturing capabilities. In fact, including aspects of the physical environment in conjunction with graphical visualisation provide a feasible approach for automation and control design. Furthermore, the architecture that was developed intended to enable the incorporation and evaluation of several control methods and distinct nanorobot shapes analyses. The automation, control, and manufacturing of nanorobots is a challenging and very new field. Realising revolutionary applications of nanorobots to health or environmental problems raises new control challenges. The design and the development of complex nanomechatronic systems with high performance should be addressed via simulation to help pave the way for future applications of nanorobots in biomedical engineering problems.

Soft Nanotechnology and Self-Assembly: Industrial Applications

The focus of this course is the design and application of nanostructured fluids and soft materials, with a particular emphasis on self-assembly processes. Many soft or fluid consumer products, such as foods, paints, detergents, personal care products, and cosmetics contain nanometer to micron scale structures. These structures are formed by the spontaneous or directed self-assembly of surfactants and polymers. In many cases complex mixtures are required to create the desired structures and performance. These materials are difficult to formulate, and challenging to characterize; but scientists in these industries, and in the academic groups focusing on these areas, have developed considerable expertise in this area. New insights and techniques have the potential to revolutionize product design. This knowledge is being applied in new areas such as drug encapsulation, solubilisation and delivery, in oil recovery, environmental remediation, and in new "active" cosmetics, nutraceuticals, antimicrobials and smart materials.

Members of the more broadly defined "nanotechnology community" are also increasingly interested in self-assembly and nanostructured fluids because of their potential to provide robust and inexpensive strategies for creating nanoscale materials. For example, IBM recently highlighted the potential for block-copolymer self-assembly to create nanometer size structures for chip fabrication, and soft nanostructured materials can provide templates for synthesis of nanoparticles. Graduates of this course will leave with a broad appreciation for current and future applications of soft nanostructured fluids, soft materials, and of self-assembly. They should have gained a sufficient grasp of the concepts and language to read and understand most research papers in these areas. They will be alerted to the latest developments in this fast moving field, and to approaches for designing materials for particular applications. They will have been introduced to characterisation and to computer-modeling techniques appropriate for these materials. They will have gained insight into the surprising commonality to design of some very different products, and also the way soft materials are starting to be used to create "hard" nanostructures.

Nano-Vasculoid and Self-Assembly as a Deliberate Strategy

As outlined below, at peak average male human metabolic rates, ~5400 cm^3 of circulating blood transports up to ~125 cm^3 of essential molecules and up to ~90 cm^3 of essential cellular elements (excluding erythrocytes). Consequently ~96 per cent of blood volume (mostly water) may be rendered superfluous if its salvation and suspension functions can be replaced with a more efficient and highly reliable nanomechanical transport mechanism. The resulting baseline vasculoid device conveys physiologically useful molecules and cells throughout its interior, and thence to the biological tissues, by containerising all materials and then using a ciliary transport system to distribute the containers to appropriate destinations. This architecture thus utilizes the fractal branching vascular network that is characteristic of efficient fluid transport systems, although many architectures are possible in this design space.

In natural human physiology, the circulating blood moves ~2 x 10^{26} molecules/sec.

However, most of these molecules are solvent water not strictly necessary in human metabolism. In the vasculoid, all non-cell materials transport takes place via ciliary transport of tanker vessels at a very conservative mean transport velocity of 1 cm/sec. (For comparison, blood velocity in the natural vasculature ranges from 0.02-0.15 cm/sec in the capillaries up to 117.5 cm/sec pulses in the femoral artery). Tanker vessels are (1-micron)^3 mostly-sapphire (possibly chamfered) cubes with 0.75 micron^3 useful interior storage volume and ~8 × 10^{-16} kg dry mass. Gases are stored in tankers at 1000 atm pressure; this is highly conservative, as a sapphire pressure vessel should easily withstand ~10^5 atm without bursting.. Liquids and solids are packed at normal macroscopic densities. All tanker mechanical subsystems (other than passive hull) have at least tenfold redundancy, hence are extremely reliable. The following is an approximate inventory and assessment of the minimum human physiological molecular transport requirement.

The human body consumes 1-20 × 10^{20} molecules/sec of O_2, depending on activity level. Similar numbers of CO_2 molecules are generated as waste products. The human red blood cell mass (~2.4 liters of RBCs) has a total storage capacity of 3.2 × 10^{22} O_2 molecules, but since hemoglobin operates between 70-95 per cent saturation in normal physiological conditions, the active capacity of human blood is only 8.1 × 10^{21} O_2 molecules. Blood CO_2 capacity is roughly the same, and O_2/CO_2 are loaded/unloaded reciprocally at lungs and tissues. Packing density of CO_2 molecules (1.11 × 10^{28} molecules/m^3) is slightly lower than for O_2 molecules (1.26 × 10^{28} molecules/m^3) at 1000 atm, so CO_2 packing density controls respiratory tanker population. Assuming a mean circulatory circuit of ~1.4 meter and a 1 cm/sec transport speed, continuous transport at the maximum 20 x 10^{20} CO_2 molecules/sec rate requires a circulating fleet of 33.6 trillion respiratory gas tankers. At normal basal metabolic rates, only 1.68 trillion active tankers are required; the remainder are held in reserve to support periods of more intense physical activity. Each tanker can hold 9.48 x 10^9 oxygen molecules or 8.36 x 10^9 molecules of carbon dioxide at 1000 atm pressure. Highly efficient reciprocal regenerative energy recovery is assumed during gas loading and unloading operations and unrecoverable gas compression energy losses should be minimal. Isothermal compression of ~10^{10} O_2 or CO_2 molecules requires—$P_i V_i \ln(V_f/V_i)$ ~ (10^5 N/m^2) (0.75 x 10^{-18} m^3/tanker) ln(1/1000) = 0.5 pJ/tanker. Assuming a need for 2 x 10^{20} molecules/sec for O_2 and CO_2 at basal demand implies ~2 x 10^{10} tanker fills/sec or ~0.01 watt for both sets of gases, or up to ~0.2 watts at peak demand, for compression of gases. Conservatively taking the CO_2 concentration in the vasculoid-endothelium interface fluid as equal to the normal CO_2 venous plasma concentration of C = 0.45-1.14 x 10^{24} molecules/m^3, then the diffusion limit for loading CO_2 from the tissues to one side of a square plate area of L^2 = 2 micron^2, with the diffusion coefficient D = 1.9 x 10^{-9} m^2/sec for CO_2 at 310 K, is J = ($4/p^{1/2}$) LDC = 2.7-6.9 x 10^9 molecules/sec; a 10-second fill time allows system operation at about one order of magnitude below the diffusion limit. Similarly, normal alveolar partial pressure of O_2 is >100 mmHg, or C ~ 3.1 x 10^{24} molecules/m^3; taking D = 2.0 x 10^{-9} m^2/sec for O_2 at 310 K, the diffusion limit for loading oxygen at the lungs is J ~ 2 x 10^{10} molecules/sec, which again allows system operation at about one order of magnitude below the diffusion limit assuming a 10-sec tanker fill time. Tankers are used to transport O_2 and CO_2, rather than the vasculoid interior in part because the theoretical rupture pressure of the natural cylindrical aorta is p_{max} ~ 1 atm (taking aortic wall thickness t_{wall} = 1.5 mm, aortic radius R = 12.5 mm, and working stress s_w ~ 10^6 N/m^2 and the rupture pressure of the vasculoid plate coating is p_{max} ~ 8 atm (taking vasculoid plate tube thickness t_{wall} = 1 micron, tube radius R = 12.5 mm, and working stress s_w ~ 10^{10} N/m^2. At peak exertion levels, respiratory gases stored in the tanker fleet at 1000 atm would fill the 4.2 liters of free vasculoid internal volume to 3.2 atm of pressure, too high for reliable safe operation under all foreseeable circumstances. By comparison, the rupture strength of individual tankers exceeds ~40, 000 atm (the rupture strength of individual plates).

The human body excretes 6-12 x 10^{20} molecules/sec of urinary water under normal circumstances. The kidney continuously filters ~18 gallons/hour of blood (~7 x 10^{23} H_2O molecules/sec) so much higher rates of urine generation are possible in theory, but it seems unnecessary to accommodate the vasculoid design to such extreme cases because the removal of waste molecules can be accomplished by direct molecular sorting rotor extraction rather than by bulk filtration and solvent repumping—that is, the principle blood-cleansing function of the kidneys is to pump nonaqueous molecules, not bulk water solvent molecules. With the vasculoid in place, only specific waste molecules and minimal quantities of carrier water need be discharged into the glomeruli of the ~10^6 nephrons of the kidney, with the result that gross water flows through the kidney and energy consumption in this biological organ can be significantly reduced—but not reduced so far as to induce urinary stone disease (urinary lithiasis or urolithiasis) or renal cytotoxicity from excessive pH, salinity (e.g., hyperosmotic stress), or toxin concentrations. The precise volume and concentration of urine generated by the biological organ is regulated by hormones that control kidney function, and these hormones may in turn be regulated by the vasculoid appliance. Additionally, 2-186 x 10^{20} molecules/sec of water are perspired through the skin, depending upon air temperature, humidity, activity level and mental state. Assuming the same transport parameters as before, the maximum requirement of 2.0 x 10^{22} H_2O molecules/sec (~0.6 cm^3/sec) requires 110.5 trillion water tankers to be available, of which only 4.5 trillion will normally be active with the rest held in reserve for periods of peak exertion. Each tanker can hold 2.51 x 10^{10} water molecules in the liquid state.

Typically 1.6 x 10^{22} molecules of glucose are present in the human bloodstream. A human on a standard 2000 kcal/day diet metabolizes 3.08 x 10^{19} glucose molecules/

sec throughout the body. However, during strenuous exercise this requirement can briefly rise to as much as 33, 000 kcal/day, a 4.93 x 10^{20} molecule/sec rate. Assuming the usual ciliary transport parameters, we require 17.6 trillion glucose tankers, providing storage of 6.91 x 10^{22} glucose molecules in the system, four times the number present in normal blood. (Slightly smaller capacity might be realised if the glucose must be stored as a 70 per cent aqueous solution concentrate to facilitate molecular mobility for sorting rotors. Only 1.1 trillion glucose tankers are needed to service basal metabolic needs; the remainder are held in reserve for periods of extreme physical activity. The average human cell is a ~30 picowatt (basal) device, so each glucose tanker, with its full complement of 3.93 x 10^9 glucose molecules, yields ~400 basal cell-seconds of energy (assuming 65 per cent efficiency for the natural cellular metabolic pathways using glycolysis and tricarboxylic acid cycles). Note that at higher power levels, the delivery rate increases correspondingly, so that even with random delivery at peak demand all cells should receive a sufficient fuel supply. Additionally, the average tissue cell normally contains an internal inventory of ~3 x 10^{10} glucose molecules, an in cyto buffer supply equivalent to ~8 full tanker loads.

Normal bloodstream concentrations of urea, uric acid, creatinine, creatine, ammonium salts, cerebroside, amino acids, etc. imply 1.73-2.49 gm present in the circulation and a maximum of 2.3 trillion tankers for full encapsulation and transport.

All plasma proteins including most notably albumen, globulins (including antibodies), complement, fibrinogen and prothrombin constitute ~7 per cent of adult blood plasma, or 210 gm, which would require 215 trillion tankers to encapsulate. Fortunately, on a transactional basis this figure is a gross overestimate. The RDA (Recommended Daily Allowance) active requirement for protein is ~70 gm/day for a 70 kg male human, an 8.1 x 10^{-7} kg/sec transport demand requiring only 0.12 trillion protein tankers for continuous ciliary transport, reducing mandatory instantaneous system storage capacity to ~0.12 gm of plasma proteins. However, almost the entire RDA protein intake is hydrolysed to amino acids and dipeptides, so the actual protein transport needs should be far less. With no need to maintain osmotic pressure, there would be almost no reason to transport albumin within the system, although the dosage of protein-bound medication and hormones might need to be corrected. Finally, there is sufficient tanker capacity to increase protein transport capacity at least 100 fold above the RDA requirement if necessary, so the design is not particularly sensitive to this transport requirement. Note also that the five antibody classes can be efficiently transported using just five types of sorting rotors having binding pockets complementary only to the constant Fc regions without regard to the numerous variable domains these antibodies may encode on their antigen-binding Fab regions, thus vastly reducing the required vasculoid rotor specificities needed to accomplish this task. Complement (immune system) components may be transported with similar efficiency. The limited free enzyme activities that occur in natural blood can still take place in the fluids at the vasculoid-endothelial interface, with due care taken to avoid inducing localised hyperenzymemia (which is often benign, as in hypertransaminasemia), although higher concentrations may be expected because of the lower total volume of the vasculoid-endothelial interface compared to the natural blood volume.

The mass of lipids including fatty acids, cholesterol, triacylglycerides (transported in the plasma lipoproteins as triacylglycerols), and phospholipids present in the human bloodstream totals roughly 35-42 gm, which would require 55-66 trillion tankers to encapsulate. However, the RDA for fat is 23 per cent of caloric intake. (Many people consume several times this amount, but others survive on fat-free diets, so the figure represents a reasonable compromise estimate of the typical human active requirement.) This implies a demand of 6.9 x 10^{-7} kg/sec for a 2000 kcal/day diet, requiring only 0.15 trillion lipid tankers for continuous ciliary transport and reducing the required vascular storage capacity to ~0.1 gm. As with plasma proteins, there is sufficient tanker capacity to increase lipid transport capacity to at least 100 times the RDA requirement, so once again the design is not particularly sensitive to this transport requirement.

Human blood plasma contains ~40 gm of salts (e.g., electrolytes) including sodium, potassium, calcium, magnesium, chloride, phosphate, and sulfate; ~1.2 gm lactic acid and other non-nitrogenous waste products; ~0.23 gm of RNA and DNA; ~0.21 gm of enzymes (protein); ~0.15 gm testosterone in the male, ~0.03 gm progesterone in the female, ~0.02 gm of corticosteroids, and a total of ~0.002 gm of more than fifty other major biologically active hormones; ~0.04 gm of vitamins including A, B2, niacin, pantothenic acid, biotin, pteroylglutamic acid, choline, inositol, C, D, and E; ~0.03 gm of essential trace elements including iron, zinc, copper, manganese, iodine, and cobalt; and smaller masses of various cell-signaling (e.g., cytokines, autocrines, neuropeptides) and numerous other specialty proteins. Other nonprotein regulatory molecules affecting numerous signaling pathways must also be locally regulated with precision by the vasculoid appliance. For example, in the present context the simple molecule NO is of particular importance as a neurotransmitter and vasoconstrictor, though it is normally present only in exceedingly small quantities. Other molecules that mediate certain regulatory pathways may become unnecessary to transport after the vasculoid is installed, as for example the renin-angiotensin pathway (since there is no blood pressure to regulate). The capacity demand totals ~42 gm, which would require ~43 trillion tankers for full encapsulation. However, the RDA for all minerals, electrolytes, vitamins and trace elements is just 6.8-16.1 gm/day for healthy adults, a maximum of 1.87 x 10^{-7} kg/sec, requiring only 0.03 trillion tankers assuming ciliary transport parameters as described above. (Sodium and potassium ion transport in individual neurons is ~10^6 ions/discharge; assuming

~10^{10} active neurons firing at near-maximum ~100 Hz, whole-brain ion transport rate is ~10^{18} ions/sec or ~7×10^{-8} kg/sec, well within the aforementioned mass flow budget for mineral and electrolyte transport. However, Na^+ and K^+ are very quickly recycled after each discharge, so the true transport requirement is more accurately measured by the small amounts of these electrolytes that are lost in the gut, sweat, and lymph.) Transport of lactic acid, RNA/DNA, enzymes (e.g., those present as a result of normal tissue breakdown such as CPK and transaminases), hormones, cytokines and so forth (~1.8 gm) requires 1.9 trillion additional tankers. In some cases these "other" molecules may be shipped mixed in a single container.

A total of 166.2 trillion tankers are required to transport all-important physiological molecules at maximum human metabolic rates. At basal metabolic rates, only 11.78 trillion tankers will typically be active. If, contrary to our assumption, RDA levels are found to be an inadequate substitute for natural blood storage of plasma proteins, lipids, minerals and vitamins, or if production and demand occur at different times and the blood volume performs a significant storage function, then vasculoid design may need to include up to ~300 cm^3 of auxiliary bulk storage or caching to extend the "just in time" inventorying of these vital substances.

In addition to conveying molecules, the vasculoid must also transport physiologically important cellular species throughout the body. Cells to be transported (comprised mainly of white cells, platelets, and the few remaining red cells) generally range in size from 2-20 microns. These are shipped in cylindrical mostly-sapphire "boxcars" 100 microns long and 6 microns in diameter. These containers, of which 75 per cent (2120 $micron^3$) is usable storage volume (~2.4×10^{-12} kg dry mass per boxcar), will just fit through average sized capillaries (~8 microns diameter, ~1000 microns in length) and all larger vessels, after accounting for plate thickness, and thus the boxcars have access to the vast majority of human tissues. The largest transportable cells (20-micron monocytes) must be temporarily dehydrated 50 per cent by volume prior to loading; such desiccated cells quickly rehydrate by natural osmosis upon release into the aqueous intercellular environment. Spherical cells 17.5 microns in diameter or smaller need not be dehydrated during loading, as they can fit into the available volume via change of shape. Typical transit time from point of entry to point of debarkation is ~70 sec (i.e., ~half the mean circuit length of 1.4 m, at 1 cm/sec), well within cellular ischemic survival times except possibly for oxygen, which may need slight supplementation during transit. Onboard computers and sensors embedded in interior boxcar walls allow the detection and interpretation of surface antigens on passenger cells to ensure that all transported cells are non-hostile, and if not, to refuse transport, convey them to a disposal site, or destroy them after entry. The following is an assessment of the minimum human cell transport requirement for the vasculoid appliance.

Under normal physiological conditions, 27-54 billion leukocytes circulate in the human blood volume, although in cases of chronic myelogenous leukemia the total count may increase to 0.13-2.7 trillion white cells. These 54 billion leukocytes consist approximately of 35-41 billion neutrophils (10-12 microns in diameter), 11-14 billion lymphocytes (8 microns), 1.6-4.3 billion monocytes (12-20 microns, average 15 microns), 1.1-2.7 billion eosinophils (12 microns), and 0.3 billion basophils (10 microns). Based on maximum container packing densities for each type of cell, 12-28 billion boxcars are required to encapsulate and transport all normal populations of bloodstream leukocytes. This will frequently include more than one cell per boxcar—multiple cells should be sedated for transport (using reversibly rotored molecular agents) to avoid unwanted interactions such as clonal expansion or induction of apoptosis. Principal origination sites are the bone marrow, the spleen, and the lymph nodes, but leukocytes may be distributed to almost any location in the body. Giant macrophages (e.g. >30 microns) and fibroblasts already present in the tissues (that normally do not enter the bloodstream) cannot be relocated by the appliance, although fresh stem cell precursors of fibroblasts with appropriate activating cytokines can be relocated from their natural source site, as required. A comparatively small population of circulating stem cells must also be transported by the vasculoid. For example, vasculogenesis is assisted by hematopoietic stem cells (surface marker CD34+) that originate in the bone marrow and enter the peripheral blood in response to colony stimulating factors, with 1-3 per cent expressing the common leukocyte surface antigen CD45. Upon reaching the vascular surface and in the presence of the cytokine IL-3, these cells may differentiate either into leukocyte precursor (CD34+/CD45+) or into endothelial precursor (CD34+/CD45–); with the cytokine VEGF present, the latter cells further differentiate into new endothelial cells. The vasculoid must provide transport for stem cells of these and other types, along with the required cytokines, to ensure vascular repair of damaged endothelium and for other purposes. Fortunately there are only a limited number of such cell types, and the total population of stem cells requiring transport is many orders of magnitude smaller than the population of leukocytes. As a result, stem cell transport should be a comparatively minor—if physiologically essential—task for the previously-specified boxcar fleet.

Although substitution of vasculoid transport for liquid blood eliminates much of the conventional health risk due to bleeding, blood vessel breaches that penetrate the endothelium but do not penetrate the vasculoid will still exude large amounts of extracellular serous fluid and other critical substances if the leakage is not promptly staunched. Furthermore, platelets are storehouses for a variety of molecules that affect vascular tone, fibrinolysis (subsequent clot dissolution), and wound healing. These substances are released during the clotting process. Platelets also communicate via chemical messengers with other blood cells (e.g. macrophages and fibroblasts) and with the endothelial cells that coat the interior of all blood vessels. Hence, platelets still have an important role in human physiology even after the vasculoid has been installed. Approximately 2 trillion platelets (~2 microns in diameter) circulate in human blood. Each boxcar can hold about 500 platelets, giving a requirement of 4 billion additional containers for platelet transport. As with white cells, multiple cells could easily activate and aggregate, hence their function should be heavily suppressed during transport (again, using reversibly rotored molecular agents).

In the vasculoid, all respiratory gases are transported in bulk and at high pressure. Boxcars could in principle be used to transport red cells, but RBCs have an effective storage pressure <<1 atm which is far less efficient than the 1000 atm storage pressure of the respiratory gas tankers. Consequently, the ~30 trillion red cells normally present in the human bloodstream are superfluous and may be permanently removed from circulation. Beyond their respiratory function, red cells also are mechanically involved in the clotting process. However, the vascular plug typically contains mostly platelets, plasma fibronectin and factor XIII-crosslinked fibrin, plus small amounts of tenascin, thrombospondin, and SPARK (secreted protein acidic and rich in cysteine) with a relatively minor RBC contribution, so RBCs probably are not essential to the clotting process or are necessary only in very small numbers. A sufficient oxygen concentration maintained in kidney peritubular cells should reduce erythropoiesis to at most 1 per cent of the normal rate of red cell production , requiring only ~34, 700 red cells/sec to be conveyed from the erythroid marrow directly to the liver or spleen for disposal. Alternatively, erythropoiesis can be reduced by directly regulating the amount of erythropoietin that is allowed to reach the bone marrow. With few erythrocytes remaining to be destroyed in the liver, hepatic heme catabolism drops significantly, greatly reducing the production of bilirubin, a yellow bile waste pigment that plays no role in fat digestion in the gut but imparts the brown color to feces; as with hepatitis patients, the stool of the envasculoided user may become chalky white. Ciliary transport of boxcars containing the few remaining red cells over a ~1.4 meter circuit at a conservative mean velocity of 0.1 cm/sec (allowing plenty of time for docking and cell unloading) requires a circulating subfleet of just 34.7 million boxcars (only ~0.1 per cent of the entire boxcar fleet).

A maximum of 32 billion boxcars are required to convey all essential bloodborne cells during normal physiological conditions, although at times as few as 16 billion boxcars may be actively involved in cellular transport. Since a principal benefit of vasculoid installation is a significant decline in susceptibility to microbial invasion, the above figures can probably be further reduced in actual clinical practice.

The interior volume of the vasculoid is lined with a sapphire surface (comprised of watertight adjacent installed plates) from which protrude trillions of mechanical cilia (arranged in patterns designed to maximize transport speed and reliability. A secondary role is to assist the vasculocytes (small mobile nanorobots) in cleaning up after accidental internal spills, component malfunctions, or other pathological events. Even by 1997, prototype MEMS ciliary arrays of up to 1024 cilia had already demonstrated transport speeds up to 200 microns/sec with ~3 micron positional accuracy.

While a highly efficient specialised ciliary mechanism can undoubtedly be designed, for the present study each vasculoid cilium is conservatively assumed to be similar in size, shape, and performance characteristics to the vacuum-sealed robotic manipulator arm described by Drexler. In brief, each cilium is a cylindrical assembly 100 nm long and 30 nm in diameter, having a transverse travel of 100 nm and a lateral speed of 1 cm/sec, consuming 0.1 picowatt during continuous operation at low load. Note that even if the vasculoid interior contained gas at 1 atm, viscous drag would contribute an additional power loss of only ~0.001 pW per cilium (taking viscosity h_{air} = 1.83 x 10^{-5} kg/m-sec at 293 K). Mean cilium/cargo contact time is ~10 microseconds. This implies an acoustic frequency >100 KHz, well above the conventional human threshold of hearing; however, ultrasonic hearing up to 108 KHz via otolithic conduction has been reported in humans, so further study is required to ensure restriction of transport motions to sufficiently high frequencies to preclude any perceptible hum directly due to ciliary operations, and possibly involving a variety of noise cancellation, anti-resonance (e.g., out-of-phase operation) and acoustic dampening techniques. Additionally, physical discomfort, intra-articular pain, and a lowering of electrical pain sensation threshold due to ultrasound exposure have also been reported in humans. The possibility of indirect audible or infrasonic vibrations due to beat frequencies, spatial acoustic interference patterns, or various resonances should be investigated further, along with the possibility of undesired vibrational energy losses into the tissues. Each cilium can apply ~1 nanonewton at the tip. This is sufficient force to accelerate a 1 micron^3 block of water (weight ~0.01 piconewton) at 10^5 G, or a 100-micron long, 6-micron wide water-filled sapphire-shell cylindrical boxcar at ~30 G. The human body is normally subject to accelerations far less than 30 G; with tenfold grasping redundancy, the probability that a container will be shaken loose from the cilia and tumble uncontrolled into the vascular lumen, due to normal macroscale accelerations of the patient's body, is extremely remote. In the rare instance where such detachment occurs, the container will eventually be caught and returned to the flow; kinetic impact energy in such cases is relatively small. Whether such cross-luminal impacts can trigger a detachment avalanche should be investigated using computer simulations. Changeable cilium tool tips may permit rapid, reversible, "blind" container grappling using van der Waals adhesive forces, which forces may be controlled by varying adhesive pad surface corrugations at the nanometer scale, or by other means. The cilia might also have a reversing function (e.g., for loosening container jams). Molecular dynamics simulations of these processes would be useful.

The principal task of the ciliary distribution subsystem is to reliably transport a maximum of 166.2 trillion 1-micron cubical tankers and a maximum of 32 billion cylindrical 6 mm x 100 mm boxcars through a physical circuit ~1.4 meters in length. Because of their smaller size and greater number, ciliary subsystem design is driven almost entirely by tanker, not boxcar, requirements. Although a single cilium can apply sufficient force to grapple and manipulate containers of either size, tenfold grappling is employed in the baseline design to achieve firm contact, to avoid physical escape of containers from the cargo traffic stream, to ensure massive redundancy hence high subsystem reliability, and to allow extremely precise steering. If a tanker occasionally encounters a patch of 9-contact cilia, or the even rarer 8-contact patch, steering or grappling capability should not be significantly reduced. Fail-safe grappling modes must be designed for cilium end-effectors. Each tanker is normally in continuous contact with 10 cilia at all times during transport, so maximum ciliary spacing is 0.32 micron. A minimum of 3000 trillion cilia are needed to achieve 100 per cent service over the entire ~300 m^2 vascular (mostly capillary) surface. A set of 3000 trillion cilia physically occupies ~2.1 m^2 (~7 per cent) of the vascular surface. Assuming natural background radiation damage causes 1.5×10^7 fatal hits/kg/sec for nonredundant structures, the mean time to failure of a ~10^{-19} kg cilium is ~10^{12} sec, which implies a mean failure rate of only ~2000 cilia/sec (~20 nanograms/day) throughout the entire vasculoid. Using redundant structures mean failure rate may be considerably reduced. Under resting metabolic conditions, 11.78 trillion tankers are transported by 117.8 trillion cilia generating waste heat of ~11.8 watts, an increment of only ~12 per cent over the basal human metabolic rate. During periods of maximum exertion and thermoregulatory stress, 166.2 trillion tankers are transported by 1662 trillion cilia generating excess heat of ~166 watts, well below the maximum human metabolic rate of 1600 watts and just slightly exceeding the ~100 watt effective load error that could trigger a response from the human thermoregulatory control system.

Once molecules or cells have been transported to the appropriate site, the container in which they reside must dock with the vasculoid surface on the interior side, and pass the delivered materials through this surface to the interstitial side of the vasculoid wall, whereupon the cargo can diffuse or migrate into the tissues as required. Tankers offload their molecular cargo in docking bays whereas boxcars unload their cellular passengers at cellulocks. Possible undesired long-term chemical interactions of transported nutrients discharged into the fluids of the vasculoid-endothelial interface at above-physiological concentrations must be carefully analysed. For example, high concentrations of glucose akin to chronic hyperglycemia could enhance protein glycosylation of long-lived proteins (e.g., vascular and myocardial collagen) which undergo continual cross-linking during aging because of the formation of advanced glycosylation end-products (AGEs) , most significantly (for this paper) in endothelial cells , which may be partly responsible for arterial stiffening and higher blood pressure with age (and which drugs like ALT-711, benfotiamine , nitric oxide and other substances may reverse). AGEs can also induce apoptosis and vascular endothelial growth factor overproduction in capillary pericytes.

Physiologically relevant molecules are delivered to specific locations by docking at 2 micron^2 (~1.4-micron square) tanker docking bays embedded at appropriate spatial intervals across the vasculoid surface. After securely connecting to the docking bay structure, and after the tanker's cargo manifest is scanned by the docking bay computer and said cargo is approved for receipt (Section 5), the tanker can be pumped dry by molecular sorting rotors in ~10 seconds, then sealed, remanifested, undocked, and released empty back into the traffic. Empty tankers may be reloaded with molecular cargo by a similar process in a similar time. Note that tankers carrying liquid or gas may discharge their contents into the narrow end of a funnel-shaped manifold leading to the active rotor banks at the wider end, to minimize the required tanker connection aperture. In different operating modes, tankers might be only partially emptied at each stop, or might be partially emptied more rapidly via bulk-flow exhausting rather than molecular rotoring. Docking bays include buffer tanks so that offloaded materials can be metered out to the underlying tissues over time periods longer than 10 seconds. If the average endothelial cell (of which blood vessel walls are comprised) has a flat luminal area of 300-1200 micron^2, then the ~300 m^2 vascular surface may be comprised of up to ~1 trillion endothelial cells. This vascular surface could theoretically accommodate a maximum of 300 trillion tankers positioned side by side. At basal metabolic rates, only 11.78 trillion tankers are actively on the move. Even during peak exertion, a maximum of 166.2 trillion tankers are being transported, requiring just 55 per cent of the available vasculoid luminal surface for their passage assuming all tankers are traveling in monolayer. Other functionally equivalent modes might include operating all tankers at reduced capacity or reduced duty cycle during periods of basal demand, or reducing the transport velocity. A detailed computer simulation study of optimal traffic patterns near peak capacity would be valuable to perform. A maximum of 166.2 trillion tankers circulating through an average ~1.4 meter circuit at ~1 cm/sec implies a docking bay cargo unloading requirement of ~1.2 trillion tankers/sec. Docks operate on a 50 per cent duty cycle to allow plenty of time for maintenance and repair, although in a less conservative design the duty cycle could possibly be boosted as high as 99 per cent. Tankers are emptied in 10 sec, so ~24 trillion docking bays are needed for reliable operation at peak loads. Docking bays occupy ~16 per cent of the vasculoid surface and have a mean center-to-center separation of ~3.5 microns. Each endothelial cell is serviced by up to ~24 docking bays. Normally only a small fraction of the body's tissue cells are directly bathed in blood. Hence, delivery of molecules to each endothelial cell surface must be sufficient to supply the needs of the up to ~30 tissue cells that lie, on average, within ~3 cell widths of the nearest capillary. The maximum whole-body transport requirement of 2.25×10^{22} molecules/sec, consisting principally of respiratory gases, water, and glucose, gives an endothelial surface mass transport rate of 2.25×10^{10} molecules/sec/endothelial cell. This transport rate may be satisfied using a minimum of 45, 000 molecular sorting rotors per endothelial cell operating on a 50 per cent duty cycle. With 24 docking bays per endothelial cell, the requirement is ~0.94 billion molecules/sec per docking bay. Allowing tenfold multiplicity of sorting rotors to ensure high reliability requires 18, 750 sorting rotors per docking bay, of which only 67 will be active at the basal rate and up to 9375 will be active during maximum physical exertion; 18, 750 sorting rotors measuring 98 nm^2 on the ventral surface (facing the endothelium) cover 1.84 $micron^2$, essentially coating the entire undercarriage of the 2 $micron^2$ docking bay facility. This provides a ~7.5×10^8 molecule/sec supply per tissue cell.

At peak metabolic load, maximum molecule offloading rates as constrained by possible effervescence and crystallescence effects appear acceptable in this application for all low-volume molecules and for glucose and CO_2, but unfortunately exceed the effervescence limit for O_2 by a factor of ten, due to the relatively poor aqueous solubility of oxygen. Hence at peak load this gas must be offloaded either ~10 times more slowly than indicated above for a single release site, or at the indicated flow rate from ten discrete release sites all spatially separated by more than the largest possible O_2 effervescence bubble radius. To simplify rotor system design, there are four classes of docking bays approximately matching the anticipated four principal tanker populations. Of the ~24 docking bays overlying each endothelial cell, ~5 stations are designated to handle tankers carrying respiratory gases, and possess rotors (on the tissue side) designed to reversibly bind only oxygen or carbon dioxide molecules. Another ~14 docking bays receive only water-bearing tankers, and employ rotors specialising in water molecule transport. Another ~3 stations accept only glucose cargoes, and use only glucose sorting rotors. The remaining ~2 docking bays are general-purpose stations equipped with rotors of up to tens of thousands of different types capable of reversibly binding all remaining biologically useful molecules (or classes of molecules) that must be transported through the vasculoid surface.

In addition to gas, water, glucose, and specialty tankers, there is a fifth class called power tankers that support the docking bay oxyglucose metabolism. Power tankers transport stoichiometric parcels of oxygen and glucose, and also remove carbon dioxide byproduct for disposal at the lungs. (Vasculoid power may be generated by glucose engines, glucose fuel cells, or other energy conversion devices that consume oxygen and native glucose; water produced by glucose engines may be vented into the biological tissues) All docking bays can accept power tankers, which supply the docking bay's own internal energy needs. Tankers reload their own oxyglucose engines and fuel tanks as they make their rounds of the various docking bays. Tanker duties are not energy intensive, so tankers refuel at most once a day, at the docking bays. Rotors conveying molecules from tankers (at high concentration) through the vasculoid integument to the endothelial surface (at low concentration) generate positive mechanical energy, which may be used to largely offset the work of concentration that must be expended to transfer molecules in the reverse direction when loading tankers from the tissue space. Concentration work losses can be reduced almost to zero in an efficient system, with systemwide entropic losses in the compression-expansion cycle as low as 0.01-0.2 watts. Hence the primary energy loss is the drag power per rotor, ~10^{-16} watts or ~0.1 zJ/molecule assuming a transfer rate of ~10^6 molecules/rotor-sec. Taking the whole-body transport requirement range as 10-225 x 10^{20} molecules/sec, the net molecular transport power dissipated per docking bay is thus 0.004-0.09 picowatt depending on physical exertion level, or 0.1-2.25 watts for the entire vasculoid docking bay subsystem. (Rotor systems that do not include efficient energy recovery and must dissipate the work of concentration, perhaps up to ~20 zJ/molecule, may require different appliance architecture) To hold docking bay subsystem total power requirement for computational tasks to a 10-watt budget, nanocomputers using an appropriate mix of local and distributed computing costing ~6 x 10^{-17} watts/(ops/sec) can provide ~200 billion MIPS systemwide or ~10, 000 ops/sec per docking bay, costing ~0.6 pW per docking bay.

These rates have been enhanced by employing reversible logic. Hence the total power per docking bay for both mechanical and computational tasks ranges from ~0.604 pW (basal) to ~0.69 pW (peak), or 14.5-16.6 watts for the entire system of 24 trillion docking bays. If power is supplied by a glucose engine achieving mechanochemical power conversion at ~10^9 watts/m^3, a 16.6 watt peak energy budget for all docking bays requires one glucose engine only ~0.0007 micron^3 in volume (~0.03 per cent of bay volume), per dock, and a similar volume of fuel tankage.

Biological cells are delivered to specific locations by docking at cellulock stations embedded at appropriate intervals across the vasculoid surface, often preferentially located near specialised post-capillary blood vessels less than 30 microns in diameter found in lymphoid tissues, commonly known as high endothelial venules (HEVs). Cellulock density requirements also vary on a tissue-by-tissue basis—for instance, intestines, lungs, throat and nasal cavity probably have much higher requirements than heart or brain. After securely docking with their loading face pressed firmly against the vasculoid surface, boxcar doors dilate open to make a 30 micron^2 (~6.2-micron diameter circle) aperture. Cellulock doors (constructed of self-cleaning sliding spiral graphene segments in snug facial contact) then dilate to the same diameter, exposing passenger cells to the interstitial fluid. Passenger cells are gently pistoned out through the opening, taking care to minimize extracellular fluid entry. The cycle ends as all doors reseal and the boxcar undocks, resuming its travels. Biochemical signals of cellular distress (e.g. cytokines such as interleukin-1 or tumor necrosis factor) are received by sensors on the interstitial side of the vasculoid. Upon receipt of this chemical signal, the cilia at that site are programed to call for leukocyte activity there. Cellulock and boxcar interiors may employ a small set of presentation semaphores which expose to the interstitial fluid minute quantities of various chemoattractants and chemorepellents keyed to each major type of cell to be transported. These chemicals are permanently bound to the rotors and are not released from the vasculoid; like semaphores, each rotor position may present differing concentrations of one chemical species, or a selection of different chemical species. White cells of the type desired to be transported can be encouraged to enter or exit, as required, by positioning presentation rotor settings so as to create "chemotactic funnels"—programable patterns of varying concentration gradient to steer motile cells toward, or away from, the cellulock aperture (e.g., haptotaxis). Contact sensors around the aperture detect the presence of a motile cell "requesting entry" to the transport system. Motile cells presenting themselves for entry at the interstitial side of the cellulock are, at minimum, subjected to a rudimentary cell protein coat identification test before they are allowed admittance. Affirmatively recognised harmful foreign pathogens such as bacteria or viruses and infected native motile cells may be transported to a specific disposal site or destroyed, rather than being admitted freely for transport elsewhere in the tissues.

The cellulock mechanism covers 60 micron^2 (~7.7-micron square) of vasculoid surface. If spiral door segments slide past each other at 0.004 m/sec over an average of one-half rotation or ~15.4 microns, dilation (or resealing) of the cellulock door takes 3.85 milliseconds. Sliding friction for 16 nm^2 contact surfaces traveling 20 nm at 0.004 m/sec dissipates ~10^{-24} joules, or ~3.13 joules/m^2-meter . Thus ~30 micron^2 cellulock sliding door surfaces moving ~15.4 microns to open, then again to close, require ~3 x 10^{-15} joules to complete one dilation cycle. Assuming a maximum rate of 32 billion boxcars per (normal blood circulation time of) ~60 sec to emulate typical physiological delivery rates, the population of vasculoid cellulocks must process a total of 533 million boxcars/sec. If there are therefore also 533 million cellulock dilation cycles/sec, frictional waste heat generated by moving door segments in the entire vasculoid cellulock subsystem is a negligible ~1.5 microwatts.

Taking cellulock power as equal to the basal rate for 30 docking bays gives a requirement of ~18 pW/cellulock, or ~0.59 watts for the entire cellulock subsystem. Serum glucose and oxygen to provide cellulock power (a minor volumetric requirement) may be rotored in directly from the interstitial fluid on the external side of the vasculoid surface. Since the active tanker/boxcar ratio varies from 736 (basal) to 5194 (max) depending on metabolic load, there should be no less than one cellulock per 736 tanker docking bays, or a total of 32.6 billion cellulocks in the vasculoid surface, again allowing a 50 per cent duty cycle for maintenance and repair. Thus there is approximately one cellulock for every boxcar in circulation. Cellulocks occupy ~0.65 per cent of the vasculoid surface and have a mean center-to-center separation of ~96 microns. Each cellulock services an average of ~31 endothelial cells, though the actual spatial distribution of cellulocks is expected to be highly nonuniform (e.g., clustering near HEVs). The average capillary is ~1 mm long and is made up of ~60 endothelial cells, and there are ~19 billion capillaries in the human body, so on average ~1 cellulock resides at either end of every capillary vessel in the body. The maximum white cell transit rate through injured or diseased tissue is ~1 cell/sec/mm^3. With 32.6 billion cellulocks there are ~163 cellulocks/mm^3 of tissue. Each cellulock has ~163 sec to pass each white cell, even at maximum traffic flows. This should be sufficient, given that cellulock doors open and close in milliseconds, and natural white cell transendothelial diapedesis may occur in as little as ~3 minutes. The creation of unnecessary interior voids where unwanted opportunistic pathogens might flourish should be avoided. It is probably unnecessary to maintain voids between vasculoid and endothelium near cellulocks because: (1) the time allocated to cycle the cellulocks appears adequate to permit WBC diapedesis, and (2) diapedesis can probably be expedited using chemical factors emitted by the vasculoid.

The ideal method for container routing will be completely automatic, consuming no incremental power or compute cycles. To this end, the outer surfaces of each of the five types of tankers (Gas, Water, Glucose, Power, and Other) and the four types of boxcars (Red Cell, White Cell, Platelet, and Other) bear a specific repeating interlock pattern, crudely analogous to Braille. These patterns remain in place on the container surface unless the container is affirmatively "reset" to carry a different cargo, in response to varying physiological needs and local container availability. Cilia surrounding each docking bay or cellulock are outfitted with gripper pads that adhere adequately to all container surfaces, but strongly only to the surface pattern of the container type sought by the docking bay or cellulock with which the cilia are associated. These pad "recognition" types normally remain constant over time, but, like the container surfaces, may be modified as required in special circumstances. Efficient methods for handling other-type tankers that can avoid the need for time-consuming docking and downloading of manifests should be investigated further. Cilia that have "recognised" a desired container transmit a trigger signal to nearby cilia, activating a collective stereotypical docking sequence. Thus, although containers follow statistically random paths across the vasculoid surface, each docking bay and cellulock automatically receives the cargo it requires from the passing container stream. Docking acquisition is demand-regulated locally.

The exemplar vasculoid has been scaled to transport essential molecules and cells at maximum physiological transfer rates. However, all engineering systems involve design tradeoffs and limitations, and the present design invokes a number of potential transport bottlenecks including intestinal water absorption and renal water elimination, glucose absorption and elimination at peak loading, boxcar geometrical clearance in narrow capillaries, and geometrical limitations of tanker monolayer transport.

Water Absorption and Elimination—The ~6 meter-long intestinal tract has a simple cylindrical surface area of ~0.65 m^2, but the ~5 × 10^6 villi of the small intestines (which is most of the length) increase the effective absorbing surface to ~10 m^2. The ileum, the lower portion of the small intestine, absorbs water into the bloodstream at the maximum rate of ~0.07 cm^3/sec. The cylindrical villi, typically measuring ~200 microns wide and ~1000 microns long, are heavily capillarised; assuming ~1000 capillaries/mm^3, each villus has ~30 capillaries presenting a total ~8 × 10^5 $micron^2$ of absorbing surface—for the entire ileum, that's ~4 × 10^{12} $micron^2$, a water transport rate of ~6 × 10^8 molecules/sec-$micron^2$. If maintenance activities are temporarily suspended during peak demand, water tankers loaded with 2.51 × 10^{10} water molecules at 2 $micron^2$ docking bays can provide the same transport rate of 6 x 10^8 molecules/sec-$micron^2$ if 49 per cent of the vasculoid surface in the ileum is given over to water transport—potentially crowding out some "applications plates" at this surface. (The requirement for alimentary water can also be greatly lessened by reducing the need for water solvent in urine by reducing the need for water cooling in sweat and possibly by reducing water losses in the alveoli (which could be covered with an improved waterproofing surfactant.) Excretion of ~0.023 cm^3/sec aqueous waste at the kidneys occurs primarily through the renal glomeruli which have a total surface area of ~1.4 m^2, giving a maximum water export rate of ~5 × 10^8 molecules/sec-$micron^2$. If maintenance activities are temporarily suspended during peak demand, water tankers loaded with 2.51 × 10^{10} water molecules at 2 $micron^2$ docking bays can provide the required transport rate if 40 per cent of the vasculoid surface in the glomeruli is given over to water transport. The ~0.001 gm/sec (~4 x 10^6 molecules/sec-$micron^2$) solute waste stream is conveyed on mixed-cargo tankers carrying ~10^9 solute molecules per tanker, which may be unloaded at mixed-cargo docking bays covering only 8 per cent of the vasculoid surface in the glomeruli.

Glucose Absorption and Elimination—Glucose is absorbed in the small intestine, mostly in the duodenum and the jejunum, at a maximum rate of ~10^{20} glucose molecules/sec, somewhat less than our maximum glucose requirement of 5 x 10^{20} molecule/sec. Given a capillary absorptive area of ~3 x 10^{12} $micron^2$ in the upper small intestine, the natural glucose transport rate is ~3 x 10^7 molecules/sec-$micron^2$. As before, if maintenance activities are temporarily suspended during peak demand, glucose tankers loaded with 3.93 x 10^9 glucose molecules at 2 $micron^2$ docking bays can provide the same transport rate if 17 per cent of the vasculoid surface in the duodenum and jejunum is given over to glucose transport. A related issue is the vasculoid's response to severe gastrointestinal overloading, as for example the rapid ingestion of a large quantity of pure sugar by the user. Sugar has a tremendous irritating effect on the stomach, with 15-30 gm (60-120 kcal) producing great outpourings of stomach mucous; as little as 0.25 liter of 20 per cent sugar solution (78 gm, 300 kcal) is sufficient to cause vomiting in some patients with chronic gastritis, according to century-old laboratory experiments. A rapid 2000 kcal (1.75×10^{24} glucose molecules) ingestion would represent a bulging stomach-full (<1.5 liters) of >22 per cent sugar solution (~526 gm of sugar) which would be severely irritating to tissues and toxic to cells, almost certainly exceeds the natural absorption abilities of the small intestine, and would likely prove emetic. Urination at the physiological maximum $\sim 10^{20}$ glucose molecule/sec elimination rate unloads a 2000 kcal sugar binge in ~5 hours, a renal export rate of 7.14×10^7 molecules/sec-micron^2 that can be handled at glucose docking bays covering 36 per cent of the vasculoid surface in the renal glomeruli. Alternatively, some excess sugar can be routed to the liver or other organs for conversion to glycogen or starch.

Boxcar Clearance in Narrow Capillaries—Cell-carrying boxcars 6 microns in diameter cannot squeeze through the narrowest of capillaries, most notably those found in the human retina which may be as small as 4 microns wide. This has two operational implications. First, boxcars carrying cellular cargo destined for tissues having capillaries <8 microns must onload and offload at cellulocks located outside such tissues, relying on natural transtissue mobility to achieve the desired placement of the transported cells. In this the vasculoid mimics the natural transport process. Second, boxcars must be capable of being rerouted around narrow capillary bottlenecks to their flow, else the carriers may lodge in the narrow passage and become an obstacle to the free flow of materials as sometimes occurs with leukocyte plugging in certain pathological conditions such as diabetes. Cilia with boxcar-coded gripper pads located at the entrances of difficult passages may be programed to refuse boxcars, and may also utilize reverse ciliary flows to assist this process. Alternatively, boxcars may be designed with flexible metamorphic surfaces (and thus a limited ability to change shape) to allow tight passage when traveling empty. With metamorphic boxcars, cilia can allow boxcars to pass that will empty their contents before reaching a capillary. Another alternative is that the vasculoid could be used directly to induce changes in capillary size.

Tanker Monolayer Transport—If container traffic is strictly limited to a single monolayer at the artery or vein wall, then ciliary transport produces a tanker bottleneck in medium- and large-diameter vessels of the arteriovenous (noncapillary) vasculature. This is most clearly illustrated in the aorta. The circulation time of ~140 sec (1.4 m circuit at v_{tanker} = 1 cm/sec transport speed) implies a system flux of 8.4×10^{10} tankers/sec at basal load and 1.2×10^{12} tankers/sec at peak load. However, the luminal surface of an aorta of diameter $2R_{aorta}$ = 2.5 cm has room to accommodate only a flux of $2p\ R_{aorta}\ v_{tanker}/S_{tanker} = 7.9 \times 10^8$ tankers/sec per monolayer of coverage, assuming S_{tanker} = 1 mm^2 tankers traveling at v_{tanker} = 1 cm/sec. The operational implication is that most tankers must be allowed to transit the aorta as a multilayer averaging 0.11 mm deep (~110 tanker layers) or <1 per cent of available luminal diameter at basal load, and up to 1.5 mm deep (~1, 500 tanker layers) or ~12 per cent of available luminal diameter at peak load. This bottleneck extends to a lesser degree throughout the arteriovenous vasculature, with the number of required tanker transport layers decreasing rapidly with decreasing vessel diameter. Purely monolayer tanker transport becomes tenable in blood vessels with diameters <200 microns (e.g., arterioles and large venules) at basal load or <20 microns (e.g., metarterioles) at peak load. Tankers can employ one or more simple grapples or temporary coupling mechanisms to join up into progressively lengthening chains (of linked tankers) as they pass blood vessel bifurcations moving upstream, and then to gradually detach and shorten the chains as they move downstream. Even a single cilium can exert sufficient force to pull a fully-loaded 1000-tanker chain at ~100 G, although the tanker at the base of each chain will normally be grasped by up to 10 independent cilia. Nevertheless, chain linkage reliability and the effects of such extended chains on flow viscosity should be studied more thoroughly. Other approaches, including facultative lateral links between chains to improve transport stability, tanker bloc motions, independently ciliated tankers (allowing high flow rates in large vessels with low inter-laminar slip rates), mechanically valved free flow, extensive water and respiratory gas caching to reduce tanker count, and increasing transport speeds up to 1 m/sec in large blood vessels, are potentially useful strategies but will not be explored further here.

An additional small population of mobile legged nanorobots, or "vasculocytes, " continuously patrols the vasculoid interior. Vasculocytes are independent nanodevices several microns in size, replete with ambulatory appendages, manipulator arms, repair and assembly tools/materials, onboard computers with mass memories, communications and sensory equipment, and independent power supplies and fuel tankage. A description of a similar system is in; a fuller discussion of these capabilities is in and is beyond the scope of this paper. Like other vasculoid subsystems, to ensure high reliability vasculocytes are designed with the usual tenfold redundancy in most major components. Vasculocytes in the vasculoid appliance have many important functions, including: (1) general maintenance and repair activities; (2) plugging internal leaks and breaches, cleaning spills, leak scavenging, and maintaining an orderly particle-free internal environment; (3) repairing or replacing malfunctioning ciliary leg mechanisms (designed for modular changeout); (4) clearing of jammed docking bays or cellulocks; (5) reconstruction of the vasculoid sapphire surface plate array after a breach or other catastrophic failure; (6) physical reconfiguration and remodeling of the vasculoid microstructure, as for example to extend additional segments into new capillaries following angiogenesis, to remove vasculoid tendrils from dead capillaries, or to expand or contract existing segments (details deferred to another paper); and (7) installation and removal of vasculoid systems from the human body. With a vasculocyte mass of ~10 picograms, volume ~3 micron^3, and peak power consumption ~50 picowatts to allow bursts of more intensive computation (at up to nearly ~1 megaflop/sec) or other activity, a 10-watt vasculocyte subsystem energy budget permits the deployment of 200 billion continuously active devices, a total 2 gm mass of nanorobots. Deployment of this number of devices would allow, for example, at least one vasculocyte to visit all 24 trillion docking bays once a day, spending 720 sec visit, enough time to perform ~10^9 manipulatory operations using a 10^6 Hz robotic arm . Alternatively, this would allow 7200 visits/day per docking bay, each visit lasting ~100 msec (~10^5 operations; travel time between adjacent bays at 1 cm/sec is only 0.35 msec) which is long enough to run a simple diagnostic set and to detect plate malfunction within just ~12 seconds (on average) of its initial occurrence. It would also allow one visit per day to each of 3000 trillion cilia, spending ~6 sec/visit (~10^7 operations). This seems sufficient. Assuming ~0.1 pW per vasculocyte leg (similar to cilium) and 4 legs in motion, simple walking consumes only ~0.4 pW of mechanical power; a 10, 000 ops/sec computation budget to control simple walking, costing ~6 x 10^{-17} watts/(ops/sec), adds another ~0.6 pW, giving a total power requirement of just ~1 pW for continuous simple walking. An onboard 1 micron^3 fuel tank with energy storage density of ~2 x 10^{10} J/m^3 for externally-supplied O_2 gives each 1-50 pW vasculocyte a range of 400-20, 000 sec between refueling stops, or ~100-5, 000 sec if the O_2 is stored internally. The power draw of the entire 200 billion vasculocyte fleet is 0.2-10 watts. During vasculoid installation, the patient is injected with a population of vasculocytes adequate to construct the initial configuration in a reasonable period of time. After installation, a sufficient number of spare vasculocytes are stored internally in dormant condition in the vesicles until needed for repair, maintenance, or injury response functions. Active vasculocytes have highly redundant subsystems, hence fail infrequently. Upon such infrequent failure, they are replaced with reactivated dormant devices (self-repairing vasculocytes are possible in principle—e.g., modular repair-by-replacement—but would add significantly to overall system complexity, hence are not included in this scaling study.). Nanorobot detritus and internal waste is containerised and ciliated to an appropriate internal holding area located in the vesicles. New vasculocytes, replacement modules, and other necessary spare parts and repair materials may be exchanged into the user as consumables, when required. Vasculocytes performing general repair and maintenance on tanker docking bays have been allocated up to 10 sec docking cycle for this task a very conservative 50 per cent duty cycle for docking bays. Vasculocyte system subtasks which must be subsumed within this time budget may include: (1) Detection and analysis of system fault; (2) determination of appropriate response; (3) dispatch of appropriate repair mechanism to problem site, including travel time and docking; (4) on-site verification that the problem is as reported, and that the repair plan is correct; (5) deployment of tools and materials; (6) performance of repair work possibly including plate changeout, rotor replacements, sensor recalibration, etc.; (7) verification of correct completion of repairs and that the faulty subsystem is now performing properly; (8) retracting/repacking all tools with final verification of job completion; and (9) undocking and return to transport stream for storage or to receive instructions for the next assignment. The vasculoid configuration, including numbers of components, their locations, their accessibility for repair and their ease of disconnection/reconnection will influence the rapidity of these steps. Of course, most of the time there will be no fault at a particular docking bay, and running a diagnostic is much quicker than correcting a fault. Since there are only enough vasculocytes to allow up to 720 seconds/day of repair and maintenance activity at each docking bay, and since a 50 per cent duty cycle implies that each docking bay is allocated ~43, 200 sec/day for this activity, then it might be possible to significantly increase the docking bay duty cycle up to (86, 400—720) / (86, 400) ~ 99 per cent in a less conservative design.

Power is supplied to the vasculoid appliance via the chemical combination of oxygen and glucose, both of which are plentiful in the human body. Total body power dissipation is ~100 watts at the basal rate and ~1600 watts at peak. As a further point of comparison, the basal human heart power output is $p_{heart} \sim P_{systolic} V_{blood} / t_{circ}$ = 1.4 watts (taking pumping pressure $P_{systolic}$ = 120 mmHg (1.59×10^4 N/m^2), pumped blood volume V_{blood} = 5.4 liters, and circulation time t_{circ} = 60 sec), rising to ~8.0 watts when cardiac output reaches ~30 liters/minute at peak exertion, assuming no rise in pressure. Experiments confirm that the basal power output of a 350 gm resting heart is ~3.5 milliwatts/gm or ~1.2 watts, and patients with chronic congestive heart failure can achieve measured peak cardiac outputs no higher than ~2 watts; efficient total artificial hearts (TAH) may draw ~5 watts of power. The four most significant subsystems requiring a regular energy supply include ciliary transport (11.8-166.2 watts), docking bays (14.5-16.6 watts), vasculocytes (0.2-10 watts), and cellulocks (~0.6 watts), giving a requirements range of 27.1 watts (basal) to 193.4 watts (peak). There are also 125 trillion applications plates which are unused in the basic vasculoid model described in this paper. If each applications plate is allotted a 0.6 pW energy budget comparable to the docking bays, and if all possible applications plates are simultaneously activated, , then the applications subsystem could require an additional 75 watts, boosting the total power draw of the vasculoid appliance to 102.1-268.4 watts, which is still only at the borderline of thermogenic significance.

Besides providing systemic materials transport throughout the human body, the water component of blood also serves an important thermoregulatory function. Removing most of the circulating water from the vasculature eliminates one of the means by which the body regulates its gross thermal conductivity. In addition, the extremely high thermal conductivity of diamond imposes the requirement for a vasculoid design using mostly sapphire, rather than diamond, construction materials, even though diamondoid materials have been more extensively discussed in connection with molecular nanotechnology and medical nanorobots and may still be employed in lesser quantities (e.g., as thin coatings) throughout the vasculoid structure. Aside from blackbody radiation, sweating, and behavioural thermoregulation (including respiratory cooling), the body regulates its temperature and offloads excess heat principally via two mechanisms: *First,* there is passive conduction. Heat travels by pure conduction through fat and muscle from the body core out to the periphery. The thermal conductivity of human tissue is $K_t \sim 0.5$ watts/m-K, so for a typical L = 10 cm path length (~half-torso thickness), heat flow $H_f \sim K_t/L = 5$ watts/m^2-K, or ~10 watts/K for a 2 m^2 human body. In a cold room, the mean temperature differential between core and periphery DT ~ 11 K, so $H_f \sim 100$ watts, which is approximately the basal metabolic rate. Experiments confirm that 5-9 watts/m^2-K is the minimum heat flow in very cold conditions (the actual value depending largely upon the thickness of subcutaneous fat layers). In this case, the peripheral capillary blood flow has slowed to a trickle, producing the minimum thermal conductivity of the human body in cold conditions. On the other hand, in a warm room or during heavy exercise, DT ~ 1 K, so $H_f \sim 10$ watts. Thus, paradoxically, at warmer temperatures when the human body is generating considerable surplus heat, the body's passive heat flow is actually very low because of the smaller temperature differential between core and periphery. *Second,* heat is transported via the active blood flow. In warm rooms, not only are the peripheral capillary sphincters fully dilated, allowing more blood to flow through the peripheral capillaries relative to the core capillaries, but also the total volume of blood flow may increase. (During heavy exercise, total blood flow volume may rise by a factor of 4 or 5.) Diathermy experiments suggest that the active blood flow mechanism alone may carry off 100-200 watts of heat before core temperature starts to rise. In cold rooms and in the absence of heavy exercise, peripheral capillary sphincters are maximally contracted, thus minimising blood flow (and hence heat transport) to the periphery.

To summarize: The passive conduction mechanism can throw off ~100 watts of waste heat when the human body is in a cold room but only ~10 watts when the body is in a warm room, while the active conduction mechanism can throw off negligible heat in a cold room but up to 100-200 watts in a warm room. Thus, as the external environment warms up, the human body shifts from passive conduction to active conduction (via increased blood flow and capillary sphincter widening); if the environment becomes hotter still, or the person begins exercising, then sweating eventually comes to dominate both processes. If we now remove the active blood flow regulatory mechanism, inhibit capillary sphincter expansion, and emplace artificial tubes inside all the blood vessels and capillaries of the body (i.e., install the vasculoid), the sweating process and surface thermal radiation remain unaltered. Thus the principal changes to the human thermoregulatory system are in the passive and active conduction systems. *First,* consider the active conduction system. In a working vasculoid operating at the maximum tanker flow rate, we have ~0.133 kg (dry mass) of tankers moving around the body in ~100 sec per circuit. We assume all tankers are filled with water, for a total water mass of 0.125 kg. Diamond has a heat capacity of 519 joules/kg-K; sapphire is 728 joule/kg-K and water is 4218 joule/kg-K at 310 K. Hence, the circulating tanker fleet can transport at most ~6 watts/K of heat from the core to the periphery. Given that DT might be as small as 1 K, the "active conduction system" is essentially disabled by installation of a vasculoid that is constructed either of diamond or sapphire. *Second,* consider the passive conduction system. For a natural biological-tissue body, heat flow is $H_f = K_t / L$ = 5 watts/m^2-K. For a human body shape comprised entirely of pure diamond (K_t ~ 2000 watts/m-K at 310 K) and again taking L = 10 cm, then H_f = 20, 000 watts/m^2-K. For a diamond-envasculoided human body, taking a mass of ~1.7 kg of diamond thoroughly interwoven with 68.3 kg of mostly aqueous biological tissue mass (for a standard 70 kg male body), as a crude estimate the effective heat flow becomes H_f ~ 500 watts/m^2-K, or ~100 times more thermally conductive than before. For comparison, a pure metal human form would have H_f ~ 170 watts/m^2-K for stainless steel, ~350 watts/m^2-K for lead, ~780 watts/m^2-K for iron, or ~3800 watts/m^2-K for copper. Hence a diamond-envasculoided human body would have passive conduction properties similar to those of solid metal.

This has implications for the maximum DT that can be maintained between core and periphery. Consider a human-shaped tissue-mass with half-thickness L ~ 10 cm and surface area A ~ 2 m^2, sufficiently heated from the inside to cause P ~ 100 watts (human basal rate) to flow via passive conduction from core to periphery, establishing a temperature differential DT ~ P L / A K_t ~ 10 K for natural human tissue with K_t = 0.5 watts/m-K. Upon switching to diamondoid envasculoided tissue, mean tissue thermal conductivity would rise to K_t ~ 50 watts/m-K and so DT would fall to ~0.1 K. In effect, the entire human body would become isothermal to within 100 millikelvins; even at the peak power output of 1600 watts for the human body, DT rises to just ~1.6 K. Thus, a diamond-envasculoided human body would tend to become isothermal with its surroundings very quickly (although partially offset by subcutaneous fat), a clear hazard to normal human health especially in very hot or very cold environments. The thermal equilibration time is approximately t_{EQ} ~ $L/v_{thermal}$ ~ 0.1 millisec, where $v_{thermal}$ ~ $K_t / h_{plate} C_V$ = 1000 m/sec for neighboring vasculoid plates in good thermal contact with each other and having thickness h_{plate} ~ 1 micron, with K_t = 2000 watts/m-K and C_V = 1.8 x 10^6 joules/m^3-K for diamond at 310 K, and taking L = 10 cm as before. This is far shorter than the typical 1-10 sec thermal response time of the purely-biological human vasculature. Substitution of sapphire for diamond significantly improves thermal performance. The thermal conductivity of synthetic sapphire may be as low as K_t ~ 2.3 watts/m-K for sapphire at 310 K, roughly a thousand fold lower than for diamond, when measured in the direction normal to the symmetry or optic axis (c-axis); heat capacity (C_V = 2.9 x 10^6 J/m^3-K) and density (3970 kg/m^3) of sapphire are slightly higher than for diamond. Thus for a sapphire-envasculoided human body, taking a mass of ~1.9 kg of sapphire (at 25 watts/m^2-K for L = 10 cm) thoroughly interwoven with 68.1 kg of mostly aqueous biological tissue mass (at 5 watts/m^2-K), the total heat flow is just 5.5 watts/m^2-K, which differs insignificantly from natural biological tissue. At P = 100 watts, DT falls to 2 K compared to 10 K for natural tissue and 0.1 K for diamond-envasculoided tissue; t_{EQ} ~ 1 sec for sapphire vs. 10^{-4} sec for diamond. Two complications regarding sapphire require additional research. *First,* the thermal conductivity of sapphire may vary significantly with both composition and crystallographic orientation, a fact which might impose additional and unknown constraints on the present design. For instance, one source reports heat flow values interpolated to 310 K of 21 watts/m-K normal to the c-axis and 23 watts/m-K parallel to the c-axis; minor extrapolations of other sources to 310 K (i.e. slightly outside the exact temperature ranges measured experimentally) imply values of 2.0 and 2.3 watts/m-K for heat flows normal to the c-axis.

However, all reported values for sapphire are at least two orders of magnitude more insulating than diamond. *Second,* much like diamond, the thermal conductivity of sapphire varies with temperature. For example, at ~200 K (near dry ice temperature) sapphire's thermal conductivity rises to 5 watts/m-K. At liquid nitrogen temperature (77 K), K_t soars to ~1000 watts/m-K; the peak is ~6000 watts/m-K at 35 K. (Diamond's conductivity also rises as it cools). At the other temperature extreme, sapphire's thermal conductivity rises to 3.9 watts/m-K by 523 K. Diamond thermal conductivity also varies significantly with isotopic composition (e.g., ^{12}C vs. ^{13}C); it is unknown whether similar opportunities may exist for the engineering of desired levels or patterns of thermal conductivity in isotopically-controlled sapphire. *Finally,* we note that much burn damage occurs due to a failure to dissipate heat.

Selectively increasing the thermal conductivity of certain parts of the body at certain chosen times is a useful feature that could mitigate the effects of transient local heating. Note that blood is an excellent electrical conductor, and its replacement by an inorganic construct may reduce the susceptibility of the body to electrical burns under ordinary conditions. Pure single-crystal alumina (sapphire), roughly analogous to the core material of the vasculoid plates, is one of the best electrical insulators known. Pure bulk diamond is also an excellent insulator, but hydrogen-terminated diamond surface shows a high p-type surface conductivity, and slight impurities (e.g., p-type (boron) or n-type (phosphorus) or dislocations may also permit conduction, so a diamond-coated sapphire vasculoid plate should be somewhat more electrically conductive than bulk sapphire. The vasculoid is envisioned as a primarily mechanical system and so its electrical characteristics have not yet been investigated. The susceptibility of the appliance to electrostatic charge transfer, electromagnetic interference or electromagnetic pulses (EMP) is unknown.

Perhaps the single most important biocompatibility factor involving the vasculoid has been termed "vascular mechanocompatibility" by Freitas. Vasculoid nanorobots in all their forms must be as mechanically biocompatible with the vascular walls as are stents and must not produce destructive mechanical vasculopathies or disrupt the endothelial glycocalyx.

Modulation of Endothelial Phenotype and Function—The luminal surfaces of all blood and lymph vessels consist of a thin monolayer (the intima) comprised of flat, polygonal squamous endothelial cells (EC) covering a much thicker layer (the media) comprised of vascular smooth muscle cells (SMC). Under normal physiological conditions, both layers are subject (and respond) to tangential fluid shear stresses across the endothelial cell surface due to the bulk flow of blood, normal hydrostatic pressure stress acting radially on the vessel wall due to the propagation of the pressure wave, and cyclic stretch or strain due to blood vessel circumferential expansion in vivo, and thus might also be sensitive to similar mechanical stresses that may be applied by stationary or cytoambulatory intravascular nanorobots such as the vasculoid basic plates. Endothelial cells (EC) are randomly oriented in areas of low shear stress but elongated and aligned in the direction of fluid flow in regions of high shear stress. In vitro endothelial cells previously acclimatised to physiological fluid shear stresses respond to artificial changes in local fluid shear stress only very slowly, and in three stages. In the first stage, EC initially respond to the imposition of stress within 3 hours by enhancing their attachments to the substrate and to neighboring cells; the cells elongate and have more stress fibers, thicker intercellular junctions, and more apical microfilaments.

In the second stage, after 6 hours the EC show constrained motility as they realign, losing their dense peripheral bands and relocating more of their microtubule organising centers and nuclei to the upstream region of the cell. In the third stage, after 12 hours the EC become elongated cells oriented in the new apparent direction of fluid flow; stress fibers are thicker and longer, the height and thickness of intercellular junctions are higher, and the number and height of apical microfilaments are increased. This produces a new cytoskeletal organisation that alters how the forces produced by fluid flow act on the cell and how the forces are transmitted to the cell interior and substrate. Physiological fluid mechanical stimuli (e.g., fluid shear stresses) are important modulators of regional endothelial phenotype and function. For example, endothelium exposed to fluid shear stress undergoes cell shape change, alignment, and microfilament network remodeling in the direction of flow (though this may be blocked via microtubule disruption using nocodazole). Interestingly, the application of a steady laminar shear stress (a physiological stimulus) upregulates the human prostaglandin transporter (hPGT) gene at the level of transcriptional activation, whereas a comparable level of turbulent shear stress (a nonphysiological stimulus) or low stress (such as would be produced by a vascular surface coated with sessile nanorobots) does not . A few of the many quantitative experimental observations include:

1. shear stresses from 0.02-1.70 N/m^2 produce flow-induced membrane K^+ currents;
2. physiological shear stresses of 0.35-11.7 N/m^2 stimulates mitogen-activated protein kinase in a 5-min peak response time;
3. 0.04-6 N/m^2 shear stress increases inositol trisphosphate levels in human endothelial cells, with a 10-30 sec peak response time;
4. shear stresses from 0.5-1.8 N/m^2 regulate (in frequency and amplitude) oscillating K^+ currents known as spontaneous transient outward currents or STOC which are observed both in isolated bovine aortic endothelial cells and in intact endothelium; activation of STOC depends on the existence of a Ca^{++} influx and is blocked by 50 microM of Gd or is significantly reduced by 20 microM of ryanodine;
5. shear stress of 1.2 N/m^2 induces transcription factor activation over response times ranging from 0.3-2 hours;
6. Arterial shear stresses of 1.5-2.5 N/m^2 (but not a venous shear stress of 0.4 N/m^2) induce endothelial fibrinolytic protein secretion;
7. shear stress of 2 N/m^2 induces TGF-b1 transcription and production in a ~60 sec initial response time, with a sustaining increase in expression after 2 hours;
8. a shear stress of 2 N/m^2 suppresses ET-1 mRNA on confluent bovine aortic endothelial cell monolayers; these effects of shear may be completely blocked (thus allowing ET-1 to be expressed) using 875 nM of herbimycin to inhibit tyrosine kinases or 10 microM of quin 2-AM to chelate intracellular Ca^{++}, partially inhibited using 3mM of tetraethylammonium (TEA), or attenuated by elevated extracellular K^+ at 70 mM or completely inhibited by K^+ at 135 mM;
9. shear stress of 3 N/m^2 induces Ca^{++} membrane currents in a 30 sec peak response time;
10. shear stress of 6 N/m^2 applied for 12 hours causes endothelial cells to align with their longitudinal axes parallel to flow;
11. membrane hyperpolarisation occurs as a function of local shear stress up to 12.0 N/m^2, with an exponential approach to steady state in ~1 minute; the process is fully reversed once the artificial fluid flow stress is removed;
12. critical shear stress of 42 N/m^2 is the disruptive threshold for endothelial cells, inducing cell mobility; and

13. Shearing stresses of 5-100 N/m^2 occur at the contact interface when a leukocyte is adhering to or rolling on the endothelium of a venule.

Endothelial cells thus respond to sustained physiological fluid shear stresses from 0.02-100 N/m^2, spanning the range of normal arterial wall fluid shear stresses of 1.0-2.6 N/m^2 from the aorta through the capillaries and 0.14-0.63 N/m^2 for the venous circulation. By contrast, legged vasculomobile medical nanorobots may apply shear stresses during luminal anchorage or cytoambulation at velocities up to 1 cm/sec of at least 40-200 N/m^2 or higher. (Self-expanding aortic stents forcibly pulled from the vessel require an extraction force of ~400 N/m^2 assuming a 10-cm length, rising to ~1200-3600 N/m^2 for stents anchored with hooks and barbs; varying the radial force applied by stents against the vascular wall has little impact on the required extraction force.) Such shear forces, if imposed unidirectionally by large numbers of closely-packed co-ambulating nanorobots for time periods of >10^3 sec, may induce significant changes in shape, orientation, and physiological function in the underlying endothelial cell population. If instead these forces are applied in randomised directions varying over short time periods (<1 hour); by vasculoid nanorobots, then mechanically-induced modulation of endothelial phenotype and function should not occur. All shear forces must not be eliminated, however. A nanorobot aggregate that shields vascular cells from fluid shear for an extended time may induce those cells to revert to their flow-unstressed phenotype or to undergo apoptosis. Endothelial cells cultured in the absence of shear stress tend to become dedifferentiated. In one study, after blood shear was artificially reduced near a wound lesion for 24 hours the local endothelial cells became less elongated, contained fewer central microfilament bundles, and exhibited a slower repair process. In another study, vein grafts removed from the higher-shear arterial circulation and reimplanted in the lower-shear venous circulation of the same animal showed regression of intimal hyperplasia and medial rethickening in 14 days, apparently due to induction of smooth muscle cell apoptosis by a reduction in pressure or flow forces. Endothelial cells can also respond to persistent static overstretching in many ways, up to and including apoptosis. For instance, hypertension caused by hydrostatic edema can induce apoptosis in capillary EC. Additionally, vascular wall cells respond to lateral stretch forces due to cyclical blood vessel expansion in vivo. For example, in one experiment bovine aortic endothelial cells were seeded to confluence on a flexible membrane to which cyclic strain was then applied at 1 Hz (0.5 sec strain, 0.5 sec relaxation) for 0-60 min. After 15 minutes of this cyclic stretching, there was an increase in adenylyl cyclase (AC), cAMP, and protein kinase A (PKA) activity of 1.5-2.2 times at 10 per cent average strain as compared to unstretched cells, but there was no activity increase at 6 per cent strain—evidently, cyclic strain activates the AC signal transduction pathway in endothelial cells by exceeding a strain threshold, thus stimulating the expression of genes containing cAMP-responsive promoter elements. Stretch-activated cation channels in bovine aortic EC are inhibited by

$GdCl_3$ at 10 microM. Human umbilical EC subjected to a 3-sec stretch pulse show an intracellular rapid-increase Ca^{++} spike, followed by a (ryanodine-inhibitable) slow-decline, due to Ca^{++} entry into the cell through stretch-activated channels; Mn^{++} also permeates mechanosensitive channels (but not Ca^{++} channels) and enters the intracellular space immediately after an application of mechanical stretch. Cyclic strains of 10 per cent at 1 Hz induce intracellular increases in Ca^{++}, diacylglycerol, inositol trisphosphate and protein kinase C (PKC) in peak response times of 10-35 sec, often sustained for up to ~500 sec, and induce transcription factor activation over response times ranging from 0.25-24 hours. Several endothelial cytokines are induced by cyclic mechanical stretch and cyclic mechanical strain modulates tissue factor activity differently in endothelial cells originating from different tissues.

Similarly, bovine aortic smooth muscle cells (SMCs) seeded on a silastic membrane and subjected to cyclic strains up to 24 per cent enhanced SMC proliferation at any strain level, although SMC under high strain (7-24 per cent) showed more proliferation than SMC at low strain (0-7 per cent) in this experiment. High-strained SMC aligned themselves perpendicular to the strain gradient, whereas low-strained SMC remained aligned randomly; PKA activity and CRE (cAMP response element) binding protein levels increased for highly strained cells, compared to low-strained cells. Other experiments have found that small mechanical strains of 1-4 per cent at 1 Hz applied to human vascular smooth muscle cells can inhibit intracellular PDGF- or TNFa-induced synthesis of matrix metalloproteinase (MMP)-1; that saphenous vein SMC distention by 0.5 atm pressure subsequently elevates cell apoptosis; that cyclic mechanical strain at normal physiological levels decreases the DNA synthesis of vascular smooth muscle cells, holding SMC proliferation to a low level; that 1 Hz, 10 per cent cyclic strain on SMC activate tyrosine phosphorylation and PKC, PKA, and cAMP pathways over response times from 10 sec to 30 min and that vascular SMC exhibited stretch-induced apoptosis when subjected to cyclic 20 per cent elongation stretching at 0.5 Hz for 6 hours. Consequently, medical nanorobot aggregates which shield the vasculature from normal cyclical strains might elicit excess growth of vascular smooth muscle cells, which growth is normally held in check by the rhythmic stretching from the arterial pulse. Intravascular nanorobot aggregates that apply cyclic mechanical strains exceeding a few percent might encourage increased SMC proliferation and activate mechanosensitive and stretch-activated channels in EC, along with cellular realignment and subsequent SMC apoptosis at the highest strain levels. In 2002 it was unknown whether high frequency (>KHz) cyclic mechanical strains likely to be employed by vasculomobile medical nanorobots would have biological effects similar to or different from those described above for low-frequency cyclic strains, excepting certain specialised mechanoreceptor cells such as the cochlear stereocilia, other hair cells and somatosensory neurons, since most mechanical cell stimulation experiments have been conducted at low frequencies.

Unrecognised effects that might be triggered by high-frequency cyclic strains cannot be ruled out. However, given the relative safety of procedures involving intravascular ultrasound with its low complication rate (e.g., only 1.1 per cent, including spasms, vessel dissection and guidewire entrapment using frequencies as high as 10-20 MHz, it seems improbable that KHz or MHz acoustic waves of the intensities that might be employed by medical nanorobots for communication or power supply will damage the luminal vascular walls. Continuous low-power ul trasound exposures exceeding ~10^4 sec are considered safe but there are few studies on the safety of long-duration chronic exposures. (Interestingly, relatively high-intensity intravascular ultrasound has been used to dissolve occlusive platelet-rich thrombi safely and effectively in myocardial infarctions and in restenosed stents.) If necessary, large wave motions and pressure patterns that are characteristic of normal blood flow can in principle be simulated by the vasculoid plate array by a combination of motile ciliary activity and periodic manipulations of interplate bumpers. It is likely that the vasculoid appliance will need to control smooth muscle cell proliferation in the simplest case releasing specific cytokines into the vasculoid-endothelial space. Such factors may include known SMC proliferation promoters such as thrombin (esp. alpha-thrombin), PDGFs (esp. PDGF-AA), FGF (esp. basic FGF), HBEGF (heparin binding epidermal growth factor), TGFb (transforming growth factor-beta) at low concentrations, angiotensin II, thrombospondin-1 (stretch/tension), and known SMC proliferation inhibitors such as heparin/heparan sulfate, TGFb (transforming growth factor-beta) at high concentrations, nitric oxide, prostaglandins, calcium antagonists, agonists that activate guanylate and adenylate cyclases, inhibitors of angiotensin-converting enzyme, interferon gamma, 18-beta-estradiol, sodium salicylate, and the topoisomerase I inhibitor topotecan. Adult arterial walls contain both differentiated and immature SMCs. R. Bradbury notes that further research is needed regarding how SMCs handle conflict resolution between "divide" and "don't divide" signals that it may be receiving from both internal and external sensors. Given the large number of signals that SMCs currently respond to, it seems highly likely that the vasculoid can "manage" them, along with other secretion products of SMCs that play important roles in the prevention of vascular disease, such as extracellular superoxide dismutase.

Vascular Response to Stenting—Mechanical biocompatibility must also be demonstrated by intravascular nanorobots that are intended to remain in long-term near-contact with blood vessel walls. A related medical analog is the vascular stent—a flexible metal coil or open-mesh tube that is surgically inserted into a narrowed artery, expanded, and pressed into the vascular wall at up to 10-20 atm pressure, in order to ensure long-term local vascular patency by providing a scaffold to hold the artery open. Within 4 days, SMC begin to appear in the intima. After a few months the stent is completely overgrown with new endothelium, forming a neointima, although the media is usually compressed with smooth muscle cell atrophy in all stented regions; stenosis is prevented in vessels 10 mm or greater in diameter but is not precluded in vessels smaller than 6 mm. Histologically, in-stent restenosis appears to derive almost exclusively from neointimal hyperplasia, which appears more abundant following stent implantation than balloon angioplasty and in stents of greater stent length and smaller vessel caliber (or inadequate stent expansion). Restenosis occurs in 22-46 per cent of all stents emplaced within 6-12 months), in some cases requiring the insertion of a second stent into the first, and varies according to the material used. In one experiment, the thickness of the neointimal layer formed over wire-mesh stents placed in canine aortas was 83.9 microns thick for gold, 103.6 microns for stainless steel, 115 microns for Teflon, 209.6 microns for silicone, and 228.6 microns for silver; a copper stent produced severe erosion of the vessel wall, marked thrombus formation, and aortic rupture. Stent surface coating and texture also affects leukocyte-platelet aggregation and platelet activation. Improved prospects are reported for diamondoid stents. For instance, diamond-coated stents produced by Phytis Corp are inserted at 16 atm pressure (vs. 2-3 atm normally for stents) and yet do not dislodge surface (thrombogenic) antigens and selectins: "The results show that in none of the control systems (systems without stents) [could] a change in glycoprotein expression (i.e. all antigens) ... be detected. With the exception of one measurement also the structural epitopes (CD 41a, CD 42b) show no significant changes during the test period. The reason is most probably that the shearing strength of the system was too weak or the expression of the antigens was too strong. Whereas for the thrombocyte activation, remarkable difference in the expression of CD 62p and CD 63 could be detected. The thrombocyte activity marker CD 62p is reduced in diamond-like coated stents and diamond-like coated stents with heparin compared with uncoated stents." However, stent devices are far from ideally mechanocompatible with blood vessel walls. For example, stents placed endovascularly in dog aorta for 4-45 weeks and then examined histologically show medial atrophy, intimal hyperplasia (tissue ingrowth), and proliferation of the vasa vasorum (the microvasculature of the aorta) more prominently for covered stents than for bare stents, probably due to hypoxia in the aortic wall. Cellular proliferation is highest when the artery wall is most hypoxic; but vasculoid nanorobot aggregates (e.g., basic plates) that entirely cover the vascular endothelium can precisely regulate oxygenation of the underlying tissue (using data from oxygen sensors in the plate wall contacting the endothelium), thus largely eliminating the possibility of hypoxia. Stentlike vascular-coating structures also may be able to inhibit stenosis due to vascular smooth muscle proliferation, migration, and neointima formation, without inducing apoptosis, by releasing the topoisomerase I SMC-proliferation inhibitor topotecan in a localised 20 min exposure, or by using other similar drugs.

Nanorobotic Destructive Vasculopathies—The physical configurations or activities of medical nanorobotic aggregates could in some circumstances be destructive of vascular tissue. Owens and Clowes point out that the severity of arterial injury is important in determining the ultimate pathophysiologic response and describe a classification system based on the immediate histologic effect of the injury:

Type I injuries involve no significant loss of the vessel's basic cellular architecture, although there may be a slight change in endothelial architecture and associated cellular adhesion. Examples include the fatty streak (an early atherosclerotic lesion), hemodynamic factors and flow disturbances which produce, at most, only a modification of the established cellular architecture.

Type II injuries involve loss of the endothelial layer, perhaps inducing platelets to adhere and begin forming a thrombus at the area of loss, but the internal elastic lamina remains intact and there is little or no damage to the media. Examples include injuries incurred during simple arterial catheterisation, endovascular procedures, vein graft preparation, or gentle filament-induced endothelial denudation of the carotid artery in a rat model.

Type III injuries involve transmural damage in which the endothelium is removed, the internal elastic lamina is often disrupted, and a significant portion of the medial cells are killed; platelets deposit and a thrombus forms at the site of endothelial loss, and an inflammatory response (vasculitis) including intimal hyperplasia is initiated within the vessel wall. Examples include spontaneous vascular dissection and various forms of surgical repair or reconstruction such as balloon angioplasty, endarterectomy and atherectomy.

Medical nanorobot device and mission designs should always seek to avoid *Type II* and *Type III* injuries, although in some special circumstances the potential even for *Type III* injuries may be inescapable. Destructive mechanical vasculopathies that might be caused by vasculoid nanorobots may be classified as ulcerative, lacerative, or concussive.

Nanorobotic Ulcerative Vasculopathy—Pressure ulcers are normally caused by a prolonged mechanical pressure against epidermal tissues (e.g., the skin of a person who is lying down; decubitus ulcer, or bedsores), typically at sites over bony or cartilaginous prominences including sacrum, hips, elbows, heels and ankles. The combination of pressure, shearing forces, friction and moisture leads to tissue death due to a lack of adequate blood supply; if untreated, the ulcer progresses from a simple erosion to complete involvement of the dermal deep layers, eventually spreading to the underlying muscle and bone tissue. In rare cases, mechanical frictional stimulation of the skin can precipitate systemic cutaneous necrosis and calciphylaxis, a state of induced tissue sensitivity characterised by calcification of tissues. Stercoral ulcers are caused by the necrosis of intestinal epithelium due to the pressure of impacted feces. Fluid mattresses can greatly reduce pressure ulcers in long-duration surgeries and pressure-relieving surfaces have been investigated for surgical patients, wheelchair users, and for other circumstances) It is generally recommended that the interior surface should employ materials having roughly the same mechanical properties as the enclosed tissue, and the applied interfacial pressures should be reduced to below 1 psi or ~ 50 mmHg. Similarly, a macroscale nanorobot aggregate such as a vasculoid appliance may cause luminal vascular ulceration by prolonged mechanical pressure against intimal tissues, similar to epithelial pressure ulcers, necrotising vasculitis, or pressure necrosis. For example, mechanical stretch induces apoptosis in mammalian cardiomyocytes and hypertension caused by hydrostatic edema can induce apoptosis in capillary endothelial cells. Another example is IUD-induced metrorrhagia (nonmenstrual uterine bleeding), wherein the IUD (intra-uterine device) elicits a vascular reaction most pronounced in the endometrium adjacent to the device and includes increased vascularity, degeneration with defect formation, congestion, and poor hemostatic responsiveness to increased vascular permeability and damage, leading to interstitial hemorrhage due to vascular damage from mechanical stress transmitted by the IUD through the endometrium to its vascular network. Indwelling catheters can rest very snugly against the vascular walls without complication for brief periods. A biological-like interface may reduce ulceration in longer-term missions. In one study, a stented aortic graft was placed endovascularly in the native aorta of male sheep, and a histological examination 6 months later found good incorporation of the graft with no pressure necrosis, although there was a foreign body reaction around the graft and an organised blood clot was noted between the graft and the aortic wall (the inner, but not the outer, aortal surfaces would be expected to have clotting-suppressive properties). However, long-duration nanoaggregates (such as vasculoid plates) that must maintain close contact with endothelium should employ a mechanically compliant coating having properties similar to extracellular matrix. All such linkages should be not just immunocompatible but also mechanocompatible, possessing equivalent elasticity or mechanical compliance as the underlying tissue to which attachment must be secured. With conventional implants, compliance design may include assessments of circumferential compliance (measurement of changes in vessel diameter over a complete cardiac cycle, including pressure-radius curves, dynamic compliance, and mechanical hysteresis effects), longitudinal compliance (elasticity of selected lengths of the vascular system, including any localised stiffening), tubular compliance (imparity of elasticity between a prosthetic conduit and the native artery, elastic energy reservoiring, and pulsatile energy losses due to interfacial impedance mismatches), and anastomotic compliance (suture line anastomotic compliance mismatch and the para-anastomotic hypercompliance zone, localised regions of excessive mechanical stress, and cyclic stretch effects on replication of vascular SMC and extracellular matrix). A mismatch in mechanical properties between relatively compliant arteries and less-compliant metallic stents and tissue grafts has been thought to influence patency and pseudointimal hyperplasia. Larger more central arteries are more compliant than the distal small caliber arteries, wall shear stress from blood flow differs on either side of a curving vessel and the stress is out of phase with the pulsing circumferential stretch strain, and compliance mismatch between host artery and prosthetic graft may promote subintimal hyperplasia. Analogous compliance issues may be assessed once the static and dynamic stress patterns in the vasculoid transport system are more precisely known. Post-installation vascular conditioning can be maintained on a permanent basis because the vasculoid appliance exercises precise control over the transport of most metabolic and cellular traffic—e.g., of cholesterol, leukocytes and platelets, the principal participants in the arteriosclerogenic process, or of various circulating pluripotent stem cells and their activating factors.

Nanorobotic Lacerative Vasculopathy—Individual vasculomobile nanorobots or nanorobotic aggregates may occasionally scratch, scrape, or gouge the vascular luminal surface, causing partial or complete loss of local endothelium (*Type II* damage), a form of mechanical vasculitis or capillaritis, particularly during installation of the vasculoid appliance. Since the typical dimensions of nanorobotic vasculoid components approximate the endothelial thickness, transmural *Type III* damage to the media is unlikely. Turnover studies of rat endothelium show that (a) injured endothelium can recover an area one cell wide (~1000 micron^2) in ~3 hours, (b) the natural loss rate is ~0.1 per cent of endothelial cell area per day (~1 micron^2/day), and (c) the steady-state vascular denudation area is ~0.125 micron^2/cell.

Smooth nanorobot hulls, boundary layer effects and low fluid flow velocity throughout most of the vasculature during installation should ensure that major "sandblasting" type erosion is unlikely to occur inside human blood vessels even at the highest nanocrits consistent with continuous flow. Free-floating nanorobots that collide with blood vessel walls (given the no-slip condition at the wall) produce minimal shear forces, on the order of <~0.1 N/m^2—this is less than the 1.0-2.6 N/m^2 shear forces normally encountered in arteries and capillaries due to normal blood flow and the 0.14-0.63 N/m^2 shear forces in veins, but may be sufficient to cause a small biological response from the vascular endothelium. Applying the maximum possible bloodstream velocity of 1 m/sec to the impact-scratch relation, it is clear that particle-wall collisions should produce only harmless submicron nicks even in the most turbulent arteries. Nevertheless, some caution is warranted because natural endothelial cell wounding of ~1-18 per cent of all cells, possibly erosionally-derived, has been observed in rat aorta. Erosion of cultured fibroblast monolayers (simulating the vascular endothelium) using MHz ultrasound at acoustic pressures of ~10^6 N/m^2 is enhanced by the presence of a microbubble (particulate) contrast agent. Injection of crystalloid cardioplegic solutions into canine hearts at pressures >110 mmHg and at peak flow rates >25 ml/sec also causes a higher incidence of mechanical-physical trauma to the vascular endothelium and the endocardium. In another unusual case, intravenous self-injection by a drug abuser of dissolved tablets containing microcrystalline cellulose as filler material produced numerous microcrystalline cellulose pulmonary emboli, intravascular foreign body granulomas, focal necrosis and edema of the pulmonary parenchyma, and fatal vascular destruction. Endothelial abrasion alone may not stimulate neointimal thickening but inevitably must involve some endothelial cell loss and other biological responses. For example, mechanically scraping cultured endothelial cells causes growth factor to be released within 5 minutes, not abating for at least 24 hours thereafter, due to plasma membrane disruption. In the case of vascular dissection, a piece of the endothelium peels up, making an intimal flap that defines regions of true and false lumina, and sometimes may induce an intramural hematoma in the aortic wall. Endothelial cells mechanically damaged with a razor blade activate extracellular-signal-regulated kinases within ~300 sec, releasing fibroblast growth factor (FGF-2) which in turn induces intimal hyperplasia. Nanorobots which detect FGF-2 are alerted that mechanical endothelial injury has taken place; by absorbing the cytokine using molecular sorting rotors, the hyperplasia signal may be suppressed by a nanorobot, if desired. However, shear-induced endothelial denudation of healthy canine arterial endothelium appears not to occur at shear stresses up to at least 200 N/m^2. The role of erythrocyte collisions with vascular walls on the detachment rate of endothelial cells is just starting to be seriously investigated. K. Clements suggests that some provision might be also be made for emergency plate disconnects in certain unusual accident situations, e.g., where the user has irreversibly caught his hand in a machine applying superior force, and now risks not just amputation of the original biological limb, but also faces either (1) forcible extraction of the diamondoid appliance from the remaining biological tissues, or (2) progressive amputation of additional biological tissues because the vasculoid network has become caught in the machine.

Nanorobotic Concussive Vasculopathy—If a patient experiences significant external crushing or concussive forces, resident medical nanorobots that are present in small numbers can simply move out of the way, as described previously by Freitas in connection with the risks of dental grinding. In the case of macroscale intravascular nanoaggregates, there are several additional risks that should be avoided in specific appliance designs. *First,* there is the possibility that a sudden mechanical external tissue compression could push macroscale nanorobotic aggregates through the soft tissues, causing deep tissue penetrations, perforations, or other serious mechanical trauma. Similarly, because the vasculoid materials may have higher density than the surrounding biological tissues, very high accelerations could produce effects on those tissues that would be not unlike pushing gelatin through a metal wire strainer. Simple activities such as hand clapping or fistfights are unlikely to produce the high accelerations required for serious damage, but this risk should be quantified in future studies. Possibly relevant analogies in the medical literature include:

(1) Ulnar artery erosion, thromboemboli, digital ischemia and skin necrosis from a glass foreign body in a patient's hand.

(2) Tantalum coil stent damage that was induced or aggravated by intravascular ultrasound inside a coronary artery.

(3) A chronic indwelling catheter that led to erosion and rupture of the anterior wall of the right ventricle, producing a near-exsanguinating hemorrhage.

(4) Cardiac perforation by a subclavian catheter.

(5) Pulmonary artery catheter-induced right ventricular perforation during coronary artery bypass surgery.

(6) An ICD patch that migrated and perforated the right ventricular cavity.

(7) A stent that migrated to an oblique position across the aorta, producing a 7-cm pseudoaneurysm after 3 years.

(8) Catheter-induced pulmonary artery rupture (a well-recognised complication of invasive monitoring) that often leads to fatal hemorrhage.

(9) Femoral artery catheterisation trauma producing hematoma, pseudoaneurysms and arteriovenous fistulas of the femoral vessel.

(10) Iatrogenic subclavian artery injury due to central venous catheterisation.

(11) Repeated and prolonged vein catheterisation that led to subsequent stenosis (presumably due to luminal vascular mechanical damage).

(12) High-pressure injection injury that induced inflammation and foreign body granulomatous reaction, progressing to necrosis.

(13) Mechanical tearing of arteries due to overstretching.

(14) Spontaneous coronary artery dissection (mechanical arterial wall failure).

(Most of these cases pertain to injury from objects much stiffer and larger than vasculoid components, or are due instead to intrinsic vessel dysfunction (as in spontaneous arterial dissection.)

Second, a sudden external tissue compression could force nanorobotic aggregates into physical contact with neighboring nanoaggregates, possibly causing major structural damage or fragmentation of the devices. This risk increases as the nanodevices become more densely packed, especially along the crushing axis. Nanoorgans (as well as looser aggregates) can be crushed if sufficient force or mechanical shock is applied. Again, a few possibly relevant analogies from the medical literature include:

(1) External compression of emplaced stents that produced premature stenosis.

(2) A transabdominal teflon stent that broke intraperitoneally during tuboplasty procedure.

(3) A strongly-beating heart that sheared off a pericardial drainage catheter.

(4) A Hickman catheter that suffered rupture and embolisation during normal use.

(5) An indwelling catheter that fractured and a distal remnant embolised to the right ventricular outflow tract and main pulmonary artery, precipitating cardiopulmonary near-collapse.

(6) A catheter embolism that was produced when a catheterised patient engaged in power training exercises, externally crushing the catheter, although no symptoms or complications accompanied this event.

(7) Spontaneous fracture of indwelling venous catheter, leading to vascular leakage.

(8) Other instances of catheter fracture and embolism, including one case that led to cardiac arrest.

(Here, too, many of these cases might differ substantially from the vasculoid, since the catheters were made from a bulk material that depended on its integrity for function while the vasculoid would consist of a large number of semi-independent components capable of restoring functional connections on their own.)

Third, there is a small risk that poor device design, poor mission design, or a loss of control might cause nanoaggregates to operate in a dangerous manner, causing macroscale concussive injury to biological tissues. For example, iatrogenic vascular trauma caused by intra-aortic balloon pumps is well-known. In one case, balloon expansion during angiography ruptured the pulmonary artery. In a canine model, an aortic valve balloon dilation to 5-12 atm pressure produced valve leaflet connective-tissue injury and hemorrhage. In an experiment with canine intestine, excessive lymph pressures (>850-1630 N/m^2) produced artificially in the central lacteals caused fluid to leak out of the villi and caused intestinal epithelial cells to be shed into the intestinal lumen. These risks can be largely avoided or at least minimised by good design and experience with related systems, initially in animal models.

Mechanical Interactions with Glycocalyx—Endothelial cells are surrounded by a well-developed extracellular glycocalyx. If this outer margin is traumatised, receptor sites and fibronectin may be exposed which could then become available for bacterial adhesion. Nanorobots that rely upon absorption of local oxygen and glucose for their power supply or whose missions include extensive small-molecule exchanges with the environment may have ~10^4-10^5 molecular sorting rotors embedded in their exterior surfaces. These spinning sorting rotors are unlikely to cause direct physical damage to cell surfaces for several reasons. First, rotors are atomically smooth and recessed into the housing, reducing physical contact with colliding surfaces and eliminating potential nucleation sites that may trigger thrombogenesis, gas embolus formation, or foaming. Second, only a small fraction of all available sorting rotors may be actively spinning at one time, further reducing the likelihood of physical trauma. And such limited contact, when it occurs, should be relatively benign: Maximum rotor rim velocity of 2.6 mm/sec is less than 1 per cent of mean aortic blood velocity and lies only slightly above maximum capillary flow speed.

Many disease processes are known that involve damage to the glycocalyx, including some bacteria that phagocytose or otherwise destroy the cellular glycocalyx during an infection. Damage to the glycocalyx creates conditions that favour the binding of immune complexes, complement activation, and intravascular coagulation, with loss of gradients between blood and parenchyma; desialylated glycocalyx of endothelium also allows an increased rate of endothelial cell detachment from arterial walls. Could the glycocalyx strands present at all tissue and nontissue cells surfaces get trimmed, even by a recessed sorting rotor? Nanorobot sorting rotor binding sites for small molecules (<20 atoms) involve pockets measuring <2.7 nm in diameter, too small to physically accommodate the 10-20 nm thick plasma membrane or the main body of the glycocalyx projections typically measuring 5-8 nm thick and 100-200 nm long, consisting of glycoproteins comprised of 10, 000 atoms or more. While an occasional sugar residue may get clipped, binding sites can be designed for maximum steric incompatibility with glycocalyx glycoproteins and proteoglycans, further minimising the opportunities for trimming. Note that clipping a covalent C-C, C-O, or C-N bond probably requires a clipping energy >500 zJ/molecule, but sorting rotors designed to pump against pressures of ~30, 000 atm can only apply ~100 zJ/molecule (i.e., per binding site) so an accidentally-bound glycocalyx moiety seems more likely to jam the rotor than to be clipped off by the rotor. If this happens, the result may be a glycocalyx-tethered nanorobot, in which case a rotor-dejamming protocol will be required to free the trapped nanorobot.

Natural rates of glycocalyx damage are just starting to be quantified, and many tissue cells replace their glycocalyx or are retired in times ranging from 10^3-10^6 sec. For example: Schistosome parasites can shed some tegument-bound complexes in only ~1200 sec to 3600 sec; plasma membrane turnover rate is ~1800 sec for macrophage and ~5400 sec for fibroblast; cholesterol turnover rate in RBC membrane is ~7200 sec membrane phospholipid half-life averages ~10, 000 sec; neutrophil lifespan in blood is ~11, 000 sec; enterocyte glycocalyx is renewed in 14, 000-22, 000 sec, as vesicles with adhered bacteria are expelled into the lumen of small and large intestine; some schistosome membrane antigen turnover may require from 68, 000 to 160, 000-430, 000 sec; typical protein turnover half-life is ~200, 000 sec; cell turnover time is ~86, 000 sec in gastric body, ~200, 000 sec for duodenal epithelium, ~240, 000 sec for ileal epithelium, and ~400, 000 sec for gastric fundus; neutrophil lifespan in tissue is ~260, 000 sec; glycocalyx turnover in rat uterine epithelial cells is ~430, 000 sec; and platelet lifespan is ~860, 000 sec).

As the cell coat is a secretion product incorporated into the plasma membrane that undergoes continuous renewal, any trimmed glycocalyx glycoproteins from tissue cells would be rapidly replaced via biosynthesis in the ribosomes of the endoplasmic reticulum, followed by final assembly with the oligosaccharide moiety in the Golgi complex and subsequent export to the plasma membrane. Glycoprotein strands or stray sugar residues released into the extracellular medium as a consequence of such trimming are nonimmunogenic and would be quickly metabolised, although it is possible that nearby parasites could absorb this released material onto their surface, affording themselves some camouflage protection against natural host immune defenses but little protection against vasculoid defenses since parasite antigens should still be visible at cell surfaces.

A nanorobotic aggregate covering a macroscale area of the capillary luminal surface may reduce the normal flow of plasma water and other substances that leaves the circulation via ultrafiltration, unless the necessary water is replaced by the nanosystem. The plasma water flow helps to remove waste products from the extracellular spaces around tissue cells, a function that could be compromised by the shielding presence of the nanoaggregate unless the aggregate replaces this flow with water transported through or around the device, by various means. Consequently, both lymph volume and the gross water flows between tissues may be affected by vasculoidisation, given that the vasculoid substitutes encapsulated fluid transport in place of bulk-flow and diffusive fluid transport in the principal global materials distribution system of the human body. More specifically, ~20 liters/day (0.23 cm^3/sec) of plasma water exit the natural circulation via ultrafiltration through leaky capillaries, of which 18 liters/day are reabsorbed after passing through the lymphatic capillaries and back into the venous loops, leaving a net flow of ~2 liters/day to pass onward through the lymphatic system. The maximum vasculoid water transport rate has been scaled to 0.60 cm^3/sec or ~52 liters/day, which is the most that the water tanker subsystem can handle. Under normal conditions only ~4 per cent of the water tanker fleet is in use, which still allows the transit of at least 0.024 cm^3/sec or ~2.1 liters/day of water which if not re-imported by the vasculoid on the venous side of the capillary bed is forced to enter the lymphatic system, maintaining normal flow and volume. This flow should be sufficient because the direct removal from the intercellular fluids of pure waterborne electrolytes, solutes, and other physiological substances by the vasculoid (through the plates) can create pericellular concentration gradients sufficient to ensure injection and removal of such substances from the vicinity of the bathed cells. This should greatly reduce the gross intercellular fluid flow rate required to maintain proper physiological conditions. One minor additional concern might arise because lymph is moved primarily by peristaltic one-way valving. Envasculoided tissues adjacent to lymphatic capillaries may have slightly higher mechanical stiffness, hence may transmit somewhat less peristaltic action resulting in reduced lymphatic flow rates. But lymph fluids transport solutes at far below maximum concentration—for example, lymph normally contains ~0.003 gm/cm^3 NaCl, two orders of magnitude less than the ~0.36 gm/cm^3 maximum solubility at 310 K– and bacterial entry will be largely precluded by the appliance, so small reductions in lymph flow rates are probably tolerable. Lymphatic venules do exhibit small-amplitude pulsations on their own, and it may be desirable for other reasons to simulate the stiffness and motion (including motion resulting from blood flow) of pre-vasculoid tissue. Lymph movement also may be replaced by an optional "lymphovasculoid" appliance. Similar considerations and conclusions apply to other fluid spaces within the body that communicate, directly or indirectly, with the blood, including the choroid plexus and cerebrospinal fluid, as well as the pericardial, pleural, peritoneal, synovial, and intraocular fluid spaces.

CVD (chemical vapor deposition) diamond coatings are said to have "low immunoreactivity" and there were no reports of diamond immunogenicity in the medical literature as of 2001. Indeed, diamond is often used as an experimental control because it is so chemically inert and biologically inactive. Pure sapphire also appears fairly nonimmunogenic, although similar hydrophilic surfaces do adsorb immunoglobulin IgG; single-crystal sapphire has excellent biological inertness and chemical stability. However, both diamond nanoparticles and various soluble aluminum salts (e.g., alum, aluminum hydroxide, and aluminum phosphate) have been shown to serve as adjuvants which enhance vaccines or immune system responses to foreign antigens. Exposed to water, the polished single-crystal a-alumina (0001) surface elicits a hydration reaction, with a water vapor pressure of ~1 torr sufficient to fully hydroxylate the surface. Alumina is corrosion-resistant because it exists in the highest oxidation state that aluminum metal can possess, and has the potential for microstructural control of the interface (with tissue) without formation of toxic corrosion products. Yet it is also known that a-alumina is very slightly soluble in highly acidic or alkaline aqueous environments. Since Al^{+++} ions can produce "dialysis dementia" and are generally considered toxic, it is of interest to determine whether or not these ions can leach into the body from alumina implants or sapphire nanorobots. Early studies in the 1970s found no movement of known contaminants into the surrounding tissue from sintered alumina implants inserted into the iliac crests (hip bones) and mandibles of rabbits. During the 1980s and 1990s, small increases in blood aluminum concentrations were demonstrated in smelter workers, though the potential exposure level is several orders of magnitude greater for body uptake of more soluble aluminum compounds used as food additives, as antacid medication, or from food packaging materials and cooking utensils. In 1990, Lewandowska-Szumiel and Komender investigated aluminum release from an alumina bioceramic during standardised biocompatibility testing in an animal experiment. Alumina implants introduced into rat femurs and guinea-pig mandibles and then removed 6-8 months later were found to be well tolerated, and no changes in the surfaces of the removed implants were observed under SEM examination. The researchers decided to compare the aluminum content of the femurs of experimental and control rats using atomic absorption spectroscopy, and discovered that the level of aluminum was higher in the bones of the experimental animals. In 1991, Arvidson et al investigated the corrosion resistance of single-crystal sapphire implants with respect to the release of aluminum ions, and found no ions in the test solutions. The next year, Christel reported that alumina exhibited greater bioinertness than all other implant materials currently available for joint replacement, and that no lymphocyte or plasma-cell infiltration into joint implants is observed "because of the absence of soluble component release." However, two studies in the early 1990s found some detectable aluminum ion release, so more research is clearly required on this issue. In any case, a thin veneer of diamond on sapphire should suffice to prevent aluminum ion release in vivo. An allergic reaction or "hypersensitivity" is an acquired and abnormal immune system response to a substance, called an allergen, that normally does not cause a reaction. An allergy requires an initial exposure to an allergen which produces sensitisation to it subsequent contact with the allergen then results in a broad range of inflammatory responses. Sapphire or alumina ceramic is considered nonallergenic—ceramic coatings are used to eliminate metal allergies on implant surfaces and hypersensitivity to oral ceramic is reported only rarely. There are no reports of allergenicity for diamond, sapphire, fullerenes, or other probable diamondoid nanorobot exterior materials and such allergenicity appears unlikely, but experiments should be done to positively confirm this expectation. The immune system could also react to small subcomponents like cilia, especially if they form aggregates with proteins, possibly requiring clonal deletion or tolerisation to deal with such issues.

Could the vasculoid surface in contact with the vascular endothelium trigger general inflammation in the human body? One early experiment to determine the inflammatory effects of various implant substances placed subdermally into rat paws found that an injection of 2-10 mg/cm^3 (10-20 micron particles at 10^5-10^6 particles/cm^3) of natural diamond powder suspension caused a slight increase in volume of the treated paw relative to the control paw. However, the edematous effect subsided after 30-60 minutes at both concentrations of injected diamond powder employed. Another experiment at the same laboratory found that intraarticulate injection of diamond powder was not phlogistic (i.e., no erythematous or edematous changes) in rabbit bone joints and produced no inflammation. Diamond particles are traditionally regarded as biologically inert and noninflammatory for neutrophils and are typically used as experimental null controls. CVD diamond and DLC diamond surfaces elicit minimal or no inflammatory response, and atomically smooth diamond may perform even better. Diamond particles are said to have little or no surface charge but unmodified graphene surfaces readily acquire negative charges in aqueous suspension, so experiments are needed to determine if negatively charged fullerenes or other diamondoid substances can contact-activate Hageman factor or kallikrein and trigger an inflammation reaction. Experiments with sapphire have generally found no serious inflammation in soft tissues or bony tissues, or only mild reactions, though there are a few modest exceptions including a brief acute inflammatory response.

Blood coagulation involves a complex series of reactions in which various proteins are enzymatically activated in a sequential manner, transforming liquid blood into a gel-like clot which is then stabilised to form a thrombus (clot) consisting of platelets, fibrin, and red cells. The series of reactions is classically divided into two pathways—extrinsic and intrinsic —involving more than a dozen factors that converge on a single common final pathway, resulting in clot formation. Since, these factors are carried by the blood, the vasculoid can regulate their local concentration and thus, prevent thrombogenic pathways from proceeding to completion. Additionally, platelets must undergo adhesion and activation for coagulation to occur. The adhesion of platelets to exposed collagen in injured blood vessels is mediated by a bridging molecule called von Willebrand's factor that is secreted by endothelial cells into plasma, which prevents platelets from detaching under the high shearing stresses developed near vessel walls. The activation of normally quiescent platelets is a complex phenomenon that includes changes in cell shape, increased movement, release of the contents of their granules (containing nucleotidyl phosphates, serotonin, various factors, enzymes and plasma proteins), and aggregation. The most potent activator of platelets in vivo is thrombin, which interacts with a receptor on the platelet plasma membrane, followed by transmembrane signaling and subsequent activation of the cell. Collagen is the other most important platelet activator; ADP can stimulate aggregation but not granule release. In principle, the blood-contacting surfaces of a nanoorgan, or of nanorobots in sufficient bloodstream numbers and concentrations, could activate platelets (and thus either of the two coagulation pathways), but careful choices of materials and of allowable mechanical motions should reduce or eliminate inherent nanodevice thrombogenicity. For example, DLC diamond-coated stents, heart valves and other blood-contacting LVAD surfaces or substrates generally show reduced thrombogenicity and no platelet activation. Sapphire (alumina ceramic) has low thrombogenicity and both platelet adhesion and activation are low. Hemolysis is near-zero for diamond and alumina powders. Additionally, in the vasculoid plateletogenesis and release of platelets could be actively regulated and reduced to minimal levels. Future experiments must determine if ordinary diamondoid surfaces will have to be supplemented with additional antithrombogenic coatings in order to achieve vasculoid performance objectives. If such coatings are required, one simple possibility is surface-immobilised heparin, a ~15 kD straight-chain anionic (acidic) mucopolysaccharide (glycosaminoglycan) that forms polymers of various lengths. Heparin, first discovered in 1916, is produced naturally by human liver mast cells and basophil leukocytes, and inhibits coagulation primarily by accelerating the interaction between antithrombin and thrombin. Nanorobot exteriors can be "heparinised", and thereby rendered thromboresistant by immobilised heparin on all blood-contacting surfaces at ~monolayer surface concentration (e.g., 7-10 pmol/cm^2). Cellulose membranes coated with 3.6 pmol/cm^2 of endothelial-cell-surface heparan sulphate show complete inhibition of platelet adhesion. If satisfactory passive nonthrombogenic surfaces cannot be found, nanorobots might employ any of several active strategies to prevent iatrogenic coagulation.

During development and after physical injury, new blood vessels may originate from pre-existing blood vessels by angiogenesis or from endothelial cell precursors (angioblasts) by a process called vasculogenesis; both processes are mediated by paracrine growth factors. In certain circumstances, such as wound healing, post-ovulation capsule repair, and exercise-related capillary formation to support development of new skeletal muscle, angiogenesis is critical in the adult human body and its lack has been associated with chronic renal failure and other pathological conditions. Without the ability to incorporate vasculoid plates into new capillary vessels, masses of new non-envasculoided tissue would accumulate around the original (envasculoided) tissues over time, reducing the effectiveness of the appliance. In principle, vasculoid can support angiogenesis and vasculogenesis by extending itself into the newly-formed space. Watertight multicomponent metamorphic surfaces have been described by Freitas), but special techniques will be required to extend a new vascular branch while maintaining continuous watertightness. An angiogenesis event can be detected by monitoring concentrations of angiopoietin-2, VEGF, and other relevant factors. Conversely, detection of tumor angiogenesis factors could induce a vasculoid-mediated reduction in oxygen supply, increased local delivery of antibodies, refusal to deliver angiogenesis factors, and optional intervention such as delivery of chemotherapeutic drugs or external signaling to call for targeted therapeutic intervention. It is envisioned that self-repair capabilities via active replacement of damaged components (from onboard inventories of spare parts including spare plates will be an important feature of the complete vasculoid system design. These capabilities may be extended to support angiogenesis. When angiogenesis is not desired (for example, near tumors), local concentrations of angiogenic factors can be regulated by the vasculoid appliance to minimize vessel formation, and the vasculoid can refuse to deliver nutrients to the undesired tissues even if ersatz capillaries form. Attention must also be paid to the removal and disposal (or temporary onboard warehousing) of damaged plates. In rare cases, traumatic events may damage plates (rather than simply causing them to separate temporarily), and radiation damage may cause rare failures. A capillary damaged beyond repair (e.g., in a severe contusion) may be abandoned and dismantled by the body; in such a case the vasculoid should cease delivery to that region and remove the plates. Damaged endothelial cells can be detected based on sensory data indicating localising chemical imbalances, and new stem cells or endothelial precursor cells can be transported to the vicinity of the damaged site to help restore the natural vascular integrity. Such restoration in some cases may require repositioning of a few plates using the motive cilia.

One important function of fluid flow in blood vessels is to ensure vascular patency. But replacing bulk fluid with a vasculoid appliance is unlikely to cause the smaller vessels to collapse because the buckling force is increased by up to several orders of magnitude by the addition of the plates in the capillaries and related small vessels, even in the absence of bulk fluid. The largest vessels such as the aorta will lose significant luminal support (even if filled with gas), but—in addition to the external muscular and elastic support from the surrounding tunica media and tunica adventitia layers—such vessels also possess an extensive capillary vasculature which, greatly stiffened and at typical capillary densities, should provide adequate replacement support. Intravascular (cross-luminal) spring-loaded diamondoid scaffolding could also be added at need. Arterioles in the natural vascular system perform extensive regulatory functions. These and related functions (including vasoconstriction and vasodilation) can be controlled, in turn, by the vasculoid (e.g., by monitoring local NO concentrations). The underlying vascular musculature will probably be maintained in the relaxed state if this maximizes tissue stability and simplicity of control, although further research of this question is warranted.

Numerous local, intermediate-scale, and global control systems and protocols will be required to ensure the proper autonomic operation of the vasculoid appliance. Special communications subsystems for local plate configuration maintenance, cargo traffic control, and fast systemic response to large-scale external stimuli must be designed. An interplate packet-switching network for long-range communications, and to support real-time autogenous user control, should be considered. System control might be simplified using local computing centers throughout tissues, directing immune, angiogenetic, and trophic functions. Besides the interplate network, trans-tissue sonic or optical communication in principle could be employed between spatially close but topologically distant portions of the vasculoid network. Precise knowledge of the functions of various cytokines and other substances will enable the vasculoid to respond appropriately to sensor data reporting local extravascular concentrations of signaling molecules. Complexity of stimulation patterns must mirror physiological conditions as much as possible. Highly responsive and user-friendly graphic user interfaces may be useful to allow the patient to communicate quickly and easily with his or her appliance; several aspects of such interfaces have been described elsewhere, but considerably more work remains to be done. The computational requirements of directed container switching/routing by the ciliary system should be investigated further. A great number of software and computational architectural issues are as yet unresolved, including a detailed molecular routing logic for each substance to be transported, comprehensive details of nanorobot control protocols, container traffic flow patterns (e.g. flow lensing, lane definition, congestion waves and gridlock, flow viscosity, etc.), and software requirements including software complexity for large-scale control issues, a discussion of which are beyond the scope of this paper. (For example, it might be interesting to apply existing vehicular traffic simulations to transport in fractal structures.) Only since the mid-1990s have multirobot control issues begun to be seriously addressed by the broader research and technical community. Computation for vasculoid accident recovery and repair will be considerably more intensive than computation for maintenance or installation, and must also be deferred to another paper. However, a few basic observations may be made here regarding vasculoid materials transport computation requirements. Concentrations of water and glucose are normally maintained by the appliance near typical serum levels, using simple feedback loops driven by concentration sensors located on the tissue side of the vasculoid device. Oxygen and carbon dioxide are similarly controlled using sensors capable of measuring partial pressures, though the precise triggering thresholds may be modified in special circumstances (e.g. high altitudes, deep sea diving, hyperbaric chambers). A simple rate-control mechanism driven by sensor data compiled once every second, employing ~1000 distinguishable concentration levels requires ~10 bits/sec per molecular type, or ~40 bits/sec for all four molecules. Most of the time, entire rotor banks are engaged using a single command. Ciliary motions, largely stereotypical, likely require a similarly small computational budget.

Establishing or maintaining appropriate concentrations of mixed-cargo molecules is far more computationally complex. These other molecules include minerals, vitamins, lipids and waste products, but the most numerous and difficult to coordinate are the blood proteins and sterols including control molecules such as insulin, leptin, gastrin, resistin, estrogen, or testosterone; synaptic transmitters and other neural biochemicals such as acetylcholine, adrenalin, dopamine, nitric oxide, or serotonin; cytokines and other signaling molecules such as follicle stimulating hormone, epidermal growth factor, or tumor necrosis factor; and many thousands of specific antibodies that reflect the body's unique learned recognition of specific antigens. (Embedded analytical units scattered throughout the vasculoid surface can test for rising concentrations of hitherto unrecognised nonmetabolizable molecules, and configure a set of programmable binding sites to rotor these molecules out of the tissues as well, though this should be a relatively rare occurrence.) Since the human genome contains ~40, 000 genes, most of which may encode several proteins each, we assume there may be on the order of ~100, 000 distinct biomolecules that must be recognised, monitored, transported, and regulated at the docking bays. This is likely a generous estimate, given that most of the proteins produced remain intracellular, but our conclusions do not depend sensitively upon the precise figure chosen. Upon receipt of any tanker, each docking bay scans the tanker manifest before offloading the contents. In the case of gas, water, and glucose tankers, the manifest is extremely brief, since only one or two molecules are involved. This content information can be encoded in as little as 15 bits. A precise molecule count of up to 3 x 1010 molecules per tanker requires an additional 35 bits, but in most cases a precise molecule will not be necessary and a range figure requiring fewer bits should suffice. At the maximum rate of one tanker unloading every 20 seconds, the manifest-reading bit rate is at most ~2.5 bits/sec. In the case of mixed-cargo tankers, 17 bits uniquely specify the identity of each of 100, 000 distinct molecules; each tanker can hold a total mixed-cargo molecule count of ~109, requiring up to 30 bits per item to ensure an exact count. Based on existing human physiology, we posit a specification that each vascular cell must have access to at least 5 per cent of all ~100, 000 molecular types during one blood circulation time, ~60 sec. Since each cell is supplied by two mixed-cargo docking bays, then restricting tankers to a maximum of 834 different molecules per load fulfills our specification, requiring in the worst case a 39, 198 bit manifest to be read in 20 seconds, a maximum bit rate of ~2, 000 bits/sec. Again assuming ~10 bits/sec per molecular type for rate-control of rotor mechanisms, rotor control will require a maximum ~8, 340 bits/sec. (One design alternative is to allow stochastic mixing and transport of the low-concentration molecules which would mimic the natural process, reduce the computational needs, and impose only a very small burden on the transport capacity.) In sum, docking bay process control appears to require ~10, 000 bits/sec of computation. Cellulock control requires reading manifests of up to ~1000 cells of at most ~1000 different types, which is ~20 bits/cell type or ~20, 000 bits/manifest. Cellulocks unloading one boxcar every 60 seconds will require a bit rate of ~333 bits/sec. Other control tasks will add only fractionally to the ~10, 000 ops/sec budget previously estimated for individual docking bays, a capacity that can be used for other purposes between docking events.

A detailed assessment of vasculoid system reliability also lies beyond the scope of this paper, especially given the possibility for prompt and severe damage if the robot is subjected to extreme accelerative loads or unplanned segmentations. However, if we assume normal gravitational loading and ignore catastrophic accidents, we find that the vasculoid promises an annual survival probability well in excess of "six nines" (failure probability < 10-6) at least against radiation damage if all major subsystems incorporate tenfold redundancy. Adopting Drexler's radiation damage model for the first part of this analysis and applying it to a system comprised of N components, with each component comprised of n redundant parts, the probability P that the system remains operational after T years is approximated by: P = exp [-N (1—p)n], where p = the probability that an individual part remains operational; p ~ exp(-1015 Dm), where m = mass of the part in kg and D = radiation dose in rads ~ 0.5 T for normal background radiation in the terrestrial environment. As a further simplifying assumption for this analysis, we shall presume that all vasculoid modular "parts" are similar in mass to the ~4, 000, 000-atom robotic manipulator device, after which the vasculoid cilium is patterned. From the above formulae, p ~ 0.999960 for each such cilium-like part, per year, even with no internal redundancy within the part. The vasculoid ciliary subsystem includes a minimum requirement of N = 300 trillion cilia incorporating a redundancy of n = 10, giving a total of 3000 trillion cilia. From the above, the annual probability of failure of the ciliary subsystem is < 10-29; assuming a redundancy of only n = 5, failure probability is still < 10-7. Each vasculoid plate is ~2 micron^3 in volume. Given that there are ~150 trillion plates, with zero redundancy among them, and then to be assured of a complete plate subsystem annual failure rate < 10-6, the annual probability of failure per plate must be reduced to ~10-20. This is achieved for plates comprised of a total of ~88, 000 "parts" as described above, with a redundancy of ~6 among such parts. However, in this study we have specified a more generous (i.e., safer) redundancy of 10 among such parts. Each vasculocyte repair robot is ~10-14 kg in mass. Given that there are ~200 billion active vasculocytes, and assuming zero redundancy among them, then to be assured of a vasculocyte subsystem annual failure rate <10-6, the annual probability of failure per robot must be reduced to ~10-17. This is achieved for robots comprised of a total of ~125, 000 "parts" as described above, with a redundancy of ~5 among such parts; we have again assumed a more generous redundancy of 10 among such parts. Non-radiation failure mechanisms have not been examined in the above analysis.

Software failure modes have been ignored, and even plate computers may be complex enough to host intentionally disruptive programmes such as computer viruses. Many other more subtle or second-order difficulties also have been ignored here, such as the potentially destabilising effects of dynamic oscillations and resonances among the various moving components which would require a detailed design before undertaking a comprehensive analysis. For instance, we should try to avoid Per Bak's notion of self-organised criticality when complex systems are pushed too close to their design limits, as in car traffic simulations. Numerous specific failure modes have not yet been exhaustively analysed, including complete single-plate failures and the seizing or locking of individual cilia while a transfer operation is in progress, e.g., near a capillary entrance where such failures could prove most troublesome.

However, the vasculocyte fleet has been scaled to accommodate a reasonable repair mission requirement. Vasculoid tankers are micron-size compressed-gas containers which may rupture explosively—though this event is extremely improbable, since tankers with rupture strength exceeding 40, 000 atm are loaded to only 1000 atm pressure. If this unlikely rupture event occurs in a large vessel such as the aorta, the possible failure modes may resemble those involved in a detachment avalanche, but the low tanker mass should preclude any serious direct infrastructural damage. However, if the explosive event occurs within a smaller vessel such as a capillary, the risk of breach is considerably greater. However, the maximum overpressure that can be explosively applied to the appliance walls is limited to the ~1000 atm compression pressure of the tanker contents, which likely does not exceed the theoretical rupture strength of the vasculoid walls in capillaries which may be crudely estimated as pmax ~ 3300-33, 000 atm, taking vasculoid plate tube thickness twall = 1 micron, capillary tube radius R ~ 3 microns and working stress sw ~ 1010 N/m^2 for solid tube walls or maximum interbumper working stress sw ~ 109 N/m^2 for more realistic plated walls jointed with bumper interconnects. Whether such explosive events will occur with sufficient frequency to warrant explicit corrective protocols (beyond conventional vasculoid self-repair activities) deserves further study. A shockwave cascade failure (such as recently occurred at the Super Kamiokande neutrino detector facility, wherein the initial implosion of a single photomultiplier tube (PMT) during refilling of the water tank triggered a runaway cascade, destroying 6, 665 of the 11, 146 PMTs in a few seconds seems unlikely in the vasculoid but also should be investigated theoretically.

In practice, one should expect that a few contaminant molecules will make their way into the vasculoid, perhaps from local strain that exceeds the adjustment capabilities of the metamorphic bumpers, accidental spills, and so forth. A molecule that is small enough to be mobile at room temperature is small enough to be captured by a sorting rotor. Larger molecules can be physically handled by vasculocytes. A specialised class of vasculocyte, also measuring several microns in size and moving at ~1 cm/sec, can employ a 0.25-mm-wide chemotactic detector bank on each mobile nanorobot to search for, bind, and remove large molecules—a complete sweep of the entire vasculoid luminal surface once a minute requires the activity of only 1 per cent of the total active vasculocyte fleet. The motion of the tankers creates a net motion of the nitrogen atmosphere that can sweep molecules toward centrally placed filters. Thus a combination of internal sorting rotors, filters, and vasculocyte activity can quickly extract contaminants from the vasculoid volume. Once contaminants are trapped, they may be dealt with in ways suggested previously. The volumetric capacity of the appliance to deal with internal contamination must be scaled according to anticipated applications and event scenarios.

Installation of the vasculoid involves, at the least, complete exsanguination of a sedated patient and an intricate vascular plating operation. Two hypothetical installation scenarios are presented. The first scenario is a detailed description of a procedure that would be feasible using the technology available currently. This is to demonstrate that vasculoid installation can in principle be carried out without violating any well-established medical or physical limits. However, the authors are aware that by the time a vasculoid-class device can be built, medical technology will have advanced significantly. We therefore also briefly sketch out a highly speculative second scenario which, if practicable in some future era, might be considerably more convenient and up to 100 times faster. This second procedure would be unduly aggressive by today's standards but might be feasible and safe, given the supporting technology available in a nanotechnology-rich medical environment. The principal installation scenario involves complete exsanguination of a sedated and hypothermic patient, replacement of the natural circulatory fluid with various installation fluids, followed by mechanical vascular plating, defluidisation, and finally activation of the vasculoid and rewarming of the patient. Installation takes ~6.5 hours from start to finish and requires a peak ~200-watt power draw midway through the procedure. (By comparison, present-day kidney dialysis treatments require 4-12 hours and the equipment also draws a few hundred watts.) Our discussion first considers the lifespan of cells temporarily denied access to external molecular transport mechanisms. We next describe the details of vasculoid installation including patient preparation (~24 hours), vascular washout (~4 hours), vascular plating (~1 hour), defluidisation (~0.3 hour), and initialisation and cold start (~1.2 hours). We conclude with a brief discussion of vasculoid removal. The hypothetical installation protocol described below has been selected for maximum comfort, reversibility, and reliability. The correct performance of the system is verified at every step, providing a plateau of safe operation before moving on to the next step. Since the entire procedure is intended to be fully reversible at each step, any step which fails or does not proceed in a manner acceptable to the installing physician may be quickly abandoned with a retreat to the previous plateau of established safety, after which a decision may be made to try again or to abort the installation. Patients must be made aware that they are about to undergo a major medical procedure which involves replacing ~8 per cent of their body mass with complex nanomachinery. They must be psychologically prepared to deal with the personal implications of this. From the turn-of-the-century perspective, vasculoid installation appears massively intrusive especially when compared to other superficially related procedures with which we are commonly familiar such as intubation, kidney dialysis, blood transfusion and blood replacement therapies, installation of pacemakers or artificial organs, and coronary angioplasty or fiberoptic endoscopy. However, in a future era when nanomedical systems are widely employed and generally accepted as standard treatment, vasculoid installation may be regarded with considerably less trepidation than it would today.

In classical medicine, "ischemia" refers to a local inadequate blood supply which is usually caused by a mechanical arterial obstruction (e.g. clot), a spasm wherein a blood vessel pinches shut, arterial narrowing (e.g. arteriosclerosis), or cessation of cardiac activity. The principal clinical outcome of local ischemia is a shortage of oxygen. Because of the high continuous biological power density of nerve cells, neurons and the brain succumb most rapidly to oxygen starvation. The precise survival limit of human brain tissue under hypoxic conditions (after which time some decline begins to occur) depends on many factors. Traditionally this has been reported as ~4 minutes (240 sec), although more recently it has been shown that 10 minutes of warm ischemia (even followed by another 10 minutes of trickle-flow CPR) is reversible without neurological deficit if followed by mild hypothermia when normal blood flow is restarted, and even modest advances in technology will likely extend this still further. Much longer periods of cold ischemia are tolerable—for example, survival with complete functional and histologic cerebral recovery has been achieved with brain temperature at 5-10°C during 1 hour of circulatory arrest, and stable spontaneous circulation has restored after water ice submersion of up to 90 minutes. In nanomedicine, where it is possible to achieve precise control of human physiology at the cellular and molecular levels, ischemia may refer to an interruption in the extracellular mass transport of any vital molecular species within the human body. Failure to adequately import a particular nutrient, or to adequately export a specific harmful waste product, or to properly regulate the transport of a particular signaling molecule, may produce some of the classical symptoms of cellular ischemia. Since the entire bloodflow must be interrupted during the vasculoid installation process (producing whole-body ischemia), it is useful to estimate how long tissue cells can maintain normal metabolism and avoid toxemia after cellular access to the bloodflow is denied. This time period—the ischemispecific limit—varies with each important cytometabolic molecule, as summarised below. (The limit also varies by cell type; a more detailed study would likely reveal specific tissues where ischemic sensitivity is higher than average for a given metabolite.)

Oxygen—The average (20 micron)^3 tissue cell consumes ~107 molecules/sec of O_2 at the basal (resting) rate of ~30 picowatts. The cytosol of such a cell can dissolve up to ~6 x 108 O_2 molecules at 310 K. If suddenly cut off from all external supply, the average cell has only enough O_2 in inventory to survive ~60 sec at the basal metabolic rate.

Glucose—Assuming ~50 per cent energy conversion efficiency and ~10-3 gm/cm3 glucose in the cytosol, then ~3 x 1010 glucose molecules are available in the average tissue cell. For a power demand of 30 picowatts, ~4800 zJ/glucose molecule, and ~50 per cent energy conversion efficiency, then the basal glucose consumption rate is 1.25 x 107 glucose molecules/sec and the basal cellular ischemispecific limit for glucose is ~2400 sec.

Carbon Dioxide—The metabolic "combustion" of one molecule of glucose with six molecules of O_2 produces six molecules of CO_2, hence the average tissue cell produces ~107 molecules/sec of CO_2 at the basal rate. The cytosol of an average cell should normally dissolve up to ~1010 molecules of CO_2 at 310 K (assuming working tissue PCO2 = 54 mmHg and applying the Henry's law constant for CO_2, an estimate which compares favourably with the measured ~5 millimoles/kg intracellular CO_2 concentration (2.4 x 1010 molecules/cell assuming an 8000 micron3 cell) in various frog tissue cells at 302 K. Hence the cellular ischemispecific limit for CO_2 is ~1000 sec, consistent with maximum physiological cellular and plasma pH levels. It may be possible to dissolve more CO_2 but this will lower pH, leading to toxic acidosis of the cell. (Acidosis has two major components, CO_2 and lactic acid. The absence of oxygen blocks electron transport and stalls the Krebs cycle, halting CO_2 production, so glucose is shunted to lactic acid in anoxia.) Additional amounts of CO_2 may be stored in combination with plasma proteins as carbamate.

Lactate—Under anaerobic conditions, lactic acid is produced during the metabolism of glucose (glycolysis) in most cells, glycogenolysis in muscle and glucolysis; two molecules of lactic acid are produced per molecule of glucose metabolised. Thus at the basal rate the average anaerobic tissue cell can produce at most 2.5 × 107 lactate molecules/sec—though this is an overestimate by perhaps a factor of 2 because in most cells the pyruvate (the immediate precursor of lactic acid) formed at the end of glycolysis enters the TCA cycle and is further oxidised by mitochondria. The maximum normal blood concentration of lactate equates to ~1.1 × 1010 molecules/cell, potassium channel openings are induced by 2-20 nM lactate (0.96-9.6 × 1010 molecules/cell) applied to the cytosol of rabbit ventricular myocytes, and the brain lactate threshold for cerebral ischemic damage is 17 mmol/gm or ~8.1 x 1010 molecules/cell, so the ischemispecific range for lactate is ~380-3800 sec.

Nitrogenous Waste Products—Urea is the principal mammalian waste product due to protein, purine, and pyrimidine nitrogen metabolism, constituting ~85 per cent of human nitrogen excretion. However, urea is formed in the human liver through reactions of the Krebs ornithine cycle, a process not available to tissue cells denied access to the circulation. Ammonia is the chief byproduct of protein and amino acid metabolism in the cell, with one molecule of ammonia produced per molecule of amino acid broken down. The daily RDA for protein represents an upper limit on the breakdown rate of ~5 × 105 amino acid molecules per cell-sec, assuming mean amino acid MW ~ 100 Da, implying a maximum ammonia generation rate of ~5 × 105 molecules/cell-sec. Given the maximum observed concentration of ammonia of ~6 × 108 NH3 molecules/cell in whole blood cells, the ischemispecific limit for nitrogenous wastes is ~1200 sec.

Acetate—Acetic acid (MW = 60 Da) is the simplest possible fatty acid having an even number of carbon atoms. It is most notably produced during the breakdown of acetaldehyde (the second step in the metabolism of alcohol), bacterial fermentation in the gut, and in other circumstances. The typical human intracellular production of acetic acid has been roughly estimated from rat studies as ~8 x 106 molecules/cell-sec. This chemical is highly metabolizable in vivo, but assuming a maximum nontoxic limit of ~10-4 gm/cm3 (vs. ~4 × 10-3 gm/cm3 for all esterified fatty acids in cells) or ~1010 molecules/cell, then the ischemispecific limit for acetic acid is ~1300 sec.

Ketones—Absent circulatory removal and in cases of low glucose levels, the metabolism of fats may produce abnormal amounts of toxic ketones including primarily beta-hydroxybutyric and acetoacetic acids (and their decarboxylation product acetone) which may be present up to ~6 × 108 molecules/cell assuming MW ~ 86 Da for ketones. Given the RDA for lipids of ~7 × 10-7 kg/sec and since one palmitic acid (typical lipid) molecule (MW = 256 Da) would convert to ~3 molecules of acetoacetic acid (MW = 86 Da), the maximum natural ketone generation rate is ~5 × 105 molecules/cell-sec and the minimum ischemispecific limit for ketones is >1200 sec.

Summary of Cellular Ischemispecific Limits—If the average human tissue cell is denied access to all extracellular molecular transport systems, the major byproducts of normal cellular metabolism may build to near-toxic levels from a near-zero initial concentration in ~103 sec. Glucose reserves last >103 sec, but oxygen runs short in ~102 sec without external resupply. These theoretical limits appear crudely consistent with the concept of the "Golden Hour" in traditional trauma care. For instance, ischemia induced by aortic cross-clamping in dogs has been survived for up to 20-60 minutes without inducing paraplegia (though decline of spinal cord electrical function is detected in the ~20-30 minute range for dogs and rabbits). Cardiac ischemia induced by aortic cross-clamping in dogs is survived for 30 minutes during normothermic (37°C) cardioplegia, 45 minutes during mild hypothermic (28-30°C) cardioplegia, and 90 minutes using potassium verapamil during profound hypothermic cardioplegia at 8-10°C. The duration of cold ischemic tolerance is enhanced by the administration of excess insulin and other pretreatments, possibly involving gene products that will be well-known in a future nanomedical era in which vasculoid installation is commonplace. (Interestingly, wood frogs can endure freezing for >2 weeks with no breathing, no heart beat or blood circulation, and with up to 65 per cent of their total body water as ice)

Patient Preparation (~24 hours): Before the vasculoid may be installed, the patient must be prepared as follows:

1. *Vascular Conditioning*—The day before the installation, the patient receives an injection containing a treatment dosage (~1 cm3) of vascular repair nanorobots or ~70 billion individual devices. These 8-picogram, 7 micron3

mobile legged artery-walking nanodevices clean out all fatty streaks, plaque deposits, complex atherosclerotic lesions, infections, vascular wall tumors, and parasites, and repair all other vascular lesions as required in less than 24 hours, in a multistep process described elsewhere in detail. As one additional task, these nanorobots inject each of the ~1012 endothelial cells lining the human vascular tree (a) with cytokine blockers to at least partially inhibit the cells' natural surface pressure and shear force responses, and (b) with adhesion-inducing glycoproteins to forestall cell migration during the installation procedure by encouraging firm anchoring. All vascular repair devices are then exfused before vasculoid installation begins; the results of their vascular reconnoiter protocol may be downloaded to an external computer and used to prepare a detailed map of the patient's vascular tree to improve efficiency during plating and plate initialisation. The patient may be started on a diuretic (e.g. furosemide) to reduce blood volume a few hours before the installation procedure begins.

2. *Sedation*— On the day of the operation, the patient arrives at the installation facility and is administered a preoperative sedative such as sodium pentobarbital an hour before the procedure is to begin, to encourage drowsiness.
3. *Cannulation*—A standard closed-chest cardiopulmonary bypass (aka. heart-lung machine support) or CPB is employed. Usually this involves a combination of femoral and thoracic vessel cannulation, but in this case a fem-fem bypass (cannulation of the vena cava and aorta via the femoral vein and artery) might suffice. This procedure, by definition, continuously supplies the equivalent of resting cardiac output through the femoral vessels and the blood pump/oxygenator circuit—typically a 50-70 cm3/sec flow rate during hypothermic CPB, allowing the entire human blood volume can be exchanged about once every few minutes. Fresh oxygenated blood flows retrograde up the femoral artery into the aorta, and oxygen-depleted blood flows retrograde down the vena cava to return to the heart lung machine via the femoral vein. The leg inferior to the cannulation point is supplied by collateral vessels. At this point in the procedure, the patient's own blood continues to circulate through the body. (Note that the above "manual cannulation" scenario is included for illustrative purposes only. In an era when advanced medical nanotechnology is available, self-directing nanocannulators will make it easy to quickly and safely establish flow-regulated channels into any desired artery or vein, up to and including the aorta and vena cava, so much higher flow rates than the conservative figures used here are theoretically available if required.)
4. *Heparinisation*—Heparin and streptokinase are injected (a) to prevent

clotting and promote lysis of hemostatic fibrin, (b) to break up axial red cell rouleaux that form spontaneously at low blood flow shear rates, and (c) to help ensure patency of the catheters throughout the procedure. The conventional cocktail of drugs given to patients placed on bypass leading up to a cold circulatory arrest procedure is rather complicated and will not be elaborated further here, but there are established procedures in conventional medical practice for cooling patients and diluting their blood for work at very cold temperatures.

5. *Cytoactivity Restraint*—The installing physician next administers: (a) an erythropoietin antagonist to maximally suppress erythrocyte production; (b) agents to temporarily suppress all leukocyte production, reversibly depress leukocyte intracellular metabolism and cytokine sensitivity, and briefly reduce or block antigen reactivity; (c) agents to temporarily suppress platelet production, reversibly depress platelet intracellular metabolism and cytokine sensitivity, and temporarily toughen the platelet cytomembrane to reduce the likelihood of shear damage; (d) minute traces of normal bloodstream hormones, cytokines, and other control biochemicals designed to minimize production of natural secretions such as insulin, adrenalin and testosterone, glucose and cholesterol, sebum and semen, etc., as well as angiogenesis inhibitors to halt the development of new capillaries; and (e) broad-spectrum antibiotics or programmable nanobiotics such as microbivores designed to prevent microbial attack during the brief period of immunological vulnerability. For example, one element of this complex and as yet incompletely specified biochemical cocktail might be dexamethasone, the most powerful anti-inflammatory and immunosuppressive adrenocortical steroid; another element might be anti-CD18, an antibody known to prevent leukocytes from sticking to blood vessel walls in the brain. (Some of these and related restraints may not be absolutely essential, since the processes in question are quite slow compared to the time course of installation.)

Vascular washout (to prepare the patient for vasculoid plating) may be performed as follows:

1. *General Anesthesia*—A general anesthetic such as propofol is administered by the installing physician at sufficient dosage (~90 mg) to establish a condition of surgical anesthesia.
2. *Respirocyte Infusion*—Over a period of 3 hours, the patient's entire blood volume is replaced with a suspension of fully charged respirocytes (1 micron^3 spherical O_2/CO_2 1000-atm pressure vessels) in an isotonic aqueous 0.01 M glucose solution (double the natural serum concentration) also

containing most of the cytoactivity restraint substances described earlier and additionally a mixture of appropriate electrolytes and other components commonly found in hypothermal blood substitutes. A 5 per cent (by volume) respirocyte suspension, containing ~1011 respirocytes/cm^3, is sufficient to provide oxygen and carbon dioxide transport equivalent to the entire human red cell mass for ~104 sec (almost 3 hours) after the cessation of respiration. The respirocyte fleet generates ~17 watts of waste heat while supplying oxygen at the human basal rate, producing at most a negligible ~0.1 oF rise in core body temperature. Typically ~20 liters of infusant is required to completely clear the vessels of all natural blood components. Exfused blood cells are removed, collected and saved for emergency reperfusion, if required.

It is true that the reduction of cerebral blood flow by 50 per cent can produce marked disturbances in brain metabolism, and at 20 per cent of normal flow the neurons can depolarize with rapid loss of intracellular potassium into extracellular spaces. But the infusant composition can be adjusted to maintain physiological electrolyte (esp. Na+, K+, and Ca++) concentrations and osmotic balances, and to extract any excitotoxins that might be released—or else respirocyte-class pharmacyte nanorobots can be employed to similar effect. Since blood cells have been removed, leukocyte plugging and other forms of neutrophil-related damage cannot occur. The basal metabolism remains sufficiently active throughout the procedure so that tissue cells suffer minimal ischemic damage. The infusant constitutes an adequate temporary replacement for the natural blood which is being removed. This is based on an established medical technology: It is well-known that the entire blood supply of a human being can be replaced with cell-free blood substitute at a low temperature, with the patient then reperfused with blood and recovered, as was first demonstrated clinically 30 years ago as a treatment modality for hepatic coma.

3. *Cooldown*—Once the entire blood volume has been completely exchanged with the respiro-infusant described in (2) in the unconscious patient on CPB (as in conventional medical settings), control of perfusate temperature will suffice to control body temperature. Core temperature follows the perfusate temperature fast and close, with any perfusate/core temperature difference decaying with a ~10 minute half-life. The heat capacity of the body is so large that the 200 watts of power dissipated by vasculoid installation over an hour or so would heat the patient by only a couple of degrees, which is tolerable. (By comparison, external cooling of a body in circulatory arrest is so inefficient that even placing the patient in an ice water bath would yield surface-to-core temperature differences decaying

with a half-life of hours.) The patient's average core temperature is reduced from 310 K (37°C) to 280-290 K (7-17 °C) in ~1 hour, a cooldown rate of ~0.5°C/minute. This cooldown schedule is qualitatively similar to hypothermic schedules commonplace in current resuscitation medicine and hypothermic CPB procedures. Mivacurium, rocuronium, or older agents such as vecuronium or metubine (recently withdrawn) may be administered to inhibit shivering and as a reversible nondepolarising muscle relaxant; the effect of vecuronium bromide may be reversed by acetylcholinesterase inhibitors such as neostigmine or pyridostigmine.

B. Wowk notes that there are three broad categories of clinical hypothermia used in medicine. First, there is mild hypothermia (a few degrees below normal body temperature). This is very commonly used during cardiopulmonary bypass for open heart surgery, and often occurs as an incidental effect of general anesthesia during other surgeries. Second, there is deep hypothermia (15-20°C). This is used when significant periods of circulatory arrest (up to 1 hour) are required for complex bloodless surgeries such as repair of cerebral aneurisms or congenital defects of the aortic arch. Hemodilution (partial blood substitution) is used during these surgeries, in part to avoid red blood cell agglutination at low temperatures. Third, there is profound (or ultraprofound) hypothermia (0-10°C). Profound hypothermia does not yet have routine clinical use but is the subject of active investigation in relevant animal models with an eye toward human use, and preliminary human clinical results recently have been reported. Total blood substitution, as proposed in the vasculoid installation protocol, is the norm for profound hypothermia. Specialised perfusates such as hypothermosol are necessary to overcome physiological problems with this temperature regime that ordinary plasma can't handle. Dogs have been held for three hours in profound hypothermia during perfusion of a blood substitute, and the record for profound hypothermic perfusion without neurological deficit is said to be ~6 hours; the record for profound hypothermic circulatory arrest (as distinct from continuous perfusion) in dogs at <5°C is ~3 hours.

4. *Cardioplegia*—After the cooldown process has begun, the heart (already in bradycardia from the muscle relaxant) may be stopped by direct infusion of cold potassium chloride solution (reversible with atropine or digitalis) or other standard cardioplegic solution. However, if the target is ultraprofound hypothermia there is no need to stop the heart with drugs during this process because the heart will stop itself when the temperature falls below 15-20 °C.

Cytometabolic processes continue, but, at 280 K, with reduced metabolic requirements and hence reduced oxygen demand. This potentially allows infused respirocytes to provide complete respiratory gas maintenance for up to ~105 sec (~1 day) before they would be exhausted after complete cessation of circulatory flow via the catheters, giving a significant safety margin for the next phase of the installation procedure. However, B. Wowk notes other risks of profound hypothermic circulatory arrest beyond mere hypoxia, including the failure of ion pumps (causing growing intra- and extracellular imbalances of calcium ions), the alteration of cell membrane permeability (allowing normally extracellular solutes to slowly leak into cells), and the decoupling of certain cellular metabolic processes (which does not occur during mild hypothermia). The vasculoid installation process may fit within the circulatory arrest times achievable with the standard off-the-shelf deep hypothermic (not profound hypothermic) protocols of conventional medicine. Further study is required to determine the optimum temperature for hypothermic installation.

The anesthetised patient is finally ready for intravenous deployment of vasculoid components.

The original 5 per cent respirocyte suspension is replaced by a new suspension containing 1 per cent fully-charged respirocytes and 10 per cent cargo-bearing vasculocytes (plus cytoactivity inhibitors and glucose), creating a mixture whose viscosity and flow characteristics are roughly equal to normal-hematocrit human blood. At an ~11 per cent nanocrit, partial plug flow might ensue in a few of the smallest capillaries, but complete plug flow (requiring much higher pumping power and pressure than laminar flow) can be avoided. Each 3 micron^3 vasculocyte grasps a single 2 micron^3 plate, ready for installation, thus the new vasculo-infusant contains ~20 × 109 vasculocytes/cm^3. Installing ~150 × 1012 basic plates thus demands a minimum of 7500 cm^3 of vasculo-infusant, requiring ~3800 sec or ~1 hour assuming a very gentle flow rate of only ~2 cm^3/sec. Each vasculocyte drifts quietly in the flow until it encounters a vessel wall for the Nth time (N is an integer control variable), which activates it, causing it to attempt to release its cargo in a clear space. (Slowly increasing N during the installation process produces a crudely progressive plating pattern.) If the immediate area is already fully plated, the legged vasculocyte walks across the surface until it reaches a clear area to deposit its cargo. Corner registration of adjacent plates is verified prior to plate release, to ensure a maximally dense tiling pattern. Because of their small size, vasculocytes can enter even the narrowest capillaries and install plates by tiling motions. This process may be loosely regarded as a robotically-guided variant of fluidic self-assembly, a well-known existing commercial process. Special procedures must also be devised to handle encounters with nonendothelial materials (e.g., fibrin strands) or stray cells (e.g., adherent leukocytes) that may be blocking the endothelial surface, or denuded patches of vasculature lacking full endothelial cell coverage, although most of these defects should have been remedied during preoperative vascular conditioning. Once its cargo plate is in place, the vasculocyte releases back into the flowing fluid, powers down, and is eventually exfused from the body. Approximately 42 billion plates/sec are deposited during this 1-hour process. We generously assume that each 1-50 picowatt vasculocyte requires ~100 sec of active operation up to peak power to find a clear space to deposit its cargo before resuming its fluidborne dormancy—though early arrivers will spend less time searching for gaps in the structure than later arrivers. Thus there are at most ~4 trillion vasculocytes active at any moment and the total power released as waste heat by the vasculo-infusant is under ~200 watts, thermal energy which is easily carried off by the cold infusant. Given that one plate is installed by one vasculocyte carrier that performs ~106 mechanical motions/sec during a ~10 sec install time, this implies ~107 motions/plate and allows an allocation of ~1800 motions per nanometer of plate perimeter during the installation of each plate, which seems sufficient. The plates are not passive during the installation process. Besides the 20 cilia positioned atop each plate to provide tanker and boxcar mobility as part of the ciliary subsystem, plates also possess "motive cilia" to assist in both installation and repair operations. Each motive cilium is fully retractable when not in use. Complete positional, rotational and translational control during installation (and functional redundancy) requires at least one motive cilium on each of the 4 sides and at least four on the underside to establish a stable tripod while walking. The motive cilia allow limited trans-endothelial cytoambulation by each plate (using adherent tool tips) and fine control of plate/plate jostling motions. A working cilium consumes 0.1 picowatt during continuous operation at 1 cm/sec roughly the speed of jostling motions involving motions of ~1 per cent plate width per microsecond. Even with all motive cilia operating at once, plate power is under ~1 picowatt; even with all 150 trillion jostling at once, maximum power would be under 150 watts—energy easily obtained from glucose and oxygen provided by the vasculo-infusant medium, and waste heat that is easily dissipated by thermal conduction. However, the motive ciliary power draw of the entire plate population normally should not exceed ~1 watt during installation because once properly positioned in a regular grid pattern (in <1 sec) all local plate jostling ceases and the motive cilia are stowed. The installation process may be made more efficient if plates are programed to dynamically shift holes (in the plating pattern) upstream. This ensures that gaps are unlikely to persist in the developing vasculoid structure (except in cases of plate malfunction), and also ensures that shifted holes will arrive in a known region of the upper vascular tree where they are most convenient for vasculocytes to quickly find and fill. Out-of-register plate domains that collide during this dynamic repositioning process require a simple interaction protocol to allow them to mutually align correctly (such a protocol has not yet been devised). Once plates are in place, their surface cilia can be adjusted to provide variable drag on the circulating fluid. This can be used as a crude method of directing resources for more efficient construction. Plates also assist the vasculocytes in system validation. Upon installation, plates briefly activate all subsystems to verify that everything is working properly. With all 18, 750 sorting rotors spinning, individual plates momentarily draw up to ~2 picowatts. Malfunctioning plates jettison themselves back into the fluid flow for ultimate removal, or if this is not possible, take other action to bring their condition to the attention of the circulating vasculocyte fleet so that they may be replaced at once. Following supravascular positioning and subsystem validation, each plate inflates fluidtight metamorphic bumpers along its contact perimeter with its neighbors. This provides at least ~14 per cent linear effective elasticity (1.6 microns vs. 1.4 microns, center-to-center, comparable to the usual stretch requirements of the elastic arteries during normal cardiac pumping) and permits plates to track lateral movements of the underlying tissue while avoiding any relative movement of the opposing surfaces. Ventral lipophilic anchors dropped into the lipid bilayer cytomembrane of the adjacent endothelial cells of the vascular wall serve as sensors to detect any such relative movement, providing continuous feedback to control bumper inflation—although with cardioplegia such wall movements will be quite small because infusant pressure may be held nearly constant throughout the installation procedure. Neighboring plates lock their bumpers firmly together with reversible fasteners embedded in the bumpers. Along vessels whose diameters change rapidly as a function of axial position, orthogonal plate alignment along a circumference can be maintained via bumper expansion. Once maximum bumper extension is reached, a further increase in vessel circumference is accommodated in the next row by deflating bumpers in the circumferential direction and inserting one additional plate. Incommensurate offsets are reconciled using multiple ports in the bumper structure. If bumpers can expand by up to 20-30 per cent, then the blood vessel being plated can change in width by 20-30 per cent before such a pattern discontinuity is required, so such discontinuities should be relatively uncommon in most of the plated vasculature. Computer modeling of the dynamic adjustments required to achieve continuous plating at blood vessel bifurcations and in tapered vessels would be useful. As the natural vascular surface is covered by the growing vasculoid, individual plates that have completed self-testing procedures and are locked in place activate a sufficient number of rotors to begin providing full molecular transport services to the underlying tissues.

Any plate malfunctions are corrected by the circulating vasculocytes, via plate changeout. Metabolism in the underlying cellular tissues continues normally, just as it did before these tissues were plated with the vasculoid. Rotors on the ventral plate face (away from the lumen) gain access to nutrients in the vasculo-infusant fluid via gated channels leading to the dorsal plate face (toward the lumen) which is directly exposed to the fluid; cellular waste products escape to the fluid by a similar route. Plate power requirements normally range from ~0.004 picowatt (basal rate) to ~0.09

picowatt (peak rate) plus ~0.6 pW for computation. At 280 K the patient's cells would require only a fraction of the basal rate; after plating is complete and in the absence of significant computation, the entire population of 150 trillion installed plates might generate only ~1 watt.

Specific procedures for cellulock installation have not been examined in detail because these vasculoid components are relatively few in number and require a net installation rate of only ~9 million/sec (compared to 42 billion/sec for basic plates). To the extent cellulock placement can be governed by easily detected geometric factors such as vascular diameter (e.g., near capillary entrances), cellulock placement may occur similarly to plate deposition, but using vasculocytes programed with specific geometric release criteria. Some modest number of cellulocks destined for specific vascular addresses may be installed after plate initialisation using a combination of vasculocytes and now-active cargo cilia, by replacing temporary plates with cellulocks. Nontubular vascular segments including flaps, valves, sphincters, portals, sinuses, plexuses, nodes, and discontinuous microvascular beds such as the red pulp of the spleen constitute only a tiny fraction of the vascular surface but may require special procedures by the vasculocytes. Endothelial pressure/shear responses may be managed by continuous emission of appropriate blockers, inhibitors or cytokines; the response may be reversibly disabled or eliminated using a designed vector to insert a short segment of revised DNA into the cells' nucleus. The vasculoid also includes an explicit mechanism to inhibit the electrical output of the sinoatrial node, thus maintaining a permanent (but in theory reversible) state of cardioplegia. After ~1 hour, the structure of the vasculoid is almost complete. All major components have been tested and are in good working order—indeed, the still-submerged vasculoid is already maintaining full cell metabolism in the tissue below. The patient is now ready for defluidisation.

At this stage, the 1-micron thick monolayer of nanorobotic plates forms a chemically inert, flexible sapphire liner on the luminal (interior) surface of the entire vascular tree. This liner may be made fluidtight and airtight to at least modest pressures. Vasculo-infusant fluid is purged from the body by introducing ~6 liters of oxygenated pure anhydrous acetone as a dehydrating rinse followed by pressurised dry air at 2 atm through the input catheters and allowing all lavage fluids to exit through the output catheters for ~15 minutes. If necessary, surface cilia may assist in the distribution of fluid or air flow to help ensure that no pockets of fluid remain trapped anywhere in the system, and to ensure that no respirocytes, inoperative or damaged vasculocytes, or other stray particles are left behind. The upper limit for laminar flow (Reynolds number ~ 2000) of aqueous vasculo-infusant fluid passing through a smooth 9-mm diameter exit catheter occurs at ~1 m/sec. Assuming a constant laminar outflow velocity of 1 m/sec, flow volume through each of the two cannula is ~60 cm3/sec and the vasculoid lumen is emptied of all installation and rinse fluids in

~100 sec. (This step is easily reversed by re-introducing an aqueous oxygen-rich or respirocyte-laden nutrient solution.) Acetone viscosity is ~40 per cent that of water, so pumping power may be reduced as the rinse proceeds. Oxygen needed to sustain human life is slightly restricted for only ~100 sec during the pure acetone rinse, and thereafter is continuously available at the requisite concentration from the dry purge gas. A temporary supply of dry air with 20 per cent O_2 at 2 atm puts 5×10^{22} O_2 molecules into the vasculoid lumen, which when burned with glucose at 50 per cent efficiency in tissue cells produces ~20, 000 joules, equivalent to ~2000 sec of power at the reduced metabolic rate of ~10 watts in a human body cooled to 280 K. The plate population consumes only an additional ~1 watt. As long as fresh purge gas continues to circulate, the oxygen requirements of both human and vasculoid can be satisfied indefinitely. Cellular glucose reserves will last ~10^5 sec at the reduced metabolic rate; subject to cautions noted earlier, cellular waste products may not approach toxic levels for ~10^4 sec. Vasculoid plates can also continue functioning in the temporary absence of glucose. Assuming a power requirement of ~0.6 pW/plate at the basal rate, then a ~1 hour supply of glucose may be stored in a small fuel tank inside each plate, representing only ~9 per cent of the 2 micron^3 plate volume. The final steps of vasculoid installation include:

(1) *Plate Initialisation*. To aid in injury response and medical diagnosis, each plate may be assigned a unique position and identity code which can be echoed upon interrogation. With 200 billion operational vasculocytes and 150 trillion plates to initialize, each active vasculocyte must contact and initialize ~750 plates. Traveling at a net velocity of 10 microns/sec across the vasculoid, a vasculocyte can traverse (and initialize) 750 plates in ~100 sec. The address block is 100 bits in length; 50 bits are sufficient to specify 2^{50} = 1000 trillion unique objects in the system, but additional bits are required to specify each plate's branch level in the fractal vascular tree and 10 bits are reserved for self-correcting parity checks.

(2) *Install Storage Vesicles*. The vesicles are small storage garages containing reserves of mobile and cargo-carrying nanodevices, other auxiliary nanodevices, spare parts, replacement modules and repair materials, critical biochemical consumables, and compressed refuse. Vesicles may also incorporate nanoscale bioprocess plants to manufacture small quantities of artificial biochemicals not normally produced in the body. Each vesicle is ~1 mm^3 in volume, providing enough space to store ~1 billion tankers, ~300 million vasculocytes, or 350, 000 boxcars. Vesicles may be attached directly to the vasculoid surface using sapphire struts affixed to reinforced base plates, or may be attached to each other to build convenient three-dimensional configurations (e.g. "bunch of grapes" formations).

As an order-of-magnitude estimate of the total number of vesicles that might be required, the volume required to store a complete replacement for the entire mobile container population would include 166.2 trillion tankers (1 micron^3/tanker) or 166.2 cm^3, 32 billion boxcars (2827 micron^3/boxcar) or 91.7 cm^3, and 2 trillion backup vasculocytes for tenfold redundancy (3 micron^3/vasculocytes) or 6 cm^3, totaling ~264 cm^3 of storage. Including additional space for bioprocess plants, spare parts, consumables and refuse storage may boost total vesicle volume to ~500 cm^3, requiring ~0.5 million vesicles which would fill roughly the combined internal volume of the four cardiac chambers which are no longer needed for pumping blood. However, conservative design principles would suggest that vesicles should be widely distributed throughout the body to maximize system survivability in the event of massive trauma. Infusion of 500 cm^3 through the two catheters at a modest infusion velocity of 1 cm/sec requires ~200 sec for infusion of all vesicle components and their contents. Each 3 mg (fully loaded) vesicle is installed by a team of ~3000 vasculocytes which can collectively apply ~30, 000 nanonewtons of force (using only 10 active legs among the ~100 available on the ventral side of each vasculocyte) resulting in a vesicle acceleration of ~1 g (enough to overcome gravity in a supine patient, and easily increased if needed). Simultaneously installing all 500, 000 vesicles requires the cooperative activity of ~1.5 billion 1-50 picowatt vasculocytes which consumes a total power of 1.5-75 milliwatts during the installation.

(3) *Install* sealed carotid, jugular, or navel access port, as required (~40 sec).

(4) *Activate* ciliary distribution system (~10 sec). This should probably include a brief stereotypical self-cleaning protocol to further ensure that all particles and stray nonessential nanorobots have been removed from the appliance interior.

(5) *Introduce* into the vasculoid interior the entire 257.9 cm^3 operational tanker and boxcar population, requiring ~100 sec at a modest ~1 cm/sec dry infusion velocity.

(6) *Warm* the patient (~1 hour) using heat generation from excess ciliary activation, energy from onboard thermogenerative systems or external heat sources imported via enhanced plate conductivity (or simple electrical or rf heating. Restart normal respiration as soon as possible. (The heart remains permanently inhibited, although the endocardium is coated with vasculoid plates.)

(7) *Reverse* cytoactivity inhibition that was initiated during patient preparation as appropriate.

(8) *Remove* catheters and seal all vascular and dermal breaches.

The vasculoid is now fully operational and self-contained. The patient is warm and breathing. All essential metabolic and immunological systems have returned to normal function. Healthy vascular conditioning is maintained on a permanent basis (post-installation) because the vasculoid exercises precise control over the transport of cholesterol, leukocytes and platelets, the principal participants in the arteriosclerogenic process. Skin and tissue suppleness may be controlled by adjusting the spring constants of linked bumpers between adjacent plates. A number of cosmetic issues must be addressed. For example, unpigmented tissues (e.g. the tongue and gums, fingernail dermae, the uvea and lacrimal apparatus of the eye, and brain tissue) will appear a pale, waxy translucent white owing to the complete absence of red cells; pigmented tissues (e.g. dark skin) should be largely unchanged. For more traditionalist users or for other aesthetic reasons, semi-natural coloration might be restored by retaining blue sapphire in venous channels but substituting red ruby (chemically similar to sapphire) in arterial plate materials. However, transmission and reflection properties of these substances may differ markedly, and such colors would likely require at least small numbers of potentially troublesome impurity atoms (10^{-3}-10^{-4} of Fe, Ti, or Cr atoms; to be present. Alternatively, a thin diamondoid-veneered sublayer of organic chromophores of appropriate colors might be employed. After a brief period of rest, postoperative checkout, and familiarisation with user interfaces, the patient is released from the installation facility.

The procedures outlined above for vasculoid installation are designed to be fully reversible at every step. Thus, vasculoid removal may be accomplished by reversing the installation procedure, except that patient preparation should be performed first. If the patient has worn his or her vascular appliance for a considerable time, heart muscles, venous valves, splenic filtration systems and the erythroid marrow may have significantly atrophied, requiring remedial cellular repair. Homologous formed blood elements such as RBCs must also be premanufactured for postoperative infusion prior to vasculoid removal, and any genetically modified endothelial cells must be reprogramed back to their original state.

For our highly speculative second scenario, we briefly (and only superficially) describe an aggressive installation procedure that attempts to install the vasculoid into a fully metabolising normothermic human body about 100 times faster than the hypothermic procedure detailed in the principal scenario. As discussed earlier vascular conditioning and mapping should occur before installation, in part to allow the manufacture of a folded, pre-assembled custom appliance tailored to the patient's unique vasculature which is then presented to the physician for installation. Our aggressive procedure involves three steps: (1) pre-charging the body with respirocyte infusant and removal of bloodborne cells; (2) physical installation of vasculoid plates as continuous rolling sheets, coincident with removal of infusant fluid; and (3) configuration and permanent connection of the appliance, followed by removal of support structures and system activation. It is intended that all steps will be carried out within the ~10^3 sec (~15 min) ischemic time limit imposed by non-respiratory metabolite concentrations. The aggressive procedure begins by accessing one or more large veins to permit rapid infusion and exfusion. The blood is quickly replaced, as it circulates, with an infusant solution containing electrolytes, glucose, and other essential substances, plus a sufficient nanocrit of respirocyte-class devices to provide respiratory support. The transfusion is speeded by using moderate numbers of "dragnet" style nanorobots as described in other contexts by Freitas—each such nanorobot manipulates an adjustable 1-100 micron diameter net in order to physically gather bloodborne cells that may be trapped in eddies near venous valves, vascular sinuses, and similar spaces. Within a few circulation times, perhaps ~200 sec, the vascular compartment is cleared of all free cells and most macromolecules. Exfused blood is retained until the procedure is complete, to facilitate reversal in the event of a medical emergency. The heart is now stopped. Respirocytes continue to provide oxygen and absorb carbon dioxide; even a modest 1 per cent respirocrit can provide ~25 minutes of resting metabolism without recharging, perhaps slightly less in the most energy-intensive tissues such as the brain. But because fresh infusant can be introduced continuously, the patient can be parked in this state indefinitely if need be. Next, the vasculoid is introduced into the arteriovenous vasculature directly through the heart via cardiopuncture or cardiocentesis. Cardiocentesis involves the surgical puncture of the heart, most often used today *in utero* on developing fetuses either to extract or to insert fluids into the vasculature (e.g., blood transfusion), or more rarely on adult organisms for related purposes. Cardiocentric installation of the vasculoid is required because the human circulation consists of two independent arterial circuits (pulmonary and systemic) each containing their own capillary beds. The appliance is installed as a continuously-everting concentric tube, a process called progressive fractal eversion that may be visualised as turning a glove inside out. Basic plates, docking bays and cellulocks are prefastened in the proper configuration to fit the various diameters and branchings of the blood vessels that they will coat. The necessary flexibility of this sheet of plates is provided by sophisticated watertight jointed interplate bumpers that have yet to be designed. Eversion may be powered by compressed gas, ciliary action between opposed plates, or by other appropriate means; the presence of pressurised gas (e.g., pure dry N_2) would also help to minimize the occurrence of leaks during installation. Each docking bay has an appropriate tanker already docked. As soon as the plate makes contact with the endothelial surface, the docking bay begins to work. Since the vasculoid contains a sufficient number of docking bays to accommodate peak metabolic loads, the preattached tankers can supply oxygen, glucose, and other nutrients, and accumulate wastes. Most critically, 5 trillion oxygen tankers are attached to all available respiratory-tanker docking bays, providing ~474 sec (~8 min) of oxygen at the basal rate. This sustainable duration is easily extended to ~1318 sec (~22 min) of basal oxygen supply by redesigning 13.9 trillion of the 24 trillion docking bays that are not needed for nonrespiratory transport to include the capacity for gas transfer during installation, since at the basal rate docking bays require only 67 active sorting rotors. Installation is initiated cardiocentrically as two primary segments imported via two cardiopuncture entry points through the relatively thin walls of the right atrium and the left atrium, respectively. Installation works outward from each chamber opening, usually in the natural direction of valve motion, moving towards the distal capillary beds. From the right atrial entry point, the unfolding right atrial appliance segment has four "fingers" and simultaneously plates:

(1) the superior vena cava; (2) the inferior vena cava; (3) the coronary sinus (which receives cardiac veins from heart tissue); and moves through (and plates) the right atrial and ventricular chambers, thence to plate; (4) the pulmonary artery. The right atrial segment also must plate all of the numerous Thebesian veins; these venules return blood from the myocardium without entering the venous current, and open directly into the right atrium. From the left atrial entry point, the unfolding left atrial appliance segment has five "fingers" and simultaneously plates: (1&2) the two left pulmonary veins (which frequently terminate by a common opening); (3&4) the two right pulmonary veins; and moves through (and plates) the left atrial and ventricular chambers, thence to plate (5) the aorta.

Additionally, a third vasculoid segment called the portal segment must be inserted through a third abdominal entry point in order most efficiently to plate the portal vein, which lies midway between the hepatic capillary beds and the intestinal capillary beds. Advancing at 1 cm/sec (2 cm/sec internal speed between the installed surface and the inverted tube), the vasculoid requires only 70 sec to plate the maximum ~70-cm main arterial or venous course. Fluid and respirocytes are withdrawn through the center of the inverted tube. As the vasculoid reaches the capillaries, the progression of the eversion may slow to ~10 micron/sec of travel, requiring another ~100 sec to complete all capillary plating. Human capillaries contain ~14 per cent of blood volume but comprise ~95 per cent of the surface area of the vascular system, so most of the appliance's plates are installed during this final ~100 seconds. If all 150 trillion plates are active throughout this process, the power draw is a physiologically-tolerable 150 watts. Upon first contacting the endothelium, each 2 mm^2 plate has ~0.1414 seconds—time enough for 14, 140 localised 100-nm movements at 1 cm/sec (100 KHz)—to adjust its position and bumper configurations, and to clear away any detritus trapped beneath it (e.g., respirocytes, stray cells, etc.), before the next plate is placed. This seems sufficient.

Once the vasculoid has filled the capillaries, the arterial and venous branches are joined, detaching the internal fluid transfer channels running through the vasculoid lumen which are quickly retracted from the interior. The ciliary system can then be activated and tankers will begin to be delivered to and from the capillaries, starting ~200 sec after the first transcardial introduction of the vasculoid mechanism. Interior rinse is unnecessary because the luminal surface of the appliance was installed dry. Vesicle installation is performed post-installation via component injection, and assembled using vasculocytes, at leisure. Since this advanced model of the vasculoid comes preassembled, it can continually report its status and monitor physiological indications of the patient during installation, via an interplate communication network. In the event of an unforeseen medical catastrophe during installation, the vasculoid may be evacuated using an emergency extraction protocol if the appliance itself has not suffered a major loss of integrity and the transfer channels are still attached. In the emergency extraction protocol, the vasculoid undergoes partial flattening with circumferential fission in the capillaries. Then the arterial and venous sides of the vasculoid are withdrawn as they were inserted, using a reverse-eversion process with progressively increasing velocities possibly reaching ~10 cm/sec in the largest arteries. During such an extraction, only a ~1.4-micron-wide annular ring will be moving at this high speed adjacent to tissue—the installed vasculoid does not move, and the moving vasculoid is surrounded by immobile vasculoid. As a result, installed appliance plate-sheets may be pulled out by applying an extraction force to the fluid transfer channel structures that are still attached to the terminus of each capillary tube. Continuous pressurised return of respirocyte-charged infusant to the natural vasculature through these channels prevents pulling a vacuum behind the retreating tendrils of the appliance, which would cause the collapse of blood vessels as the appliance was removed. Thus, in less than a minute, the patient has returned to the relative safety of the respirocyte-infused parked state from which the installation procedure was begun. In the context of a medical technology capable of manufacturing a vasculoid, the temporary bloodless state would not appear to be a cause for concern.

Self-Assembly as a Deliberate Strategy

"Self-assembly" is a term that by now figures prominently in the literature of nanostructured materials and nanofabrication. The term therefore carries a variety of implicit or explicit meanings, and we cite the definition given by Kuhn and Ulman: "Self-assembly is a process in which supermolecular hierarchical organisation is established in a complex system of interlocking components." The mechanisms that produce the hierarchical organisation are determined by competing molecular interactions (e.g., interactions between hydrophobic versus hydrophilic components, van der Waals, Coulombic, or hydrogen bonding), resulting in particular microphase separation or surface segregation of the component materials. Thus, the use of a hierarchy of bond strengths and/or chemical specificity can produce a hierarchy of lengths in the final nanostructured material (Muthukumar et al. 1998; Stupp and Braun 1998). As one example of such self-assembled or self-organised materials, McGehee et al. (1994) have mixed silica precursors with surfactants that have self-ordered to form various surfactant-water liquid crystals, producing various structures built from walls of amorphous silica, organised about a repetitive arrangement of pores up to a hundred angstroms in diameter. In the "natural" formation of inorganic nanostructures, the addition of organic molecules can strongly influence the resulting structure of inorganic components. Such strategies have been adopted in synthetic formation of nanostructures, such as in the formation of networks of gold clusters (Andres et al. 1998). Gold clusters 3.7 nm in diameter, formed in the vapor phase, are encapsulated in organic surfactants, such as dodecanethiol, forming a colloidal suspension. The surfactants prevent the agglomeration of the gold clusters. Addition of small amount of dithiol precipitates out a 3-D cluster network, which can in turn be deposited onto another solid substrate. A transmission electron microscope (TEM) image of a cluster array spin-cast onto MoS2. The methodology of self-assembly has even been extended to physical vapor deposition processes where it would seem more difficult to control the nucleation and growth of three-dimensional nanostructures. Utilising the strain inherent in the epitaxial growth of lattice-mismatched materials, and the expected strain-induced transition from two-dimensional (layered) to three-dimensional (islanded) growth, together with careful monitoring of the growth process through RHEED analysis, researchers have been able to form arrays of semiconductor quantum dots, ~ 200-300 Å in diameter, ~ 1011 cm-2 in density, and with a size variation of about ±7% (Leonard et al. 1993). An example of such "self-assembled" semiconductor quantum dots.

The achievement of arrays of several billions of quantum dots of these dimensions with such a size variation is beyond the capability of standard high resolution lithographic and pattern transfer processes. Moreover, the controlled formation of critical surfaces and interfaces without the intercession of ion-assisted processing that can introduce potential defects into the materials has produced a rich source of optically and electronically efficient quantum structures. A number of researchers have already incorporated such self-assembled dots into laser structures (Bimberg et al. 1998). Chemical specificity may provide the most robust means of ensuring control of size and placement of nanostructured building blocks, and recent work in the synthesis of compound semiconductor quantum dots from chemical precursors have provided even tighter distributions of size variation (±5%) than those shown in the strain-induced self-assembled dots (Katari et al. 1994; Murray et al. 1993).

Great Opportunities in Self-assembly

In self-assembly large molecular structures are obtained from the organisation of a large number of molecules or atoms into a given shape, typically through specific interactions of the molecules among themselves and with a template. The interaction of the different bonding mechanisms is an area of strong fundamental research interest. Only two areas will be highlighted here: zeolites and carbon materials. Both of these materials exhibit characteristics of self-assembly, namely novel and reproducible structures that can be fabricated in industrially significant quantities.

Carbon Materials

The carbon-based materials of interest from a molecular self-assembly point of view include fullerenes and their relatives, including endohedral fullerenes and metal-coated fullerenes, carbon nanotubes, carbon nanoparticles, and porous carbons. Since 1990 with the discovery of techniques to produce soluble carbon in a bottle (for examples, see Krätschmer et al. 1990 and references therein), research on and with carbon materials has skyrocketed (Dresselhaus et al. 1996; Dresselhaus and Dresselhaus 1995). Not only can the molecular forms of carbon (the fullerenes and their derivatives) be synthesised, characterised, and studied for applications, but many other new carbon materials such as multi- and single-walled carbon nanotubes can now be produced in macroscopic quantities.

The broad variety of carbon nanotube structures whose properties are now being examined both theoretically and experimentally. A rich literature on these new carbon materials now exists. This report will only attempt to highlight a few important recent examples in the area of high surface area materials. Of particular interest for future catalytic applications is the recent report that not only can C60 be coated with metal atoms, but that the metal coating can consist of a precise number of metal atoms. For example, C60Li12 and C60Ca32 have been identified mass spectroscopically (Martin, Malinowski, et al. 1993; Martin, Naher, et al. 1993; Zimmerman et al. 1995). C60 has been coated with a variety of different metals, including Li, Ca, Sr, Ba, V, Ta and other transition metals. Interestingly, addition of more than 3 Ta atoms to C60 breaks the C60 cage. Replacement of one carbon atom in C60 by a transition metal atom such as Co or Ir is being studied for possible catalytic applications. The future technological challenge will be to discover techniques to fabricate large quantities of such materials, so that such catalyst materials can be put in a bottle and not just in molecular beams.

Examples of carbon nanotube structures, including multiwalled and metal-atomfilled nanotubes. Carbon nanotubes have the interesting property that they are predicted to be either semiconducting or conducting (metallic), depending on the chirality and diameter of the nanotube. Such materials are being studied as conductive additives to plastics and for use in electrochemical applications where the uniformity of the nanotube diameter and length is not overly critical (Dresselhaus 1998). Another approach is to use the carbon nanotube as a template for a nanotube of an inorganic oxide. Hollow nanotubes of zirconia and yttria-stablilised zirconia have been prepared by coating treated carbon nanotubes with a zirconium compound and then burning out the carbon template (Rao et al. 1997). Finally, large scale production of singlewalled nanotubes has recently been demonstrated, so one may anticipate a strong upsurge in the characterisation and potential usage of single-walled carbon nanotubes in the future (Jounet et al. 1997). Porous carbons are of interest as molecular sieve materials, both as sorbants and as membranes, or as nanostraws for filtration.

One of the major research objectives is to develop materials or structures with exceedingly high storage capacity per unit volume and weight for gases such as H2 or CH4. H2 or CH4 could become an economic source of combustion fuel or a means to power fuel cells for ultralow-emission vehicles or for electric power generation. Microporous hollow carbon fibers have exhibited high permeance and high selectivity as hydrogen selective membranes, and development is now underway to scale up these membranes to commercial levels (Soffer et al. 1987; Jones and Koros 1994; Rao and Sircar 1993). Carbon fiber materials produced via catalytic decomposition of hydrocarbon vapors have also recently been reported to exhibit exceptionally high hydrogen adsorption capacity (Baker 1998). More mundane uses of nanotubes are as nanometer reinforcing rods in polymers or even in concrete. Incorporation of conducting carbon nanotubes in construction materials such as concrete or structural plastics opens opportunities for real time monitoring of material integrity and quality.

Microporous and Dense Ultrathin Films

Research and development of microporous thin films for use as molecular sieving membranes using inorganic crystalline materials such as zeolites or porous silica is another area of active research around the world. For molecular sieving membranes, one critical challenge rests on discovering ways to create large scale, thin, nearly defect-free membranes. One recent example is the fabrication of mesoporous conducting thin films grown from liquid crystal mixtures (Attard et al. 1997). Transmission electron microscopy (TEM) reveals an ordered array of 2.5 nm diameter cylindrical holes in a 300 nm thick Pt film. The hole diameter can be varied either by changing the chain length of the surfactant molecule or by adding an alkane to the plating solution. It is interesting that this technique produces a continuous thin film with nanoscale porosity in an electrically conducting material. Dense ultrathin films such as single monolayer films would be of significant importance in the semiconductor industry. Thin films of specialised coatings for corrosion, thermal, and/or chemical stability should be valuable for the chemical and aerospace industries. Novel chemical sensors may be anticipated through use of ultrathin films composed of specialised clusters. Typical techniques for production of thin films are physical vapor deposition, chemical vapor deposition, and Langmuir- Blodgett processes.

Zeolitic Materials

Aluminosilicates (e.g., zeolites) are crystalline porous nanostructures with long range crystalline order with pore sizes which can be varied from about 4 Å to 15 Å in conventional zeolites. A 3-dimensional (e.g., MFI) zeolite cage structure together with a depiction of the straight and ziz-zag channels and a 2-dimensional zeolite with channels only in 2 directions. The vertices in the stick drawing denote position of the O atoms in the crystalline lattice. This particular zeolite has 10 atoms in the zeolite "window." The size of the window is determined by the number of oxygens in the ring. Typical zeolite structures together depicting the positions of the O atoms and two different zeolitic structures one (lower left) with a three dimensional structure and (lower right) a zeolite with a two dimensional channel structure.

For example normal hexane with a kinetic molecular diameter of about 5.1 Å can pass through a 10 ring or larger, whereas cyclohexane with a kinetic molecular diameter of 6.9 Å would be hard pressed to pass through a 10 ring. Thus all other things being equal, a 10-ring zeolite could be used to separate mixtures of normal hexane and cyclohexane. It is this property together with the ability to chemically modify the acidity of zeolitic materials that makes them extremely valuable as selective sorbants, as membranes and for use in selective catalytic reactions. In 1992, a new family of aluminosilicates (M41S) with pores sizes between 20 and 100 Å in diameter were reported by Mobil researchers (Beck et al. 1992; Kresge et al. 1992). One of particular interest is MCM-41, which consists of hexagonal arrays of uniform 2 to 10 nanometer-sized cylindrical pores. Not only can such materials be synthesised, but novel structures such as "tubules-within-a-tubule" have been fabricated as mesoporous molecular sieves in MCM-41 (Lin and Mou 1996).

Of particular interest is the possibility of expanding the so-called "liquid crystal templating" mechanism (Chen et al. 1993) to non-aluminum dopants within the silicate MCM-41 framework (Tanev et al. 1994) and to derive nonsiliceous MCM-41 type of materials (Braun et al. 1996). Another approach to synthesising large pore and large single crystals of zeolytic materials is being pioneered by Geoffrey Ozin and his group at the University of Toronto, who have demonstrated that crystals as large as 5 mm can be synthesised (Kupperman et al. 1993). The ability to synthesize such large crystals has important implications for discovery of new sensors (selective chemical adsorbants) and membrane devices (selective transport of molecular species), since large single crystals can now be available to the laboratory researcher to carry out fundamental studies of adsorption and diffusion properties with such materials. These materials are expected to create new opportunities for applications in the fields of separations science, for use directly as molecular sieves or as new molecular sieving sorbant materials; in catalysis, as heterogeneous catalysts; and as supports for other catalytic materials as well as other novel applications (Bowes et al. 1996; Brinker 1996; Sayari 1996). The ability to synthesize zeolitic materials of precise pore size in the range between 4 and 100 Å continues to expand the possibilities for research and technological innovation in the catalytic, separations, and sorption technologies (Ruthven et al. 1994; Karger and Ruthven 1992).

Biomimetic Self-Assembly and Self-Assembled Monolayers

Langmuir-Blodgett (LB) films are formed when synthetic or naturally occurring surfactants are placed at the air-water (or liquid gas) interface where they can organise into a layer that is one molecule thick (a monolayer), two molecules thick (bilayer) or several molecules thick (multi-layer). The polar part of the surfactant molecule tends to be attached to the surface of the water while the nonpolar hydrophobic regions of neighboring molecules will stick together and orient themselves away from the surface. As the lateral pressure (2 dimensional pressure created by compressing the interface area) is changed, the film undergoes phase transitions corresponding to various levels of self-organisation.

Thiol self-assembly *can be achieved by reacting thiol containing compounds with clean gold surfaces. Sulfhydryl groups on molecules will covalently bind to gold, thus allowing molecules to be arranged two dimensionally over a gold surface. This is extra useful because gold conducts electricity and makes for excellent electrical contacts, thus Scanning Tunneling Microscopy (STM) measurements can be made on such samples. For example, redox proteins can be mutagenically modified with cysteine residues (an amino acid that contains a thiol group). Exposing a gold surface to such engineered molecules results in self-assembled monolayers of protein.*

Electrons Microscopy and Lithography

The electron is a most convenient particle (or wave) for microscopy and lithography, not to mention electronic devices such as your computer. The mass of this subatomic particle is so small that only a negative exponent can do it justice: about 10^{-27} g. That is about one billionth of a billionth of a billionth of a gram. See what I mean? Too small for words. Compared to this super-small mass, it has a quantised negative charge of (elementary charge = $1.6*10^{-19}$ Coulomb). As this charge is exactly the same for all electrons, it is possible to precisely predict and control the movement of electrons. This tendency for very small objects such as electrons to be identical to one another is one of the motivating factors behind nanotechnology. That is, unlike components used in larger scale manufacturing, there need not be an inspector to check the mass and charge of each electron in a nanodevice.

The small and uniform *mass to charge ratio* of an electron makes its momentum extremely sensitive to electric fields, which exert a force on charged particles. Such fields can be used to accelerate, direct, and focus electrons in the form of an *electron beam*. Scanning an electron beam over a surface, and measuring its interactions with the surface is a method known as *Scanning Electron Microscopy (SEM)*, which allows nanoscale resolution of conducting surfaces. Furthermore, if a surface is chosen that reacts in response to exposure to electrons then an electron beam can be used as a writing tool for the rapid patterning of surfaces with ultra-high resolution, a method known as *Electron Beam Lithography (EBL)*.

Now, consider the wave properties of this beam. The *De Broglie wavelength* of a particle can be related to its momentum by the relatively simple formula $l = h/p$, where l is the wavelength, h is Plank's constant and p is the momentum of the particle. Electrons differ from photons in that their mass is much larger, which results in a larger momentum and smaller wavelength. While a photon has a wavelength much larger than the size of one atom, the wavelength of an electron is far smaller than an atom. Thus, unlike optical microscopy or lithography, the *far-field diffraction limit* is not a barrier to the resolution of electron microscopy or lithography. Rather, other difficulties must be (and are being) overcome, such as our ability to focus the beam on a spot so small as a single atom. In Feynman's founding speech of 1959, he explained how electron microscopy might one day see individual atoms:

"I would like to try and impress upon you while I am talking about all of these things on a small scale, the importance of improving the electron microscope by a hundred times. It is not impossible; it is not against the laws of diffraction of the electron. The wave length of the electron in such a microscope is only 1/20 of an angstrom. So it should be possible to see the individual atoms."

It is remarkable how prophetic Feynman was in his speech, seeing as this exact feat was was accomplished 22 years later with the invention of *Scanning Tunneling Microscopy (STM)*. STM is indeed invaluable for determining atomic precision within conducting samples and even selectively positioning individual atoms. However, even with its sub-atomic resolution, STM relies on slow mechanical actuators and is not currently useful for industrial *nanolithography*. Thus, it is still entirely possible that continued improvements in the focusing of electron beams (or other particle beams) might result in more rapid atomic resolution.

Alchemy and Atomic Precision and Universal Assemblers

Nanotechnology can be qualitatively understood as an approach to alchemy that is based on the current model of matter. For more than 2000 years, the Aristotelian model of matter went relatively unquestioned in the western world. In this model, matter consisted of four elements: earth, air, fire and water. The variety of materials in existence was explained by the idea that the ratios of these elements within a material determined the form that the matter took. Based on this theory, the idea that one might rearrange the elements to create a desired material naturally followed. The arts of alchemy were early attempts to achieve this goal. In the last centuries, alchemy became more of a philosophy while its techniques were divided between chemistry, nuclear physics and various other disciplines.

In the 18th and 19th century, a far more accurate and productive model of the elements was finally developed. Lavoisier started the change with the discovery of a material that he called "oxygen" and began to arrange the other known elements into four groups. Dalton then proposed the importance of atomic masses, after which Mendeleyev noticed a pattern that resulted in the periodic table of the elements. Early in the 20th century, Pauling provided an accurate model of how these elements *bond* together to form molecules. These molecules can in turn interact by means of *weak interactions*. The chemical bonds and molecular interactions seem to be flawlessly explained by *quantum mechanics*, making our model of matter relatively complete.

With the new model being far more complex than the original one, the alchemist's goal of arbitrarily rearranging the elements became too confusing to pursue directly, until recently. During the last century, physicists and mathematicians have been ironing out the kinks in quantum mechanics. Chemists have found numerous ways to synthesize and characterize a plethora of molecules. Biologists have made enormous progress in understanding the molecular mechanisms and weak interactions that allow life to function. Meanwhile, the industrial revolution has provided us with machines, computers and other tools enabling further progress.

Atomic Precision

Our technology has now reached a point where the idea to more or less arbitrarily arrange the elements to create materials ranging from *nanomachines* to food is beginning to seem achievable. In 1959, Feynman translated the goals of alchemy into more modern terms, proposing that we miniaturize our tools for observing and manipulating matter until atomic precision is reached. In 1991, Drexler coined the term *molecular nanotechnology* referring to thorough, inexpensive control of the structure of matter. In the last decade, the concept of *nanotechnology* has been accepted by the scientific community and numerous organisations have been formed.

Unlike alchemy, nanotechnology is based on a more modern model of matter. Nanotechnology does not try to transmute elements as it is far cheaper to extract and purify elements or molecules from existing materials. Arranging the atoms in a material to get the desired elements where we want them is a far more economical approach but still requires significant effort and cooperation. Thus, nanotechnology is an *interdisciplinary* subject drawing from biology, chemistry, physics, engineering and computer science, to name a few. Each discipline contains an important piece of the puzzle, and as the puzzle is assembled we are witnessing remarkable advancements in our ability to economically control the structure of matter.

The latest examples include *molecular imprinting*, *single molecule detection*, *DNA arrays* and *better sunscreens*. However, with nanotechnology just now beginning, what might result is still a matter of imagination. For instance, the Drexlerian nanotech community is looking forward to a future where a nanorobotic *assembler* could be programmed to build just about any stable arrangement of atoms, including more assemblers. Others imagine *quantum computers* or long term *life extension*. Only time will tell whether these dreams will be realised or face the same fate as the philosopher's stone and the elixir of life. Regardless of whether the ultimate goal of atomic precision is reached, novel *materials*, *methods* and *models* are resulting from this second attempt at alchemy.

Universal Assemblers

These second-generation nanomachines—built of more than just proteins—*will do all that proteins can do, and more*. In particular, some will serve as improved devices for assembling molecular structures. Able to tolerate acid or vacuum, freezing or baking, depending on design, enzyme-like second-generation machines will be able to use as "tools" almost any of the reactive molecules used by chemists—but they will wield them with the precision of programmed machines. They will be able to bond atoms together in virtually any stable pattern, adding a few at a time to the surface of a workpiece until a complex structure is complete. Think of such nanomachines as *assemblers*. Because assemblers will let us place atoms in *almost any reasonable arrangement (as discussed in the Notes)*, they will let us build almost anything that the laws of nature allow to exist. In particular, they will let us build almost anything we can design—including more assemblers. The consequences of this will be profound, because our crude tools have let us explore only a small part of the range of possibilities that natural law permits. Assemblers will open a world of new technologies.

Advances in the technologies of medicine, space, computation, and production—and warfare—all depend on our ability to arrange atoms. With assemblers, we will be able to remake our world or destroy it. So at this point it seems wise to step back and look at the prospect as clearly as we can, so we can be sure that assemblers and nanotechnology are not a mere futurological mirage.

9

Fundamentals and Applications of Carbon Nanotubes, Nanotube Theory and Nanotube Rings

Carbon Nanotubes

Here, we focus on our recent work on *carbon nanotubes*; their structure, properties and uses in nano-electronic devices. Carbon nanotubes are extremely thin (their diameter is about 10, 000 times smaller than a human hair), hollow cylinders made of carbon atoms.

This study consists of the following sections (to access them you may also use the navigation bar on the left):

- *Nanotube manipulation:* We can manipulate the nanotube positions, change their shape, cut them and place them on electrodes.
- *Molecular mechanics:* We can simulate the mechanical behaviour of nanotubes by calculating the forces acting between nanotubes and other objects such as the substrate.
- *Nanotube Field-Effect Transistor:* We have successfully used semiconducting single and multi-walled nanotubes as channels of field-effect transistors.
- *Nanotube rings:* While normally nanotubes are straight, we have devised ways to prepare them in a ring form.
- Nanotube theory: Computation and theory of the electrical and mechanical properties.

Nanotubes, depending on their structure, can be metals or semiconductors. They are also extremely strong materials and have good thermal conductivity. The above characteristics have generated strong interest in their possible use in nano-electronic and nano-mechanical devices. For example, they can be used as nano-wires or as active components in electronic devices such as the field-effect transistor shown in this study.

Potential Applications

(i) Conductive plastics; (ii) Structural composite materials (iii) Flat-panel displays' (iv) Gas storage; (v) Microm and nano-electronics; (vi) Radar-absorbing coating; (vii) Technical textile; (viii) Ultra-capacitors; (ix) Atomic Force Microscope (AFM) tips Batteries with improved lifetime; (x) Biosensors for harmful gases; and (xi) Extra strong fibres.

Market

According to market report in 2004, "a total tonnage of 65 tons of nanotubes`and fibers was produced in 2004 which equates at manufacturer's current prices to`•144 million worth. By 2010 this figure will surpass euro 3 billion. This`represents a compound annual growth rate of well over 60%." Nanocyl's basic technology had its foundation in the research work carried out at the Universities of Liège and Namur in Belgium in the late 1990's. After becoming a company in 2002, Nanocyl continued to extend its own IP portfolio in carbon nanotube technology. Nanocyl's Intellectual Property covers carbon nanotube synthesis technology; modifications to the carbon nanotubes; as well as products utilising carbon nanotubes for their unique properties—such as polymer nanocompositesn Nanocyl's method of synthesis is the catalytic carbon vapour deposition (CCVD), which is based on the mechanism shown below. This reaction is carried out in a continuous reactor technology enabling the production of industrial quantities of carbon nanotubes with uniform product characteristics and quality. Nanocyl owns and develops original processes for modifying surface characteristics of carbon nanotubes such as imparting chemical functionality to the surface of the nanotube. This treatment makes CNTs more suitable for applications such as reinforcement of material compositesl interconnections, surface graftingl metal plating and nanocomposite manufacturing.

Carbon nanotubes have exceptional properties which can potentially be used in many applications ranging from macroscopic material composites down to nanodevices.

Following is a list of some potential applications:

- Polymer additives for adhesives, coatings, fibers, films and engineered plastic parts
  - Flame retardants
  - Electrical conductive and antistatic polymers
  - EMI/RFI shielding

- Organic photovoltaic cells
- Mechanical reinforcement
- Thermal conductive polymers

- Additives for inorganic materials
  - Ceramics
  - Metals
- Optical films
  - Optical Filtering devices
- Transparent electrodes
  - Flexible Display screens
- Gas storage
  - Hydrogen storage medium
- Catalyst support
  - Precious metals support in fuel cells
  - Polymerisation
- Electron field emitters
  - Field Emission Displays
  - Flat lamps
  - Gas-discharge tubes
  - LCD backlighting
  - X-ray sources
- Electrode materials
  - Lithium-ion batteries
  - Bipolar plates of fuel cells
  - Supercapacitors
- Filters
  - Water and gas filtration
- Chemical and biological sensors
- Nanoprobes
  - Atomic force microscope
  - Scanning probe microscope
- Nanolithography

- Circuit Interconnect
- Nanotransistors
- Actuators
  - Artificial muscles
  - Electro-mechanical systems

Nanocyl was established in 2002 as the first European company with the goal of manufacturing carbon nanotubes (CNT) in commercial quantities.

The foundation of Nanocyl stems from the research work of Professor Janos B.Nagy of the University of Namur, and Professor Jean-Paul Pirard of the University of Liège. Today these researchers along with Nanocyl scientists and engineers continue to push the envelope in developing high-quality grades of carbon nanotubes for commercial applications and to support research activitiesn.

The scope of Nanocyl's current activities includes:

- Refinement of continuous-process industrial CNT production
- Improved methods for CNT dispersion in materials
- Global commercialisation of CNTs in a variety of sectors
- Partnerships for the development and commercialisation of new products
- Research programmes and technology transfer agreements with Universities
- Active participation in the European Union sixth framework programme.

Jean Kluyskens, member of the Board of Directors of Nanocyl, specialist in recruitment and staff management, founder of Mercuri Urval Europe and board member of several companies including Namur Invest. With his daughter is a main shareholder of Inbev, the largest brewery in the word. Dominique Jozeau, member of the Board of Directors of Nanocyl, former CEO of Nanocyl. MBA from University of Dallas |/ST1:CITY> TX, Dean of the European University |/ST1:PLACE> and Director of the MBA programme. CEO of ADM Group and European Managing Director of Franklin Electronic Publisher. Jean-Claude Jungels, member of the Board of Directors of Nanocyl, Electrical Engineer (ULg), MBA from Columbia Business School. Manager of Start-It, venture capital fund specialised in High Tech Start-ups. Luc Geuten, member of the Board of Directors of Nanocyl has a chemical engineering degree from KU Leuven and a MBA from Harvard Business School. Began his career at McKinsey & Company, is comfounder of Mitiska nv, a holding company active in retailing (Brantano, AS Adventure) and quoted on Euronext Brussels stock exchange. He is also comfounder and director of Innogeneticsl director of AXA-Royale Belge and VPK Packaging Group".

Dr Michel Morant, Board Observer of Nanocyl is Managing Director of Gesval, the technology transfer company of University of Liege. He graduated as Electrical Engineer, PhD and MBA from University of Liege, spent 12 years in the AEG Daimler-Benz Company. He is a member of the board of several start ups, and also member of international associations (Proton, ASTP, CURIE, TII, ...).

Jean Huybrechtsl representing Strafincol member of the Board of Directors of Nanocyl, graduated in lawl pure economics and holds a MBA from the Johnson School, Cornell University, NY. After a banking career and a management consulting experience, he launched Strafinco SA, a corporate finance consulting firm.

Jean Stéphenne, member of the Board of Directors of Nanocyl, Chemical Engineer, is CEO of GlaxoSmithKline Biologicalsn He is a former president of the Walloon Association of the Industry. Member of the board of several Belgian companies including Fortis. Belgian Manager of the year in 1996.

Vincent Doumier, member of the Board of Directors of Nanocyl, CEO of Compagnie du Bois Sauvage, a Belgian Holding company, listed on the Brussels Stock Exchange having investments in the Banking Sector, Real Estate, Industry (mainly Plastici and Luxury Consumer Sector, both in Belgium and abroad.

Baudouin Dubuisson, Chairman of the Board of Nanocyl, graduated with a degree in economics. After a career in audit and industry, he became advisor of the regional Minister of economy and he founded Franfin SA, a holding company specialised in start ups and middle sized companies in difficulties.

Dr. Janine Al-Asswad, member of the Board of Directors of Nanocyl, PhD in Chemistry, experience in R&D, Production, Technical Marketingl and Sales in SME's companies.

Frédéric Pouchain, is Investment Manager with the "Compagnie Nationale à Portefeuille", shareholder of Nanocyl for which he follows several equity participations, was consultant for McKinsey & Company and holds a degree of "Commercial Engineer" from UCL (Belgium) and a MBA from the Chicago Graduate School of Business.

Robert Van den broeckl Shareholder of Nanocyl, board member of several companiesl Managing Partner of Privast Capital Partners.

Eric de Lamotte, member of the Board of Directors of Nanocyl, director of several technology companies and investor in start-ups, former managing director and CFO of IBA, held management positions in the banking sector.

Dr. Frédéric Luizi, shareholder and Quality Manager of Nanocyl, has worked in various positions ranging from researcher in Universities (Portugal, UKl Denmark), project manager at BP (Norway), technology transfer officer for the University of Namur (Belgium), civil servant in Canada to European Project Leader (Belgium), Board member of 2 other start-up companies.

Dr. Antonio Fonseca, shareholder and adviser to Nanocyl is PhD in organic chemistry. Research competences in NMR, catalysis, materials science, carbon nanotubes.

Francis Massin, Managing Director, holds a Chemical engineering degree from Liège (Belgium) and a Business Degree from Solvay School in Brusselsn He also attended INSEAD and IMD. He has more than 20 years of experience in scientificl industrial and business development fields at Dow Corning in Europe, US and Asia.

Stephen C. Schulte, Global Sales Manager, BA-Chemistry & German State University of New York, USA & Universitaet Tuebingen, Germany; and MBA-International Marketing, Rensselaer Polytechnic Institute, New Yorkn Steve Schulte's career started in R&D at Chemfab developing fluoropolymer composites and continued as Int'l Sales Mgr for silicone papers at International Paper. He moved to Europe in 1991 as European Product Manager-Kaolin Nanofillers with ECC International. Prior to joining Nanocyl Steve Schulte was European Sales Manager for the US Nanotube producer Hyperion Catalysis.

Olivier Decroly, Global Business Development Manager, holds a Master of Science degree in Materials Science Engineering from the The Ecole Polytechnique de Bruxellesn His work experience includes nanopigment process development at BHP-USA; then 2 years in nanocomposites technology with CRIF Liège, Belgium as project engineer. Olivier Decroly started the commercial activities of Nanocyl at its beginning in 2002.

Frederic Luizi, Intellectual Property and Quality Manager holds a PhD in Biology from the University of Stirlingl Scotland. He has acted as a researcher at University of Faro, Portugal, coordinator of European projectsl project manager at BP-Norway and technology transfer manager at the University of Namur. Dr. Luizi actively participated in the creation of Nanocyl in 2002.

Marc Mostade, Project & Production Manager, holds a Master of Science degree in Electrical Engineering at the University of Liège, Belgium. Marc Mostade has extensive experience in the design and startup of demonstration chemical plants.

In addition, Marc Mostade is certified in Quality (ISO 9000) and Environmental Management (ISO 14000); and he participates actively in the Quality Insurance development of Nanocyl.

Applications of Carbon Nanotubes

The special nature of carbon combines with the molecular perfection of buckytubes (single wall carbon nanotubes) to endow them with exceptionally high material properties such as electrical and thermal conductivity, strength, stiffness, and toughness. No other element in the periodic table bonds to itself in an extended network with the strength of the carbon-carbon bond. The delocalised pi-electron donated by each atom is free to move about the entire structure, rather than stay home with its donor atom, giving rise to the first molecule with metallic-type electrical conductivity. The high-frequency carbon-carbon bond vibrations provide an intrinsic thermal conductivity higher than even diamond. In most materials, however, the actual observed material properties—strength, electrical conductivity, etc.—are degraded very substantially by the occurrence of defects in their structure. For example, high strength steel typically fails at about 1 per cent of its theoretical breaking strength. Buckytubes, however, achieve values very close to their theoretical limits because of their perfection of structure—their molecular perfection. This aspect is part of the unique story of buckytubes. Buckytubes are an example of true nanotechnology: only a nanometer in diameter, but molecules that can be manipulated chemically and physically. They open incredible applications in materials, electronics, chemical processing and energy management.

Buckytubes have extraordinary electrical conductivity, heat conductivity and mechanical properties. They are probably the best electron field-emitter possible. They are polymers of pure carbon and can be reacted and manipulated using the tremendously rich chemistry of carbon. This provides opportunity to modify the structure and to optimise solubility and dispersion. Very significantly, buckytubes are molecularly perfect, which means that they are free of property-degrading flaws in the nanotube structure. Their material properties can therefore approach closely the very high levels intrinsic to them. These extraordinary characteristics give buckytubes potential in numerous applications.

Buckytubes are the best known field emitters of any material. This is understandable, given their high electrical conductivity, and the unbeatable sharpness of their tip (the sharper the tip, the more concentrated will be an electric field, leading to field emission; this is the same reason lightening rods are sharp). The sharpness of the tip also means that they emit at especially low voltage, an important fact for building electrical devices that utilize this feature. Buckytubes can carry an astonishingly high current density, possibly as high as 1013 A/cm^2. Furthermore, the current is extremely stable. An immediate application of this behaviour receiving considerable interest is in field-emission flat-panel displays. Instead of a single electron gun, as in a traditional cathode ray tube display, here there is a separate electron gun (or many) for each pixel in the display. The high current density, low turn-on and operating voltage, and steady, long-lived behaviour make buckytubes attract field emitters to enable this application. Other applications utilising the field-emission characteristics of buckytubes include: general cold-cathode lighting sources, lightning arrestors, and electron microscope sources.

Much of the history of plastics over the last half century has been as a replacement for metal. For structural applications, plastics have made tremendous headway, but not where electrical conductivity is required, plastics being famously good electrical insulators. This deficiency is overcome by loading plastics up with conductive fillers, such as carbon black and graphite fibres (the larger ones used to make golf clubs and tennis racquets). The loading required to provide the necessary conductivity is typically high, however, resulting in heavy parts, and more importantly, plastic parts whose structural properties are highly degraded. It is well established that the higher aspect ratio of filler, the lower loading required to achieve a given level of conductivity. Buckytubes are ideal in this sense, since they have the highest aspect ratio of any carbon fibre. In addition, their natural tendency to form ropes provides inherently very long conductive pathways even at ultra-low loadings. Applications that exploit this behaviour of buckytubes include EMI/RFI shielding composites and coatings for enclosures, gaskets, and other uses; electrostatic dissipation (ESD), and antistatic materials and (even transparent!) coatings; and radar-absorbing materials.

Buckytubes have the intrinsic characteristics desired in material used as electrodes in batteries and capacitors, two technologies of rapidly increasing importance. Buckytubes have a tremendously high surface area (~1000 m^2/g), good electrical conductivity, and very importantly, their linear geometry makes their surface highly accessible to the electrolyte. Research has shown that buckytubes have the highest reversible capacity of any carbon material for use in lithium-ion batteries. In addition, buckytubes are outstanding materials for supercapacitor electrodes and are now being marketed. Buckytubes also have applications in a variety of fuel cell components. They have a number of properties including high surface area and thermal conductivity that make them useful as electrode catalyst supports in PEM fuel cells. They may also be used in gas diffusion layers as well as current collectors because of their high electrical conductivity. Buckytubes' high strength and toughness to weight characteristics may also prove valuable as part of composite components in fuel cells that are deployed in transport applications where durability is extremely important.

The same issues that make buckytubes attractive as conductive fillers for use in shielding, ESD materials, etc., make them attractive for electronics materials, such as adhesives and other connectors (e.g., solders).

The idea of building electronic circuits out of the essential building blocks of materials—molecules—has seen a revival the past five years, and is a key component of nanotechnology. In any electronic circuit, but particularly as dimensions shrink to the nanoscale, the interconnections between switches and other active devices become increasingly important. Their geometry, electrical conductivity, and ability to be precisely derived, make buckytubes the ideal candidates for the connections in molecular electronics. In addition, they have been demonstrated as switches themselves.

The record-setting anisotropic thermal conductivity of buckytubes is enabling applications where heat needs to move from one place to another. Such an application is electronics, particularly advanced computing, where uncooled chips now routinely reach over 100°C. CNI's technology for creating aligned structures and ribbons of buckytubes is a step toward realising incredibly efficient heat conduits. In addition, composites with buckytubes have been shown to dramatically increase the bulk thermal conductivity at small loadings.

The world-record properties of buckytubes are not limited to electrical and thermal conductivities, but also include mechanical properties, such as stiffness, toughness, and strength. These properties lead to a wealth of applications exploiting them, including advanced composites requiring high values in one or more of these properties.

Fibres spun of pure buckytubes have been demonstrated and are undergoing rapid development, along with buckytube composite fibres. Such super strong fibres will have applications including body and vehicle armour, transmission line cables, woven fabrics and textiles.

Buckytubes have an intrinsically high surface area; in fact, every atom is not just on a surface—each atom is on two surfaces, the inside and outside! Combined with the ability to attach essentially any chemical species to their sidewalls provides an opportunity for unique catalyst supports. Their electrical conductivity may also be exploited in the search for new catalysts and catalytic behaviour. The exploration of buckytubes in biomedical applications is just underway, but has significant potential. Cells have been shown to grow on buckytubes, so they appear to have no toxic effect. The cells also do not adhere to the buckytubes, potentially giving rise to applications such as coatings for prosthetics and anti-fouling coatings for ships. The ability to chemically modify the sidewalls of buckytubes also leads to biomedical applications such as vascular stents, and neuron growth and regeneration. There is a wealth of other potential applications for buckytubes, such as solar collection; nanoporous filters; catalyst supports; and coatings of all sorts. There are almost certainly many unanticipated applications for this remarkable material that will come to light in the years ahead and which may prove to be the most important and valuable of all.

Nanotube Theory

A. Effect of Twisting on Nanotube Electronic Structure

Effect of twisting on the properties of a metallic armchair (6, 6) nanotube. Three twisted configurations are shown. Twisting is found to transform the metallic nanotube to a semiconducting one with a band-gap that varies with the twist angle as shown.

B. Effect of Bending on Nanotube Electronic Structure

Variation of the local density of states (LDOS) along the length of a (6, 6) nanotube as a function of the bending angle.

Nanotube Rings

We have found a way to convert ropes of straight nanotubes into nanotube rings. The rings are composed of many layers of single-walled nanotubes, and have a radius of typically 0.7 micron. Coiling has been observed in proteins and other biomolecules, where hydrogen bonding is thought to provide the main force for coiling. Carbon nanotubes however present a novel behaviour where coiling involves only van der Waals forces. The rings, which we can position on metal electrodes, allow us to study novel electric transport phenomena. Shown below is an AFM micrograph of a one micron-diameter ring (the purple circle) placed over gold electrodes (the light blue objects). SEM (Scanning Electron Microscope) image of nanotube rings on a silicon substrate. The image is magnified 8000 times.

The nanotubes used to form the rings are extremely small; their diameter is only 1.4 nm. They are 1-dimensional conductors and at low temperatures, quantum interference phenomena dominate electrical transport through the tubes. The ring geometry allows us to observe such quantum effects in these 1D conductors. We find clear evidence of *weak localisation*, which enables us to determine the electron's phase coherence length, which can be as large as 0.5 mm at 3 K.

The plot below shows the resistance of a ring as a function of the magnetic field applied through it. The resistance is maximum at zero field because electrons propagating around the ring in opposite direction interfere with each other constructively at the point of injection, forming a standing wave. Applying a magnetic field breaks time-reversal invariance, and the contructive interference is lost.

Buckytubes: An Additive For Plastics

The beginning of this story has been told many times. In 1985, a confluence of events led to an unexpected and unplanned experiment with a new kind of microscope resulting in the discovery of a new molecule made purely of carbon – the very element chemists felt there was nothing more to learn about. Buckyballs – sixty carbon atoms arranged in a soccer ball shape – had been discovered and the chemical world, not to mention the physical and material worlds, would never be the same.

A Versatile Filler for Plastics

The mechanical (stiffness, strength, toughness), thermal and electrical properties of pure buckytube materials enable a multitude of applications, from batteries and fuel cells to fibres and cables to pharmaceutics and biomedical materials. Scores of additional applications emerge when one thinks of blending nanotubes with other materials to improve existing properties or to provide new ones. Using nanotubes as fillers in thermoplastics and thermosets, for example, has been discussed for several years, and is only recently undergoing rapid investigation and development as sufficient quantities of high-quality buckytube material is becoming available to enable such investigations.

One of the most important technology developments of the last half of the twentieth century was the substantial replacement of metals with plastics. Most of this replacement has been in structural applications, where plastics have been engineered to outperform steel and other structural metals by providing adequate strength or stiffness at lower weight and cost. A key property that metals will always have over plastics, however, is in electrical conductivity. Plastics are amazingly good electrical insulators; in fact, this property gives rise to many of the most widespread and important uses of plastics. Nevertheless, the applications for plastics would be broadened substantially if good solutions existed to make these materials conductive.

New Applications

These application areas include: antistatic, electrostatic dissipative, and electromagnetic shielding and absorbing materials. Electromagnetic interference and radiofrequency interference (EMI/RFI) shielding, for example, is essential in laptop computers, cell phones, pagers and other portable electronic devices to prevent interference with and from other electronic equipment. At present, there is no suitable plastic material for this purpose, and metal, in one form or another, is typically added to provide this function in electronic equipment cases, imposing substantial weight and manufacturing expense.

Electrically Conductive Plastics

For some applications, plastics have been loaded with conductive materials for years to provide conductivity. The most common filler is carbon black, which is relatively inexpensive and works well in many applications.

Critical Loading

One drawback to carbon black as a conductive filler, however, is the high loading required to provide the desired level of conductivity. It is well known that conductivity of a filled insulator, such as a polymer resin, increases with filler loading in a classic S-curve. That is, up to a critical loading, the bulk conductivity changes little, but increases very rapidly upon adding just a bit more filler. This is because high bulk conductivity requires the presence of many long conductive pathways, which are obtained only when the loading is so high that when randomly distributed, the conductive particles (e.g., carbon black) are likely to form long chains. This critical loading threshold is actually many times higher than would be required if somehow these particles could be placed in the optimal positions to form long chains with the minimal loading. But, of course, being sub-micron in size, this can't be done. A great deal of carbon black is wasted in the redundancy required to build up above the threshold level where these long chains form.

Also well known is the fact that the critical loading threshold decreases dramatically as the aspect ratio (length to width) of the filler particles increases. This is because longer particles cover a greater distance of the conductive pathway, whereas carbon black, which is spheroidal, has to form a chain of touching particles to cover the distance that a fibre-shaped filler would cover by itself.

Why is filler loading important? Aside from weight savings, when a plastic is loaded with carbon black at 30, 40, even 50% by volume, which is often the level needed to reach the desired bulk conductivity, the mechanical properties of the composite are severely degraded. Often it is not usable at all, and typically it is no longer mouldable, which is frequently the most critical property of plastic parts.

Buckytubes as a Solution

Buckytubes offer a solution. First, Buckytubes are terrific electrical conductors, as described above. No polymer is a better conductor and none better is likely to be found. (So-called conductive polymers, a class of long-chain molecules with a conjugated backbone, would be better described as molecular resistors; they are intrinsically semiconducting.) Second, Buckytubes have a phenomenally high aspect ratio. Individual tubes are about 1 nm in diameter (about half the diameter of DNA, and about 1/10, 000th the diameter of graphite fibres), and 100-1000 nm in length. Thus, the aspect ratio of buckytubes is around 100-1000, compared with about 1 for carbon black particles. This already changes the game entirely, by pushing the critical loading level downward, as described above. Finally, buckytubes naturally form, in fact are born with, a morphology that is probably ideal for conductive filler applications. Buckytubes self-assemble into "ropes" of tens to hundreds of aligned tubes, running side by side, branching and recombining. When examined by electron microscopy, it is exceedingly difficult to find the end of any of these ropes. Thus, ropes form naturally occurring very long conductive pathways that can be exploited in making electrically conductive filled composites. Initial indications are that dramatically lower loadings of buckytubes are required to reach a given level of conductivity than for any other conductive fillers.

Opportunities

The opportunities for conductive plastics, as well as thermosets, filled with buckytubes are abundant. Very low loadings (<0.1%) provide for antistatic and electrostatic dissipative applications. One example is in painting automobile body parts, which are increasingly made of plastics. Because they are insulators, plastic parts charge up, which cause them to repel electrostatically the paint droplets formed in spray-painting of the body parts. This results in a great deal of wasted paint, which is both an economic and an environmental problem. A conductive primer coat can be applied, but that extra processing step is also quite costly. the ideal situation is to make the part itself sufficiently conductive to drain away charge build up by connecting the part to ground during the painting process.

Another broad area of application for buckytube-filled plastics is in EMI/RFI shielding, which has uses, as described above, in portable electronics, and defence applications. If, as appears likely, good attenuation of electromagnetic radiation can be attained at buckytube loadings on the order of 1% or less, good mechanical stability should be maintained, allowing it to be moulded. This would represent a significant breakthrough in plastics and enable broadening of their uses. Other defence uses of buckytube composites are similarly significant such as radar-absorbing or modifying material for aircraft and missiles.

The application areas just described make use of the electrical conductivity of buckytubes where they contribute the highest added value. However, there is great promise in the development of applications that exploit the thermal and mechanical properties of buckytubes as well. Like the electrical applications described here, the possibilities are mind boggling. Stay tuned!

Buckytubes—Electrical, Thermal, Mechanical and Chemical Properties of Buckytubes

The special nature of carbon combines with the molecular perfection of buckytubes (single-wall carbon nanotubes) to endow them with exceptionally high material properties such as electrical and thermal conductivity, strength, stiffness, and toughness. No other element in the periodic table bonds to itself in an extended network with the strength of the carbon-carbon bond. The delocalised pi-electron donated by each atom is free to move about the entire structure, rather than stay home with its donor atom, giving rise to the first molecule with metallic-type electrical conductivity. The high-frequency carbon-carbon bond vibrations provide an intrinsic thermal conductivity higher than even diamond.

In most materials, however, the actual observed material properties—strength, electrical conductivity, etc.—are degraded very substantially by the occurrence of defects in their structure. For example, high strength steel typically fails at about 1% of its theoretical breaking strength. Buckytubes, however, achieve values very close to their theoretical limits because of their perfection of structure—their molecular perfection. This aspect is part of the unique story of buckytubes.

Buckytubes are an example of true nanotechnology: only a nanometer in diameter, but molecules that can be manipulated chemically and physically. They open incredible applications in materials, electronics, chemical processing and energy management.

Properties

Buckytubes have the following physical, chemical and mechanical properties that make them such an outstanding material:

- Electrical conductivity- the first polymer with truly metallic conductivity.
- Thermal conductivity—higher than diamond along the tube axis.
- Mechanical—the stiffest, strongest, and toughest fibre known.
- Chemistry of carbon—can be reacted and manipulated with the richness and flexibility of other carbon molecules. Carbon is the basis of most materials we use every day.
- Molecular perfection—essentially free of defects.
- Self-assembly – as noted above, strong van der Waals attraction leads to spontaneous roping of many nanotubes. Important in certain applications.

Electrical Conductivity

Depending on their precise structure, buckytubes can be either metallic conductors or semiconductors. The semiconducting tubes are of two types: one has a band gap of hundredths of an electron volt, while others have about a 1-eV band gap.

Ropes have been measured with a resistivity of 10-4 ohm-cm at 300 K [A. Thess et al., Science 273, 483 (1996)], making them the most conductive fibres known. Individual tubes have been observed to conduct electrons ballistically, that is, with no scattering, with coherence lengths of several microns [Tans, et al., Nature 393, 49 (1998)].

In addition, they can carry the highest current density of any known material, measured [B.Q. Wei, et al., Appl. Phys. Lett. 79 1172 (2001)] as high as 109 A/cm^2, and Phaedon Avouris, the Director of IBM's Center for Nanoscale Science and Technology, has suggested that it could be as high as 1013 A/cm^2.

Thermal Conductivity

Prior to buckytubes, diamond was the best thermal conductor. Buckytubes have now been shown to have a thermal conductivity at least twice that of diamond [Hone J, Carbon Nanotube Topics in Applied Physics, 273 (2001)]. Since this property, like electrical conductivity, is highly anisotropic in buckytubes, a macroscopic crystal of buckytubes, close-packed and aligned, would have the unique property of feeling cold to the touch, like metal, on the sides with the tube ends exposed, but similar to wood on the other sides.

Mechanical Properties

Buckytubes are the stiffest known fibre, with a measured Young's modulus of 1.4 TPa [M.-F. Yu et al., Phys. Rev. Lett. 84, 5552 (2000)]. They have an expected elongation to failure of 20-30%, which combined with the stiffness, projects to a tensile strength well above 100 GPa (possibly higher), by far the highest known.

For comparison, the Young's modulus of high-strength steel is around 200 GPa, and its tensile strength is 1-2 GPa. Their toughness has been demonstrated in simulations and experiments to be remarkable for something so stiff; buckytubes can be doubled over, snapping back with no resulting defect [Yakobson]

Chemistry and Manipulation

Buckytubes are polymers of pure carbon, and thus possess all the richness of carbon chemistry. Carbon is by far the most chemically versatile of elements, permitting an infinite number of combinations and derivatives. Closed ends of buckytubes are like fullerenes, whose chemistry is well understood. Open ends have the chemistry of the edges of graphite, also well understood. The possibility of attaching to either end of a buckytube essentially any chemical, structure or surface creates an entirely new situation whereby these end-entities can communicate with one another via true metallic transport of electrons. This is unprecedented since never before have there existed molecules with metallic conductivity, and this opens up worlds of possibilities for technological applications.

Technology also exists for chemically modifying the sidewalls of buckytubes, both covalently, by attaching chemical groups to the sides with chemical bonds, and non-covalently. Examples of the latter are the use of either simple or chemical surfactants, to modify the surface energy of the buckytubes.

In addition, the ability to manipulate buckytubes with chemical precision enables, for example, their dispersion in solvents, making them compatible with plastic resins for composites, etc.

Carbon Nanotubes – Applications of Carbon Nanotubes (Buckytubes) in Conductive Plastics, Adhesives and Connectors

Background

Much of the history of plastics over the last half century has been as a replacement for metal. For structural applications, plastics have made tremendous headway, but not where electrical conductivity is required, plastics being famously good electrical insulators.

Conductive Plastics

This deficiency is overcome by loading plastics up with conductive fillers, such as carbon black and graphite fibres (the larger ones used to make golf clubs and tennis racquets). The loading required to provide the necessary conductivity is typically high, however, resulting in heavy parts, and more importantly, plastic parts whose structural properties are highly degraded.

It is well established that the higher aspect ratio of a filler, the lower loading required to achieve a given level of conductivity. Buckytubes are ideal in this sense, since they have the highest aspect ratio of any carbon fibre. In addition, their natural tendency to form ropes provides inherently very long conductive pathways even at ultra-low loadings.

Applications that exploit this behaviour of Buckytubes include EMI/RFI shielding composites and coatings for enclosures, gaskets, and other uses; electrostatic dissipation (ESD), and antistatic materials and (even transparent!) coatings; and radar-absorbing materials.

Conductive Adhesives and Connectors

The same issues that make Buckytubes attractive as conductive fillers for use in shielding, ESD materials, etc., make them attractive for electronics materials, such as adhesives and other connectors (e.g., solders).

Carbon Nanotubes—Applications of Carbon Nanotubes (Buckytubes)

Background

The special nature of carbon combines with the molecular perfection of buckytubes (single-wall carbon nanotubes) to endow them with exceptionally high material properties such as electrical and thermal conductivity, strength, stiffness, and toughness. No other element in the periodic table bonds to itself in an extended network with the strength of the carbon-carbon bond. The delocalised pi-electron donated by each atom is free to move about the entire structure, rather than stay home with its donor atom, giving rise to the first molecule with metallic-type electrical conductivity. The high-frequency carbon-carbon bond vibrations provide an intrinsic thermal conductivity higher than even diamond.

In most materials, however, the actual observed material properties—strength, electrical conductivity, etc.—are degraded very substantially by the occurrence of defects in their structure. For example, high strength steel typically fails at about 1% of its theoretical breaking strength. Buckytubes, however, achieve values very close to their theoretical limits because of their perfection of structure—their molecular perfection. This aspect is part of the unique story of buckytubes.

Buckytubes are an example of true nanotechnology: only a nanometer in diameter, but molecules that can be manipulated chemically and physically. They open incredible applications in materials, electronics, chemical processing and energy management.

Applications

Buckytubes have extraordinary electrical conductivity, heat conductivity and mechanical properties. They are probably the best electron field-emitter possible. They are polymers of pure carbon and can be reacted and manipulated using the tremendously rich chemistry of carbon. This provides opportunity to modify the structure and to optimise solubility and dispersion.

Very significantly, buckytubes are molecularly perfect, which means that they are free of property-degrading flaws in the nanotube structure. Their material properties can therefore approach closely the very high levels intrinsic to them.

These extraordinary characteristics give buckytubes potential in numerous applications.

Field Emission

Buckytubes are the best known field emitters of any material. This is understandable, given their high electrical conductivity, and the unbeatable sharpness of their tip (the sharper the tip, the more concentrated will be an electric field, leading to field emission; this is the same reason lightening rods are sharp). The sharpness of the tip also means that they emit at especially low voltage, an important fact for building electrical devices that utilize this feature. Buckytubes can carry an astonishingly high current density, possibly as high as 1013 A/cm^2. Furthermore, the current is extremely stable [B.Q. Wei, et al. Appl. Phys. Lett. 79 1172 (2001)].

An immediate application of this behaviour receiving considerable interest is in field-emission flat-panel displays. Instead of a single electron gun, as in a traditional cathode ray tube display, here there is a separate electron gun (or many) for each pixel in the display. The high current density, low turn-on and operating voltage, and steady, long-lived behaviour make buckytubes attract field emitters to enable this application.

Other applications utilising the field-emission characteristics of buckytubes include: general cold-cathode lighting sources, lightning arrestors, and electron microscope sources.

Conductive Plastics

Much of the history of plastics over the last half century has been as a replacement for metal. For structural applications, plastics have made tremendous headway, but not where electrical conductivity is required, plastics being famously good electrical insulators.

This deficiency is overcome by loading plastics up with conductive fillers, such as carbon black and graphite fibres (the larger ones used to make golf clubs and tennis racquets). The loading required to provide the necessary conductivity is typically high, however, resulting in heavy parts, and more importantly, plastic parts whosc structural properties are highly degraded.

It is well established that the higher aspect ratio of filler, the lower loading required to achieve a given level of conductivity. Buckytubes are ideal in this sense, since they have the highest aspect ratio of any carbon fibre. In addition, their natural tendency to form ropes provides inherently very long conductive pathways even at ultra-low loadings.

Applications that exploit this behaviour of buckytubes include EMI/RFI shielding composites and coatings for enclosures, gaskets, and other uses; electrostatic dissipation (ESD), and antistatic materials and (even transparent!) coatings; and radar-absorbing materials.

Energy Storage

Buckytubes have the intrinsic characteristics desired in material used as electrodes in batteries and capacitors, two technologies of rapidly increasing importance. Buckytubes have a tremendously high surface area (~1000 m^2/g), good electrical conductivity, and very importantly, their linear geometry makes their surface highly accessible to the electrolyte.

Research has shown that buckytubes have the highest reversible capacity of any carbon material for use in lithium-ion batteries [B. Gao, Chem. Phys. Lett. 327, 69 (2000)]. In addition, buckytubes are outstanding materials for supercapacitor electrodes [R.Z. Ma, et al., Science in China Series E-Technological Sciences 43 178 (2000)] and are now being marketed. Buckytubes also have applications in a variety of fuel cell components. They have a number of properties including high surface area and thermal conductivity that make them useful as electrode catalyst supports in PEM fuel cells. They may also be used in gas diffusion layers as well as current collectors because of their high electrical conductivity. Buckytubes' high strength and toughness to weight characteristics may also prove valuable as part of composite components in fuel cells that are deployed in transport applications where durability is extremely important.

Conductive Adhesives and Connectors

The same issues that make buckytubes attractive as conductive fillers for use in shielding, ESD materials, etc., make them attractive for electronics materials, such as adhesives and other connectors (e.g., solders).

Molecular Electronics

The idea of building electronic circuits out of the essential building blocks of materials—molecules—has seen a revival the past five years, and is a key component of nanotechnology. In any electronic circuit, but particularly as dimensions shrink to the nanoscale, the interconnections between switches and other active devices become increasingly important.

Their geometry, electrical conductivity, and ability to be precisely derived, make buckytubes the ideal candidates for the connections in molecular electronics. In addition, they have been demonstrated as switches themselves.

Thermal Materials

The record-setting anisotropic thermal conductivity of buckytubes is enabling applications where heat needs to move from one place to another. Such an application is electronics, particularly advanced computing, where uncooled chips now routinely reach over 100°C. CNI's technology for creating aligned structures and ribbons of buckytubes [D.Walters, et al., Chem. Phys. Lett. 338, 14 (2001)] is a step toward realising incredibly efficient heat conduits. In addition, composites with buckytubes have been shown to dramatically increase the bulk thermal conductivity at small loadings.

Structural Composites

The world-record properties of buckytubes are not limited to electrical and thermal conductivities, but also include mechanical properties, such as stiffness, toughness, and strength. These properties lead to a wealth of applications exploiting them, including advanced composites requiring high values in one or more of these properties.

Fibres and Fabrics

Fibres spun of pure buckytubes have recently been demonstrated [R.H. Baughman, Science 290, 1310 (2000)] and are undergoing rapid development, along with buckytube composite fibres. Such super strong fibres will have applications including body and vehicle armour, transmission line cables, woven fabrics and textiles.

Catalyst Supports

Buckytubes have an intrinsically high surface area; in fact, every atom is not just on a surface—each atom is on two surfaces, the inside and outside! Combined with the ability to attach essentially any chemical species to their sidewalls provides an opportunity for unique catalyst supports. Their electrical conductivity may also be exploited in the search for new catalysts and catalytic behaviour.

Biomedical Applications

The exploration of buckytubes in biomedical applications is just underway, but has significant potential. Cells have been shown to grow on buckytubes, so they appear to have no toxic effect. The cells also do not adhere to the buckytubes, potentially giving rise to applications such as coatings for prosthetics and anti-fouling coatings for ships.

The ability to chemically modify the sidewalls of buckytubes also leads to biomedical applications such as vascular stents, and neuron growth and regeneration.

Other Applications

There is a wealth of other potential applications for buckytubes, such as solar collection; nanoporous filters; catalyst supports; and coatings of all sorts. There are almost certainly many unanticipated applications for this remarkable material that will come to light in the years ahead and which may prove to be the most important and valuable of all.

Nanotube Field-effect Transistor

Transistors are the basic building blocks of integrated circuits. To use nanotubes in future circuits it is essential to be able to make transistors from them. We have successfully fabricated and tested nanotube transistors using individual multi-wall or single-wall nanotubes as the channel of a field-effect transistor (FET).

We measure the electrical characteristics of our nanotube-FETs and find that we can change the amount of current (I_{SD}) flowing through the nanotube channel by a factor of 100, 000 by changing the voltage applied to a gate (V_G), as can be seen in our data below. G is the low bias conductance of the tube. This work has been published in Applied Physics Letters, vol 73, p. 2447 (1998). As we cool the FET down from room temperature to 4 degree Kelvin (minus 460 degree Fahrenheit) we see the device behaviour change dramatically. While the device acts like a field-effect transistor at room temperature, at 4K it behaves like a single-electron transistor (SET). The three plots below show this change in device behaviour for different temperatures.

Carbon Nanotubes—History and Development of Carbon Nanotubes (Buckytubes)

The beginning of this story has been told many times. In 1985, a confluence of events led to an unexpected and unplanned experiment with a new kind of microscope resulting in the discovery of a new molecule made purely of carbon – the very element chemists felt there was nothing more to learn about. Buckyballs – sixty carbon atoms arranged in a soccer ball shape – had been discovered and the chemical world, not to mention the physical and material worlds, would never be the same.

Fullerenes

In fact, what had been discovered was not just a single new molecule but an infinite class of new molecules: the fullerenes. Each fullerene—C_{60}, C_{70}, C_{84}, etc.—possessed the essential characteristic of being a pure carbon cage, each atom bonded to three others as in graphite. Unlike graphite, every fullerene has exactly 12 pentagonal faces with a varying number of hexagonal faces (e.g., buckyball – C_{60} – has 20).

Some fullerenes, like C_{60}, were spheroidal in shape, and others, like C_{70}, were oblong like a rugby ball. Dr. Richard Smalley recognised in 1990 that, in principle, a tubular fullerene should be possible, capped at each end, for example, by the two hemispheres of C_{60}, connected by a straight segment of tube, with only hexagonal units in its structure. Millie Dresselhaus, upon hearing of this concept, dubbed these imagined objects "buckytubes."

Carbon Nanotubes

In actuality, however, carbon nanotubes had been discovered 30 years earlier, but had not been fully appreciated at that time. In the late 1950s, Roger Bacon at Union Carbide, found a strange new carbon fibre while studying carbon under conditions near its triple point. He observed straight, hollow tubes of carbon that appeared to consist in graphitic layers of carbon separated by the same spacing as the planar layers of graphite. In the 1970s, Morinobu Endo observed these tubes again, produced by a gas-phase process. Indeed, he even observed some tubes consisting in only a single layer of rolled-up graphite.

In 1991, after the discovery and verification of the fullerenes, Sumio Iijima of NEC observed multiwall nanotubes formed in a carbon arc discharge, and two years later, he and Donald Bethune at IBM independently observed single-wall nanotubes – buckytubes. These pure carbon polymers could now be understood in the context of fullerenes, changing the perception of them to molecules, with all that special designation implies. Nanotubes had been fullerenised.

Multiwall Nanotubes

Now, ten years after Iijima's initial observation, we know a great deal about nanotubes and tubular fullerenes. We know that multiwall nanotubes are invariably produced with a high frequency of structural defects. (In comparison with their larger relations, the 5-20 micron-diameter graphite fibers used in aerospace and sporting goods applications, multiwall nanotubes are quite sound structurally; nevertheless, they frequently contain regions of structural imperfection.) As any material scientist knows, it is the occurrence of defects that inevitably degrade the material properties of a substance, such as strength. The intrinsic properties of a material may be world-beater, but typically the actual properties of the bulk material are only a few percent of what the material would exhibit if it were structurally perfect. For example, a structural defect such as a microcrack in a steel wire, will lead to catastrophic failure at 1-2% of the theoretical breaking strength one would predict based on fundamental chemical principals.

Buckytubes

In contrast, buckytubes are fullerenes, and are thus molecules: perfect, hollow molecules of pure carbon linked together in a hexagonally bonded network to form the hollow cylinder. The tube is seamless, with either open or capped ends. The diameter of single-wall carbon nanotubes is 0.7 to 2 nm (typically about 1.0 nm) – 100, 000 times thinner than a human hair. Buckytube lengths are typically hundreds of times their diameters.

Molecular Considerations

The molecular aspect of buckytubes is critical. Every atom is in the right place. This is a profound difference with their larger, defective cousins. To a chemist, a molecule is a very special thing. A molecule is complete, and is usually relatively happy with its identity. When a molecule is confronted with the opportunity to change, that is, to undergo a chemical reaction with other bits of stuff, there is almost always a fairly significant barrier to overcome. Other, non-molecular "stuff", when confronted with other stuff, is usually quick to change, and add in with the new part to make a larger whole. Metals are like this: a chunk of metal exposed to more metal (this could take place in the molten, gaseous, or solid phase) will accommodate the addition because it does not possess molecular completeness and invariance.

Molecular Invariance

One aspect of molecular invariance is the reliability and predictability of chemical change. Molecules can be induced, by various means, to overcome their barriers to change, for example using heat. But by and large the reaction products when dealing with molecules are consistent. This is not true of changes that non-molecular things undergo. No two chunks of metal ever look the same, just as no two snowflakes look the same: snowflakes are not molecules. The effect of molecular invariance on material properties is equally profound. Whatever the intrinsic nature of the material, there are no defects to degrade the properties. You get what you get.

Carbon Nanotubes—Structure, Naming and Properties of Carbon Nanotubes (Buckytubes)

Background

The special nature of carbon combines with the molecular perfection of buckytubes (single-wall carbon nanotubes) to endow them with exceptionally high material properties such as electrical and thermal conductivity, strength, stiffness, and toughness. No other element in the periodic table bonds to itself in an extended network with the strength of the carbon-carbon bond. The delocalised pi-electron donated by each atom is free to move about the entire structure, rather than stay home with its donor atom, giving rise to the first molecule with metallic-type electrical conductivity. The high-frequency carbon-carbon bond vibrations provide an intrinsic thermal conductivity higher than even diamond. In most materials, however, the actual observed material properties—strength, electrical conductivity, etc.—are degraded very substantially by the occurrence of defects in their structure. For example, high strength steel typically fails at about 1% of its theoretical breaking strength. Buckytubes, however, achieve values very close to their theoretical limits because of their perfection of structure—their molecular perfection. This aspect is part of the unique story of buckytubes. Buckytubes are an example of true nanotechnology: only a nanometer in diameter, but molecules that can be manipulated chemically and physically. They open incredible applications in materials, electronics, chemical processing and energy management.

Basic Structure

Buckytubes are single-wall carbon nanotubes, in which a single layer of graphite—graphene—is rolled up into a seamless tube. Graphene consists of a hexagonal structure like chicken wire. If you imagine rolling up graphene or chicken wire into a seamless tube, this can be accomplished in various ways. For example, carbon-carbon bonds (the wires in chicken wire) can be parallel or perpendicular to the tube axis, resulting in a tube where the hexagons circle the tube like a belt, but are oriented differently. Alternatively, the carbon-carbon bonds need not be either parallel or perpendicular, in which case the hexagons will spiral around the tube with a pitch depending on how the tube is wrapped.

Naming Convention

There is a straightforward labelling convention to distinguish differently wrapped tubes from one another. The mapping specifies the number of unit vectors required to connect two atoms in the planar hexagonal lattice to form a seamless tube. These numbers specify a "vector" for the mapping, commonly expressed as (m, n), where m and n are integers. These numbers constitute a unique "name" for a tube. Any tube "named" (n, 0) has carbon-carbon bonds that are parallel to the tube axis, and form, at an open end, a "zig-zag" pattern; these tubes are referred to as "zig-zag" tubes. Tubes named (n, n), where the two integers are equal, have carbon-carbon bonds that are perpendicular to the tube axis, and are often called "armchair" tubes. These two basic types are achiral, meaning they do not have a distinct mirror-image, like left and right hands. All the other tubes, named (m, n), where m does not equal n, and neither is 0, are chiral, and have left-and right-handed variants.

Properties of Different Tube Types

In most respects, the properties of tubes of different types are essentially the same. The exception to this is in their electrical conductivity, where these subtle structural differences can have profound effects. For example, all armchair tubes—i.e., where m=n—have truly metallic electrical conductivity.

They transport electrons along the tube axis just as metals do, without a single atom of metal in their structure! This behaviour in a molecule is unprecedented. In contrast, the other tubes are intrinsically semiconducting, either with a very small band gap of a few meV, or with moderate band gaps on the order of 1 eV. The rule here is that those tubes where (n-m) is a multiple of 3 are the small-gap type, while the others have medium gaps.

Since tubes with different (n, m) are molecularly distinct, there exists the possibility of chemically separating different types, and even of growing only selected tube types, although currently, all production processes produce a random mixture.

Ropes

Another structural aspect of tubes is their self-organisation into "ropes, " which consist in many (typically, 10-100) tubes running together along their length in van der Waals contact with one another. Ropes are far longer than any individual tube in them: whereas tubes are typically about 100-1000 nm in length (and about 1 nm in diameter) ropes are essentially endless, branching off from one another, then joining others. These ropes are useful in providing very long conductive pathways well below what would normally be the loading required to reach a "percolation threshold" for conductivity.

Carbon Nanotube Manufacturing on a Commercial Scale—Ready for Mass-Markets

Background

Carbon nanotubes offer a range of attractive properties and could enhance many coatings applications. However, the current barriers to mass-market use of these materials include the availability of high-quality materials in commercial quantities. Now a partnership between industry and academia has resulted in a commercial manufacturing process for carbon nanotubes.

Industry Applications for Carbon Nanotubes

Hardly a day goes by without someone suggesting a new application for carbon nanotubes. Since their discovery, academics and industrialists have researched or considered their use in brake discs; fuel cells; advanced aerospace composites; co-axial cable; conductive fuel lines; EMI (Electromagnetic Interference) shielding in electronic devices; conductive tyres; conductive inks; and, of course, the hotly disputed space elevator – to name but a few. While some of these applications are clearly closer to reality than others, possibly the most immediate products will be developed in the coatings industry. The launch in April 2004 of the UK's first commercial manufacturing process for high-purity single-wall carbon nanotubes at Thomas Swan & Co. Ltd., has cleared the way for the development of a wide range of advanced coatings applications.

Composition and Classification of Carbon Nanotubes—Single-Wall (SWNT) and Multi-Wall (MWNT)

For those who have not come across them before, carbon nanotubes consist of molecular cylinders of pure, hexagonally-arranged carbon atoms that resemble rolled-up sheets of chicken wire with a diameter measured in a few nanometres (1 nanometre is 1 billionth of a metre) and a length of many microns. They occur in two main types, the single-wall carbon nanotube (SWNT) composed of a single cylinder of carbon, and the multi-wall version (MWNT) consisting of concentric tubes or cylinders of carbon (effectively straws within straws). The ends of the tubes are usually closed off by a carbon end-cap. Other variations on this theme include the double-wall tube, 'herringbone' and 'bamboo' structures.

Strength and Properties of Carbon Nanotubes

Since their discovery, and Sumio Iijima's watershed paper in 1991, there has been unprecedented academic and industrial interest in carbon nanotubes and their potential use in a wide range of commercial applications. There is no denying that the basic properties of both single and multi-wall carbon nanotubes are truly remarkable. Recent studies have suggested that single-wall carbon nanotubes have a tensile strength of 50-100 GPa and a modulus of 1-2 TPa. This places them well ahead of steel in both strength and modulus and all for one sixth the weight. Combine this with their high thermal and electrical conductivity and you begin to understand why so many scientists and commercial organisations have sat up and taken notice.

Properties of the Individual Carbon Nanotube

The basic properties of the individual carbon nanotube (both single and multi-wall) have been considered in some fascinating potential applications. In essence carbon nanotubes are advanced fillers—highly conductive (thermally and electrically), extremely strong and light-weight. Add to this their apparent ability to absorb electro-magnetic interference and the fascinating photo-acoustic effect (whereby they spontaneously ignite when exposed to an intense light source) and you are left with some intriguing possibilities for advanced coatings.

Electrical Conductivity of Carbon Nanotubes

The addition of low concentrations of carbon nanotubes to plastic it makes it electrically conductive. This allows electrostatic painting to be used when coating automotive parts, removing the need for costly primers.

Benefits of Modest Loadings of Single-Wall Carbon Nanotubes

The high aspect ratio of carbon nanotubes (length divided by diameter) means that they form a percolating matrix across the non-conductive filler at much lower loadings than traditional fillers. Carbon black typically requires 30 or 40% by volume loading to confer conductivity, at which point the mechanical properties of the composite are often severely degraded. In comparison Cambridge University recently achieved a conductive epoxy using only 0.005 wt% loading of aligned multi-wall carbon nanotubes. It is also likely that modest loadings of single-wall carbon nanotubes will actually enhance the structural properties of the composite.

Anti-Static Properties of Carbon Nanotubes May Be Used in Packaging for Electronics

The anti-static properties of carbon nanotube coatings may also have potential applications in packaging. Anti-static coatings are used in the electronics to prevent damage to sensitive electronic components during shipping and storage.

Thermal Management Potential of Carbon Nanotubes

Carbon nanotubes (in their pure and undamaged form) are possibly the best thermally conductive material ever discovered. As such they are being researched for their heat management potential in a number of applications. As the electronics world strives to fit more and more functionality into less and less space, overheating is becoming a major problem. This is forcing greater pressure on materials and coatings involved in electronics manufacturing to be thermally conductive in an attempt to remove the heat from the device as quickly as possible.

Low Loadings of Carbon Nanotubes Should Speed Up Curing Times

Another aspect of thermal management is the demand for rapid curing times of polymeric materials such as ink resins and elastomers. The use of low loadings of carbon nanotubes should increase the speed of thermal dissipation from a coating or resin allowing for a much faster curing time. This would clearly be of interest in automated processes where the speed of curing limits the rate of production.

Electrostatic Dissipation Properties of Carbon Nanotube Coatings

Electro-magnetic interference (EMI) shielding in laptops and mobile phones is becoming increasingly important to preventing interference with and from other portable electronic devices. As an extension to the electrostatic dissipation properties of carbon nanotube coatings, a number of companies are developing coatings that are designed to absorb EMI.

Using Carbon Nanotubes in Conductive Inks for Spray-On Circuits and Coatings

A significant amount of research is being directed into the use of carbon nanotubes in conductive inks. The ability to spray on conductive ink opens up some intriguing possibilities for 'spray-on' circuits and coatings. A leader in this field is US-based Eikos which was awarded USD 860, 000 in May this year from the US Air Force Research Laboratory to develop transparent conductive polymers specifically for military aircraft canopies.

While such contracts indicate that the carbon nanotube market is already up and running in some niche applications, there are still some major barriers to overcome.

The Three Main Restrictions in the Market-Place for Carbon Nanotubes

While there are already some products in the market-place and a whole range of applications in development, the carbon nanotube market has been constrained by three main issues:

1. A lack of commercially-available material of consistently high quality.
2. Cost—until recently single-wall carbon nanotubes were selling for EUR 400/g which is clearly some way from commercial viability as a bulk material.
3. Successfully expressing the basic properties of the raw material in the end application – the incredibly small scale of the material poses some interesting challenges for advanced material and coatings science.

Such teething problems are not new. Carbon fibres took many years to be widely accepted in the materials world from both a cost and performance aspect and have only just made a regular appearance in popular sporting markets such as golf equipment.

Early Applications of Carbon Nanotubes now on the Market

Despite this, some early applications of carbon nanotubes in products such as conductive fuel lines in the automotive industry have already made a successful appearance on the market. In addition, recent advances in the manufacture of carbon nanotubes on an industrial scale will allow vital commercial research and evaluation to be conducted and this will open the way to a wide range of fully industrialised applications.

The UK's First Production Plant for Single-Wall Carbon Nanotubes (SWNTs)

The UK's first commercial manufacturing process for high-purity single-wall carbon nanotubes was announced by Thomas Swan & Co in April 2004. The plant is the result of four years of collaboration between Thomas Swan and Cambridge University's Department of Material Science and Department of Chemistry. The unique collaboration brought the academic expertise of Cambridge University together with over 77 years of chemical manufacturing experience from Thomas Swan.

Problems Encountered When Setting Up this Plant for Carbon Nanotubes

The goal of the project was to supply high-purity single and multi-wall carbon nanotubes in commercial quantities and to back up this initial offering with the ability to scale up the process to keep up with demand as the industrial market expanded. The problem was that at the start of the programme no-one knew which production method would supply both single and multi-wall tubes and be fully scalable. It was this problem that was presented to Cambridge University.

Different Methods Used for Manufacturing Carbon Nanotubes

Carbon nanotubes can be manufactured using a variety of methods:

- Laser ablation uses a high-power laser to vaporise a graphite source loaded with a metal catalyst. The carbon in the graphite reforms as predominantly single-wall nanotubes on the metal catalyst particles.
- Arc discharge involves an electrical discharge from a carbon-based electrode in a suitable atmosphere to produce both single and multi-wall tubes of high quality but in low quantities.
- Chemical vapour deposition (CVD) is where a hydrocarbon feedstock is reacted with a suitable metal-based catalyst in a hot furnace to 'grow' nanotubes which are subsequently removed from the substrate and catalyst by a simple acid wash.

Using a System Based on Chemical Vapour Deposition (CVD) Growth

The team at Cambridge developed a novel system based on CVD growth, as it enabled the production of both single and multi-wall nanotubes of reasonably high quality and consistency while offering the greatest potential for scale up. This concept was then taken to design stage (and patented by Cambridge University). Using a team of chemical engineers and an independent consultant, the plant was built at Swan's site in North East England towards the end of 2003. Over the following months the plant was commissioned and then optimised until the material that was being produced was of consistently high quality.

Benefits of Having Access to a Regular Supply of High-Purity and Scaleable Material

A consistent supply of high-purity material that is also scaleable is a major stepping stone for the developing carbon nanotubes market. It allows companies to conduct meaningful industrial research using commercial quantities of raw material. In addition, this research can now be done with the knowledge that, should an application prove itself, there is a scalable source of the same raw material available to supply any ongoing development. Combine this with lowering prices (Thomas Swan predicts that within a few years the price of single-wall carbon nanotubes with be measured in tens of GBP per kg) and the market should start to take off.

Problems in Seeing What has been Manufactured

A major problem in achieving a saleable product that is on average 100, 000 times smaller than a human hair is that it is rather hard to 'see' what you have made. The advent of nanomaterial manufacturing brings to light the necessity for advanced industrial and cost-effective analytical techniques. While Scanning Electron Microscopes

(SEM) provide the ability to view the sample in remarkable detail, it only shows a minute fraction of the product. The analogy often used is like looking at a football pitch through a jam jar. The grass beneath you is green and healthy but you have no idea what it is like on the other side of the pitch.

Tools and Techniques Used to Help Manufacturers See what they are Making

This will not be an easy problem to solve and while RAMAN techniques (which involve studying the spectrum profile of a reflected laser from your sample) offer some hope, it will be some time before a standard industry protocol for purity measurement is identified and accepted. Thomas Swan currently bases purity measurement on a combination of SEM, (Transmission Electron Microscopy) TEM, RAMAN and Thermal Gravimetry Analysis (TGA) techniques, combined with an experienced technical eye and has so far received very encouraging feedback from early customers. As the company moves to scale the process up, ongoing customer feedback and collaboration will become increasingly vital.

Prices of Carbon Nanotubes Will Fall as Demand Increases

Carbon nanotubes are still in the very early stages of industrial development. While they are still expensive compared to a fully commoditised product, such as carbon black, the price will fall as the demand increases. Despite the problems associated with analysis and downstream processing (namely dispersion) the basic properties of the raw material offer a wide range of exciting applications that are beginning to be realised.

What the Next Ten Years Might Hold for Producers of Carbon Nanotubes

As the research into these applications moves out of the lab and into industry it is important that the supply of carbon nanotubes is consistent, of high quality and performed under the appropriate regulatory and health and safety protocols. Thomas Swan has taken the first step towards this goal and material is now available to allow meaningful industrial evaluation and development. The next ten years of development should see the creation of a whole new area of material and coatings science and associated commercial applications.

Flexural Strain Sensing Using Carbon Nanotube Film

The sensor/actuator embedded smart materials and systems are essential for structural damage detection and control. There is need for smart structural materials, which can be used to sense strains, and adjust the structural properties so as to achieve desired response. Carbon nanotube composites have the potential to be such a smart structural material. Results (Baughman *et al.*, 1999) have shown that when carbon nanotube films, also called as "buckypapers", are used as actuators, large actuator strains can be achieved by smaller operating voltages as compared with ferroelectric and electrostrictive materials. Single wall carbon nanotubes (SWCNTs) promise to be a reinforcement material for nanocomposites because of their superior mechanical properties. Researchers (Hadjiev *et al.*, 2001; Zhao *et al.*, 2001) have studied the Raman band shift in SWCNT's as a function of stress/strain in carbon nanotubes. Raman studies on multi-wall carbon nanotubes have also been carried out (Schadler *et al.*, 1998; Wagner *et al.*, 1998). Though the experiments are carried at nanoscale, the results illustrate the potential of SWCNT as strain/stress sensor. It has also been shown that the electrical band gap changes as a function of axial compression, tension stretch, torsion, and bending strain (Peng and Cho, 2002; Tombler *et al.*, 2000; Yang and Han, 2000). Strong dependence of SWCNT's band structure on mechanical deformation makes it possible to develop nano-electro-mechanical sensors. Most of the studies to date relate the mechanical deformation with the change in electrical properties at the nanoscale. In this study, change in the electronic property of carbon nanotubes due to strain at macroscale is demonstrated using the experimental results. Such carbon nanotube films have randomly oriented SWCNTs with isotropic strain sensing properties; thus, have multidirectional and multi-location strain sensing capability.

The carbon nanotube films used in this study are produced by mixing SWCNTs with 0.25?mg/ml *N*, *N*-Dimethylformamide (DMF) and filtering the mixture through a 0.2?mm teflon membrane. Freestanding carbon nanotube film (Buckypaper) is peeled from the filter after rinsing and drying of the remaining material. The carbon nanotube film is further dried for 24?h under vacuum. The SEM picture of a carbon nanotube film with ~15?ìm thickness. To explore the strain sensing potential of the carbon nanotube film, it is attached to a 0.25×2×24?in. brass specimen and subjected to pure bending moment using two-point loading. For insulation, a PVC film is attached between the brass specimen and carbon nanotube film. High strength epoxy and a vacuum-bonding method is used to attach carbon nanotube film and PVC to the brass specimen in order to ensure the perfect strain transfer – between brass, PVC, and the carbon nanotube film. A conventional electrical resistance strain gage is attached next to the carbon nanotube film to measure flexural strain for comparison purposes. Load P is applied in increments using a servo-hydraulic machine. The corresponding change in voltage across the carbon nanotube film is measured using four-point probe measurement (Smits, 1958). Input voltage across two outer probes is kept constant during the measurement and change in voltage across the two inner probes is recorded. Strain measurements are also made using conventional strain gage. The change in voltage is proportional to the change in resistivity of the film (Dharap *et al.*, 2004). The change in voltage due to the change in dimensions of the film is small as compared to the change in voltage due to the change in resistivity of the film (Li *et al.*, 2003).

The change in voltage in the carbon nanotube film as a function of measured flexural strain from the conventional strain gage. As the strain measured by the conventional strain gage increases from 0 to ~600?ìm/m the voltage change measured across the two outer probes increases from 0 to ~200?ìV. A nearly linear trend between the change in voltage and the strain is evident; although, deviation from the linear trend exists. Further study is needed to establish causes for deviation from linear trend such as temperature and gas exposure history (Bezryadin *et al.*, 1998; Collins and Avouris, 2000; Hone *et.al.*, 2002). The change in voltage is measured by moving the four-point probe to several parallel locations on a single carbon nanotube film, which also yielded a similar linear trend.

It is evident from the experimental study that the carbon nanotube film can be used for measuring flexural strains at the macroscale. Carbon nanotube films are made up of randomly oriented SWCNTs; hence, their electronic properties are independent of direction. Hence taking measurement along different directions provides corresponding strains. Carbon nanotube films can also be integrated into composites to simultaneously act as strong structural material and sensor. Such composites may find application in smart structures.

Properties and Future Applications of Nanocapsules and Dendrimers

A nanocapsule is any nanoparticle that consists of a shell and a space, in which desired substances may be placed. Technologies for microencapsulating materials have been around for several years, primarily for applications involving minimisation of hygroscopy and chemical interactions, elimination of oxidation, and controlled release of nutraceuticals.

The Use of Man-Made Liposomes

Man-made liposome's have been used in cosmetics for some years to control the release of substances or protect them from the environment. Recently many other materials, such as polymers, have been used to make nanocapsules.

The Properties of Polymeric Nanocapsules

Polymeric nanocapsules can be made in specific sizes, shapes, and in reasonable quantities. Nanocapsules can be made to function in various ways. They can be produced as monodisperse particles with exactly defined biochemical, electrical, optical, and magnetic properties. They can be tailored to suit the complexity of whatever application they are intended for, such causing the release of the contents in response to a particular bimolecular triggering mechanism in targeted drug-delivery systems.

The Use of Nanocapsules as Smart Drugs

Nanocapsules can be used as smart drugs that have specific chemical receptors and only bind to specific cells. It is this receptor that makes the drug 'smart, ' allowing it to target cancer or disease. The advantages of nano-encapsulation technologies for pharmaceutical applications include:

- Higher dose loading with smaller dose volumes
- Longer site-specific dose retention
- More rapid absorption of active drug substances
- Increased bioavailability of the drug
- Higher safety and efficacy
- Improved patient compliance

The Future Benefits of Nanocapsules in Drugs

Beyond the ability to deliver existing drugs to their target, nanocapsules would allow for as much as a 10, 000-fold decrease in drug dosages, reducing the harmful side effects of drugs used in chemotherapy. Quite often, drugs don't make it to market is because they have too many unwanted side effects. However, placing the same drug inside a nanocapsule and delivering it directly to its intended target in a reduced dosage, eliminates some of those side effects, or at least reduces them to an acceptable level.

Further Applications of Nanocapsules

Nanocapsules also have potential applications in agrochemicals, cosmetics, genetic engineering, wastewater treatments, cleaning products, and adhesive component applications. They can be used to encapsulate enzymes, catalysts, oils, adhesives, polymers, inorganic micro- and nanoparticles, latex particles, or even biological cells.

What is a Dendrimer?

A dendrimer is an artificially manufactured or synthesised large molecule comprised of many smaller ones linked together—built up from branched units called monomers. Technically, dendrimers are a unique class of a polymer, about the size of an average protein, with a compact, tree-like molecular structure, which provides a high degree of surface functionality and versatility. Their shape gives them vast amounts of surface area, making them useful building blocks and carrier molecules at the nanoscale and they come in a variety of forms, with different physical (including optical, electrical and chemical) properties.

Dendrimer As A Biologically Active Carrier

Dendrimers can act as biologically active carrier molecules in drug delivery to which can be attached therapeutic agents and as scavengers of metal ions, offering the potential for environmental clean-up operations because their size allows them to be filtered out with ultra-filtration techniques.

Benefits of Nanomaterials and Nanoparticles in Agriculture Sector

A number of projects around the world are exploring the use of nanoparticles on the farm for purposes other than pesticides—from enhanced photosynthesis to better germination and soil management.

Buckyball Fertiliser?

Researchers at Kyoto University (Japan) have discovered a method of producing ammonia using buckyballs. Ammonia is a key component of fertiliser, but it is not clear if the resulting product for use in the fields would contain buckyballs.

TiO_2 Nano Mixture

Scientists at the University of Korea have applied for a patent on a liquid after which time it dissolves in the groundwater and becomes indistinguishable from naturally occurring iron.

Seeding Iron

The Russian Academy of Sciences reports that they have been able to improve the germination of tomato seeds by spraying a solution of iron nanoparticles on to fields.

Soil Binder—Using Chemical Reactions at the Nanoscale to Bind Soil Together

In 2003, ETC Group reported on a nanotech-based soil binder called SoilSet developed by Sequoia Pacific Research of Utah (USA). SoilSet is a quick-setting mulch which relies on chemical reactions on the nanoscale to bind the soil together. It was sprayed over 1, 400 acres of Encebado mountain in New Mexico to prevent erosion following forest fires, as well as on smaller areas of forest burns in Mendecino County, California.

Soil Clean-Up Using Iron Nanoparticles

A number of approaches are being developed to apply nanotechnology and particularly nanoparticles to cleaning up soils contaminated with heavy metals and PCBs. Dr. Wei-Xang Zhang has pioneered a nano clean-up method of injecting nano-scale iron into a contaminated site. The particles flow along with the groundwater and decontaminate en route, which is much less expensive than digging out the soil to treat it. Dr. Zhang's tests with nano-scale iron show significantly lower contaminant levels within a day or two. The tests also show that the nano-scale iron will remain active in the soil for six to eight weeks, after which time it dissolves in the groundwater and becomes indistinguishable from naturally occurring iron.

Impact of Nanotechnology on Agriculture Industry

Using Nanosensor Networks to Track Livestock

Livestock tracking has been a problem for farmers since before Little Bo Peep lost her sheep. Nano-Bo-Peep, however, would have no such problems. Just as converging technologies in crop production will use nanosensor networks to continuously monitor the health of plants, so, too, will sensors monitor livestock. The USDA envisions the rise of 'smart herds'—cows, sheep and pigs fitted with sensors and locators relaying data about their health and geographical location to a central computer.

Linking Nanosensors to Drug Delivery Systems for Animals and Humans

This is a vision of precision agriculture on the hoof. The long-term aim is not merely to monitor, but also to automatically and autonomously intervene with pharmaceuticals using small drug delivery devices that can be implanted into the animal in advance of illness. The notion of linking in-built sensors to in-built smart delivery systems has been called "the fuel injection principle" since it mimics the way modern cars use sensors to time fuel-delivery to the engine. The closest applications to market are implantable insulin-delivery devices or "drug chips" that will be linked with glucose sensors for (human) diabetics to automatically regulate blood sugar levels. Over time, this could become the model for all drug delivery, in both humans and animals.

Why Biocompatible Nanomaterials Are Very Suitable for Implantable Medical Devices

One of the current barriers to implantable medical devices is that their composite materials (e.g., metal or plastics) are often incompatible with living tissue. New materials, engineered at the nano-scale to be biocompatible, seek to address this problem.

Possible Industry Uses for Animal Tracking Devices

Implanting tracking devices in animals is nothing new—either in pets, valuable farm animals or for wildlife conservation. Injectable microchips are already used in a variety of ways with the aim of improving animal welfare and safety—to study animal behaviour in the wild, to track meat products back to their source or to reunite strays with their human guardians. In the nanotech era, however, retrofitting farm animals with sensors, drug chips and nanocapsules will further extend the vision of animals as industrial production units.

Nano-Eugenics: Possible Dangers in Using Microfluidics for Breeding Animals

Animals also are likely to be used as the testing ground for less savoury or more risky applications that could later be extended to human beings. Using microfluidics for breeding is likely to accelerate genetic uniformity within livestock species and also opens the possibility of applying new nano-eugenic technologies to humans in the future. The ability to remotely regulate animals may have adverse affects as livestock go longer periods without direct human care.

Implantable Microchips in Humans—A Threat to Civil Liberties or an Innovative Healthcare Product?

The same technologies transferred to humans raises profound concerns about quality of life and civil liberties. In October 2004, the US Food and Drug Administration approved the use of implantable microchips in humans to provide easy access to an individual's medical records—the first approval of microchips for medical uses in the United States.

Benefits and Drawbacks of Automated Drug Delivery for Humans

As healthcare is driven more and more by the bottom line, the future use of implantable chips for automated drug delivery may become economically preferable to nursing. When dealing with the elderly or those with different cognitive abilities or with any condition requiring regular treatment, ethical questions may arise about who decides to make an individual 'fuel injected.' Automated drug delivery could allow some people to live independently who would otherwise be institutionalised. However, the absence of human caretakers is also a factor.

Nano-Aquaculture—Fish Farming Looks to Incorporate Nanotechnology into its Future Products

The world's fastest growing area of animal production is the farming of fish, crustaceans and molluscs, particularly in Asia. According to the FAO there were 45.7 million tonnes of aquaculture production in 2000 and it is growing at a rate of more than 9% per year. With a strong history of adopting new technologies, the highly integrated fish farming industry may be among the first to incorporate and commercialise nanotech products.

Emerging Nanotechnology Applications in the Aquaculture Sector Cleaning Fish Ponds with Nanotechnology Devices

Nevada-based Altair Nanotechnologies makes a water cleaning product for swimming pools and fishponds called 'NanoCheck.' It uses 40 nm particles of a lanthanum-based compound which absorbs phosphates from the water and prevents algae growth. NanoCheck is currently undergoing large-scale testing in swimming pools and Altair is expected to launch a swimming pool cleaner in early 2005. Altair has its eye on a potentially large demand for NanoCheck for use in thousands of commercial fish farms worldwide where algae removal and prevention is costly at present. According to Altair, the company plans to expand its tests to confirm that its nanoparticles will not harm fish, but no mention is made of the tests that will be undertaken to examine the impacts of nanoparticle-laden run-off on human health or on the environment.

DNA Nano-Vaccines Using Nanocapsules and Ultrasound Methods

The USDA is completing trials on a system for mass vaccination of fish using ultrasound. Nanocapsules containing short strands of DNA are added to a fishpond where they are absorbed into the cells of the fish. Ultrasound is then used to rupture the capsules, releasing the DNA and eliciting an immune response from the fish. This technology has so far been tested on rainbow trout by Clear Springs Foods (Idaho, US)—a major aquaculture company that produces about one third of all US farmed trout.

Using Iron Nanoparticles to Speed Up the Growth of Fish

Scientists from the Russian Academy of Sciences have reported that young carp and sturgeon exhibited a faster rate of growth (30% and 24% respectively) when they were fed nanoparticles of iron.

Carbon Nanotube Transistors

Scientists have long been intrigued by the electrical properties of carbon nanotubes, those tiny straws of pure carbon measuring less than a hair's width across. Now one research team has unveiled an improved design for transistors based on the minuscule tubes, one that could substantially reduce the lower limit on transistor size.

The size of conventional transistors made of metal oxide semiconductors (MOS) has shrunk over the years and can now be as small as about 100 nanometers across. This reduction has allowed for an impressive increase in both the speed and power of a variety of electronics. By employing carbon nanotubes, a team led by Prabhakar R. Bandaru of the University of California at San Diego hopes to shrink transistor dimensions even more—down to just a few nanometers thick. "The small size and dramatic switching behaviour of these nanotubes makes them candidates for a new class of transistor, " he explains.

The novel transistors were first grown as regular, straight nanotubes. Using catalyst particles containing iron, the team then coaxed a second branch to grow out of the first tiny straw. The resulting y-shaped nanotubes carry current when attached to electrical contacts, and varying the amount of voltage applied to the stem controls the flow of electrons through the junction. The new approach improves on previous carbon nanotube-based transistors, in which separate components had to be added in. "We can synthesize functionality at the nanoscale, in this case to include the three elements of a circuit—the gate, source and drain, " Bandaru says, "and we don't have to go to the trouble of making them separately and assembling them." A report detailing the results was published online yesterday by *Nature Materials*.

Nanotechnology and Nanotubes

The term nanotube is normally used to refer to the carbon nanotube, which has received enormous attention from researchers over the last few years and promises, along with close relatives such as the nanohorn, a host of interesting applications. There are many other types of nanotube, from various inorganic kinds, such as those made from boron nitride, to organic ones, such as those made from self-assembling cyclic peptides (protein components) or from naturally-occurring heat shock proteins (extracted from bacteria that thrive in extreme environments). However, carbon nanotubes excite the most interest, promise the greatest variety of applications, and currently appear to have by far the highest commercial potential. Only carbon nanotubes will be covered in this chapter. Carbon nanotubes are often referred to in the press, including the scientific press, as if they were one consistent item. They are in fact a hugely varied range of structures, with similarly huge variations in properties and ease of production. Adding to the confusion is the existence of long, thin, and often hollow, carbon fibers that have been called carbon nanotubes but have a quite different make-up from that of the nanotubes that scientists generally refer to. To distinguish these we will refer to them as carbon nanofibers.

Carbon nanotubes were 'discovered' in 1991 by Sumio Iijima of NEC and are effectively long, thin cylinders of graphite, which you will be familiar with as the material in a pencil or as the basis of some lubricants. Graphite is made up of layers of carbon atoms arranged in a hexagonal lattice, like chicken wire. Though the chicken wire structure itself is very strong, the layers themselves are not chemically bonded to each other but held together by weak forces called Van der Waals. It is the sliding across each other of these layers that gives graphite its lubricating qualities and makes the mark on a piece of paper as you draw your pencil over it.

Now imagine taking one of these sheets of chicken wire and rolling it up into a cylinder and joining the loose wire ends. The result is a tube that was once described by Richard Smalley (who shared the Nobel Prize for the discovery of a related form of carbon called buckminsterfullerene) as "in one direction . . . the strongest damn thing you'll ever make in the universe". In addition to their remarkable strength, which is usually quoted as 100 times that of steel at one-sixth of the weight (this is tensile strength—the ability to withstand a stretching force without breaking), carbon nanotubes have shown a surprising array of other properties. They can conduct heat as efficiently as most diamond (only diamond grown by deposition from a vapour is better), conduct electricity as efficiently as copper, yet also be semiconducting (like the materials that make up the chips in our computers). They can produce streams of electrons very efficiently (field emission), which can be used to create light in displays for televisions or computers, or even in domestic lighting, and they can enhance the fluorescence of materials they are close to. Their electrical properties can be made to change in the presence of certain substances or as a result of mechanical stress.

Nanotubes within nanotubes can act like miniature springs and they can even be stuffed with other materials. Nanotubes and their variants hold promise for storing fuels such as hydrogen or methanol for use in fuel cells and they make good supports for catalysts. But let's look at some of the different types of nanotubes, and nanotube pretenders. One of the major classifications of carbon nanotubes is into single-walled varieties (SWNTs), which have a single cylindrical wall, and multi-walled varieties (MWNTs), which have cylinders within cylinders.

The lengths of both types vary greatly, depending on the way they are made, and are generally microscopic rather than nanoscopic, i.e. greater than 100 nanometers (a nanometer is a millionth of a millimeter). The aspect ratio (length divided by diameter) is typically greater than 100 and can be up to 10, 000, but recently even this was made to look small. In May 2002 SWNT strands were made in which the SWNTs were claimed to be as long as 20 cm. Even more recently, the same group has made strands of SWNTs as long as 160 cm, but the precise make-up of these strands has not yet been made clear.

A group in China has also found, purely by accident, that packs of relatively short carbon nanotubes can be drawn out into a bundle of fibers, making a thread only 0.2 millimeters in diameter but up to 30 centimeters long. The joins between the nanotubes in this thread represent a weakness but heating the thread has been found to increase the strength significantly, presumably through some sort of fusing of the individual tubes. These are the stars of the nanotube world, and somewhat reclusive ones at that, being much harder to make than the multi-walled variety. The oft-quoted amazing properties generally refer to SWNTs. As previously described, they are basically tubes of graphite and are normally capped at the ends, although the caps can be removed. The caps are made by mixing in some pentagons with the hexagons and are the reason that nanotubes are considered close cousins of buckminsterfullerene, a roughly spherical molecule made of sixty carbon atoms, that looks like a soccer ball and is named after the architect Buckminster Fuller (the word fullerene is used to refer to the variety of such molecular cages, some with more carbon atoms than buckminsterfullerene, and some with fewer). The theoretical minimum diameter of a carbon nanotube is around 0.4 nanometers, which is about as long as two silicon atoms side by side, and nanotubes this size have been made. Average diameters tend to be around the 1.2 nanometer mark, depending on the process used to create them. 8 SWNTs are more pliable than their multi-walled counterparts and can be twisted, flattened and bent into small circles or around sharp bends without breaking.

Discussions of the electrical behaviour of carbon nanotubes usually relate to experiments on the single-walled variety. As we have said, they can be conducting, like metal (such nanotubes are often referred to as metallic nanotubes), or semiconducting, which means that the flow of current through them can be stepped up or down by varying an electrical field. The latter property has given rise to dreams of using nanotubes to make extremely dense electronic circuitry and the last year has seen major advances in creating basic electronic structures from nanotubes in the lab, from transistors up to simple logic elements. The gulf between these experiments and commercial nanotube electronics is, however, vast. There are various ways of producing SWNTs, which are briefly discussed later. The detailed mechanisms responsible for nanotube growth are still not fully understood and computer modeling is playing an increasing role in fathoming the complexities. The ambition of SWNT producers is to gain greater control over their diameters, lengths, and other properties, such as chirality (explained below). The volumes of SWNTs produced are currently small and the quality and purity are variable. Carbon Nanotechnologies Inc. of Houston, Texas, are currently ramping up production to a half a kilogram a day, which is actually huge in comparison to amounts of SWNTs that have been made historically. Various companies pursuing specific nanotube applications produce their own material in house.

Imagine again the chicken wire that we roll up to make the nanotube. In fact, imagine a chicken wire fence out of which will be cut a rectangle to roll into a tube. You could cut the rectangle with the sides vertical or at various angles. Additionally, when joining the sides together, you can raise or lower one side. In some cases it will not be possible to make a tube such that the loose ends match and hexagons are formed, but in other cases it will, and these represent the possible permutations of SWNTs. The possibilities are two forms in which a pattern circles around the diameter of the tube, often called zigzag and armchair (not the most intuitive of names, unfortunately, but they are now widely used), and a variety of forms in which the hexagons spiral up or down the tube with varying steepness, these being the chiral forms. There is theoretically an infinite variety of the latter, if you allow for infinite diameters of nanotubes.

Which of these forms a nanotube takes is the major determinant of its electrical properties, i.e. whether the tube is semiconducting or conducting. For a long time, the fact that all known production methods created a mix of types has been considered one of the hurdles to be overcome if the electronic properties are to be exploited. Claims have now been made that it is possible to produce only the semiconducting kind (specifically, the zigzag form). Additionally, there are approaches that can yield only semiconducting nanotubes from a mix of semiconducting and conducting ones. One such approach relies on vaporising the conducting nanotubes with a strong electric current, leaving only the semiconducting kind behind. A more recent approach is simply to leave the mix of nanotubes lying around for a while—the metallic ones are oxidised and become semiconducting (the process can, of course, be speeded up). In the early part of 2001 there were also reports that nanotubes had been induced to form crystals, each of which contained just one type of nanotube. This would have been a nice separation method but the silence on this approach since then suggests that these results have not been duplicated.

Multi-walled carbon nanotubes are basically like Russian dolls made out of SWNTs—concentric cylindrical graphitic tubes. In these more complex structures, the different SWNTs that form the MWNT may have quite different structures (length and chirality). MWNTs are typically 100 times longer than they are wide and have outer diameters mostly in the tens of nanometers. Although it is easier to produce significant quantities of MWNTs than SWNTs, their structures are less well understood than single-wall nanotubes because of their greater complexity and variety. Multitudes of exotic shapes and arrangements, often with imaginative names such as bamboo-trunks, sea urchins, necklaces or coils, have also been observed under different processing conditions. The variety of forms may be interesting but also has a negative side—MWNTs always (so far) have more defects than SWNTs and these diminish their desirable properties. Many of the nanotube applications now being considered or put into practice involve multi-walled nanotubes, because they are easier to produce in large quantities at a reasonable price and have been available in decent amounts for much longer than SWNTs. In fact one of the major manufacturers of MWNTs at the moment, Hyperion Catalysis, does not even sell the nanotubes directly but only pre-mixed with polymers for composites applications. The tubes involved typically have 8 to 15 walls and are around 10 nanometres wide and 10 micrometers long. Other companies are moving into this space, notably formidable players like Mitsui, with plans to produce similar types of MWNT in hundreds of tons a year, a quantity that is greater, but not hugely so, than the current production of Hyperion Catalysis. This is an indication that even these less impressive and exotic nanotubes hold promise of representing a sizable market in the near future.

These are single-walled carbon cones with structures similar to those of nanotube caps that have been produced by high temperature treatment of fullerene soot. Sumio Iijima's group at NEC has demonstrated that nanohorns have good adsorptive and catalytic properties (i.e. desired substances stick to them and they enhance chemical reactions), and the company is working on using them in a new generation of fuel cells for personal electronics.

We use this term to refer to hollow and solid carbon fibers with lengths on the order of a few microns and widths varying from some tens of nanometers to around 200 nanometers. These materials have occasionally been referred to as nanotubes. However, they do not have the cylindrical chicken wire structure of SWNTs and MWNTs but instead consist of a mixture of forms of carbon, from layers of graphite stacked at various angles to amorphous carbon (lacking any large-scale regular structure). Because of this variable structure they do not exhibit the strength of pure nanotubes but can still be quite strong and possess other useful properties. The US company Applied Sciences Inc. is already producing tons of such material a year and there are several major producers in the Far East.

Production processes for carbon nanotubes, crudely described, vary from blasting carbon with an electrical arc or a laser to growing them from a vapor, either en masse (usually in tangled bundles) or on nanoparticles, sometimes in predetermined positions. These processes vary considerably with respect to the type of nanotube produced, quality, purity and scalability. Carbon nanotubes are usually created with the aid of a metal catalyst and this ends up as a contaminant with respect to many potential applications, especially in electronics. IBM have very recently, however, grown nanotubes on silicon structures without a metal catalyst. Scalability of production processes is an essential commercial consideration—some of the approaches use equipment that simply cannot be made bigger and the only way to increase production is to make more pieces of equipment, which will not produce the economies of scale required to bring down costs significantly.

By now we hope you have an idea of the different types of nanotubes and how their qualities and ease of production vary. Reports of one company producing tons of the material, alongside reports of researchers not being able to get enough to get meaningful results from their research, should no longer be confusing. It would, of course, be useful if commentators got into the habit of making clear the type of material they are talking about. Understanding these differences is essential for understanding the commercial potential of the various applications of nanotubes and related structures that already exist or have been proposed. The variety of these is vast, and the commercialisation timelines involved vary from now to ten years from now or more. Some of the potential markets are enormous. We will leave you with a taste of the possibilities. The materials markets are already seeing applications for composites based on multi-walled carbon nanotubes and nanofibers. In many ways this is an old market—that of carbon fibers, which have been around commercially for a couple of decades. The benefits of the new materials in these markets are the same as those of carbon fibers, just better; the main properties to be considered being strength and conductivity. Carbon fibers are quite large, typically about a tenth of a millimeter in diameter, and blacken the material to which they are added. MWNTs can offer the same improvements in strength to a polymer composite without the blackening and often with a smaller amount of added material (called the filler). The greater aspect ratio (i.e. length compared to diameter) of the newer materials can make plastics conducting with a smaller filler load, one significant application being electrostatic painting of composites in products such as car parts. Additionally, the surface of the composite is smoother, which benefits more refined structures such as platens for computer disk drives.

When thinking about structural applications such as these, it should be remembered that, in general, as the fibers get smaller so the number of defects decreases, in a progression towards the perfection of the single-walled nanotube. The inverse progression is seen in terms of ease of manufacturing—the more perfect, and thus more structurally valuable, the material, the harder it is to produce in quantity at a good price. This relationship is not written in stone—there is no reason that near-perfect SWNTs should not be producible cheaply and in large quantities. When this happens, and it might not be too far off, the improvements seen in the strength-to-weight ratios of composite materials could soar, impacting a wide variety of industries from sports equipment to furniture, from the construction industry to kitchenware, and from automobiles to airplanes and spacecraft (the aerospace industry is probably the one that stands to reap the greatest rewards). In fact a carbon nanotube composite has recently been reported that is six times stronger than conventional carbon fiber composites. This is an appropriate point at which to introduce a note of caution, and an area of research worth keeping an eye on. Just because the perfect nanotube is 100 times stronger than steel at a sixth of the weight doesn't mean you are going to be able to achieve those properties in a bulk material containing them. You may remember that the chicken wire arrangement that makes up the layers in graphite does not stick at all well to other materials, which is why graphite is used in lubricants and pencils. The same holds true of nanotubes—they are quite insular in nature, preferring not to interact with other materials. To capitalize on their strength in a composite they need to latch on to the surrounding polymer, which they are not inclined to do (blending a filler in a polymer is difficult even without these issues—it took a decade to perfect this for the new nanoclay polymers now hitting the markets).

One way of making a nanotube interact with something else, such as a surrounding polymer, is to modify it chemically. This is called functionalisation and is being explored not just for composite applications but also for a variety of others, such as biosensors. For structural applications, the problem is that functionalisation can reduce the valuable qualities of the nanotubes that you are trying to capitalize on. This is an issue that should not be underestimated. Of course, in theory you don't need to mix the nanotubes with another material. If you want to make super-strong cables, for example, the best solution would be to use bundles of sufficiently long nanotubes with no other material added. For this reason, one of the dreams of nanotube production is to be able to spin them, like thread, to indefinite lengths. Such a technology would have applications from textiles (the US military is in fact investigating the use of nanotubes for bullet-proof vests) to the 'space elevator'. The space elevator concept, which sounds like something straight out of science fiction (it was, in fact, popularised by Arthur C. Clarke) involves anchoring one end of a huge cable to the earth and another to an object in space. The taut cable so produced could then support an elevator that would take passengers and cargo into orbit for a fraction of the cost of the rockets used today. Sounds too far out? It has, in fact, been established by NASA as feasible in principle, given a material as strong as SWNTs. The engineering challenges, though, are awesome, so don't expect that 'top floor' button to be taking you into orbit any time soon.

The materials market is a big one, and there are others, which we'll come to, but smaller ones exist too. Nanotubes are already being shipped on the tip of atomic force microscope probes to enhance atomic-resolution imaging. Nanotube-based chemical and biosensors should be on the market soon (they face stiff competition from other areas of nanotechnology). The thermal conductivity of nanotubes shows promise in applications from cooling integrated circuits to aerospace materials. Electron beam lithography, which is a method of producing nanoscale patterns in materials, may become considerably cheaper thanks to the field emission properties of carbon nanotubes. Recent developments show promise of the first significant change in X-ray technology in a century. Entering more speculative territory, nanotubes may one day be used as nanoscale needles that can inject substances into, or sample substances from, individual cells, or they could be used as appendages for miniature machines (the tubes in multi-walled nanotubes slide over each other like graphite, but have preferred locations that they tend to spring back to). Big markets, apart from materials, in which nanotubes may make an impact, include flat panel displays (near-term commercialisation is promised here), lighting, fuel cells and electronics. This last is one of the most talked-about areas but one of the farthest from commercialisation, with one exception, this being the promise of huge computer memories (more than a thousand times greater in capacity than what you probably have in your machine now) that could, in theory, put a lot of the $40 billion magnetic disk industry out of business. Companies like to make grand claims, however, and in this area there is not just the technological hurdle to face but the even more daunting economic one, a challenge made harder by a host of competing technologies.

Despite an inevitable element of hype, the versatility of nanotubes does suggest that they might one day rank as one of the most important materials ever discovered. In years to come they could find their way into myriad materials and devices around us and quite probably make some of the leaders in this game quite rich.

Nanotechnology Being Used to Create the Worlds Stickiest Material

Scientists in Britain and USA are taking inspiration from the gecko to try and develop the ultimate in adhesive tape. The gecko has evolved with toe surfaces covered in the perfect adhesive that allows them to stick in the most difficult of circumstances—even upside down on wet glass. Geckos have micro and nano-scale hairs on their toes. Each main hair is a thousand times thinner than a human hair. On the end of each hair are more hairs—each a thousand times smaller again. Molecular attraction then occurs by way of Van der Waals forces that allow the gecko to cling on. Synthetic versions of these hairs have now been developed. Applications include incorporating the new adhesive material into lightweight rescue robots that can readily climb walls in search and rescue missions. NASA has plans to use the material on space flights as early as next year. The theory is that by being able to have astronauts stick to the exterior of a space station or the space shuttle, then spacewalks would become far less dangerous.

REFERENCES

Baird, D.; Nordmann, A. & Schummer, J. (eds.): 2004, *Discovering the Nanoscale*, Amsterdam: IOS Press.

Baker, R.T.K. Synthesis, properties and applications of graphite nanofibers. In *R&D status and trends*, ed. Siegel et al.

Ball, P.: 2002, 'Tutorial: Natural Strategies for the Molecular Engineer', *Nanotechnology*, 13, 15-28.

Bate, R., Frazier, G., Frensley, W., Reed., M., "An Overview of Nanoelectronics," *Texas Instruments Technical Journal*, July-August 1989, pp. 13-20.

Baum, R.; Drexler, K.E.; Smalley, R.: 2003, 'Nanotechnology: Drexler and Smalley Make the Case For and Against 'Molecular Assemblers', *Chemical and Engineering News*, 81, 37-42.

Beck, J.S., J.C. Vartuli, W.J. Roth, M.E. Leonowicz, C.T. Kresge, K.D. Schmitt, C.T.-W. Chu, D.H. Olsen, E.W. Shepard, S.B. McCullen, J.B. Higgins, and J.L. Schlenker. 1992. *J. Am. Chem. Soc.* 114:10834.

Bensaude-Vincent, B.: 2001, 'The Construction of a Discipline: Materials Science in the United States', *Historical Studies in the Physical and Biological Sciences*, 31, 223-248.

Bensaude-Vincent, B.: 2004, 'Two Cultures of Nanotechnology?', *Hyle: International Journal for Philosophy of Chemistry*, 10(2), 65-82.

Berube, D.M.: 2004, 'The Rhetoric of Nanotechnology', in: D. Baird, A. Nordmann & J. Schummer (eds.), *Discovering the Nanoscale*, Amsterdam: IOS Press, pp. 173-192.

Bijker, W.E.: 1995a, *Of Bicycles, Bakelite, and Bulbs: Toward a Theory of Sociotechnical Change*, MIT Press, Cambridge, MA.

Bijker, W.E.: 1995b, 'Sociohistorical Technology Studies', in: S. Jasanoff, G.E. Markle, J. Petersen, T. Pinch (eds.), *Handbook of Science and Technology Studies*, Sage, London, pp. 229-256.

Bijker, W.E.; Law, J.: 1992, 'Do Technologies Have Trajectories?', in: W.E. Bijker, J. Law (eds.), *Shaping Technology/Building Society: Studies in Sociotechnical Change*, MIT Press, Cambridge, MA, pp. 1-16.

Bijker, W.E.; Pinch, T.: 1987, 'The Social Construction of Facts and Artifacts: Or How the Sociology of Science and the Sociology of Technology Might Benefit Each Other', in: W.E. Bijker, T.P. Hughes, T. Pinch (eds.), *The Social Construction of Technological Systems: New Directions in the Sociology and History of Technology*, MIT Press, Cambridge, MA, pp. 17-50.

Bimber, B.: 1994, 'Three Faces of Technological Determinism', in: M.R. Smith, L. Marx (eds.), *Does Technology Drive History?: The Dilemma of Technological Determinism*, MIT Press, Cambridge, MA, pp. 79-100.

Bowes, C.L., A. Malek, and G.A. Ozin. 1996. *Chem. Vap. Deposition* 2:97.

Braun, P.V., P. Osenar, and S.I. Stupp. 1996. *Nature*. 368: 2.

Braun, T.; Schubert, A. & Zsindely, S.: 1997, 'Nanoscience and nanotechnology on the balance', *Scientometrics*, 38, 321-325.

Brinker, C.J. 1996. *Curr. Opin. Solid State Mater. Sci*. 1:798.

Brooks, L.J.: 2003, *The Institute for the Future: A Site for Mapping, Maintaining, and Transforming Images of the Future*, (Ph.D. dissertation) University of California, San Diego, La Jolla.

Brown, N.: 2004, 'Needle on the Real: Technoscience and Poetry at the Limits of Fabrication', in: N.K. Hayles (ed.), *Nanoculture: Implications of the New Technoscience*, Bristol, UK: Intellect Books, pp. 173-190.

Bueno, O.: 2004, 'The Drexler-Smalley Debated on Nanotechnology: Incommensurability at Work?', *Hyle: International Journal for Philosophy of Chemistry*, 10(2), 83-98.

Bueno, O.: 2004, 'Von Neumann, Self-Reproduction and the Constitution of Nanophenomena', in: D. Baird, A. Nordmann & J. Schummer (eds.), *Discovering the Nanoscale*, Amsterdam: IOS Press, pp. 101-115.

Bush, V.: 1945, *Science, the Endless Frontier*, US Government Printing Office, Washington, DC.

Carroll, J.S.: 2001, 'Social Science Research Methods for Assessing Societal Implications of Nanotechnology', in: M.C. Roco and W.S. Bainbridge (eds.), *Societal Implications of Nanoscience and Nanotechnology*, Dordrecht: Kluwer, pp. 188-192.

Chen, C-Y., S.L. Burkett, H.-X. Li, and M.E. Davis. 1993. *Microporous Mater*. 2:27.

Chianelli, R.R. 1998. Synthesis, fundamental properties and applications of nanocrystals, sheets, and fullerenes based on layered transition metal chalcogenides. In *R&D status and trends*, ed. Siegel et al.

Chianelli, R.R., M. Daage, and M.J. Ledoux. 1994. *Advances in Catalysis* 40:177.

Claeson, T., and Likharev, K.," Single Electronics," *Scientific American,* June 1992, pp. 80-85.

Clausen-Schaumann, H.; Seitz, M.; Krautbauer, R.; Gaub, H.E.: 2000, 'Force Spectroscopy with Single Bio-Molecules', *Current Opinion in Chemical Biology*, 4, 524-530.

Coenen, C.: 2004, 'Nanofuturismus: Anmerkungen zu seiner Relevanz, Analyse und Bewertung', *Technikfolgenabschätzung ' Theorie und Praxis*, 13 (2), 78-85.

Constant, E.W., II: 1980, *The Origins of the Turbojet Revolution*, Johns Hopkins University Press, Baltimore.

Crandall, B.: 1999, 'Molecular Engineering', in: B. Crandall (ed.), *Nanotechnology: Molecular Speculations on Global Abundance*, MIT Press, Cambridge, MA, pp. 1-46.

Crandall, B.C. and Lewis, J., *Nanotechnology: Research and Perspectives*, MIT Press, Cambridge, 1992.

Crow, M. & Sarewitz, D.: 2001, 'Nanotechnology and Societal Transformation', in: M.C. Roco and W.S. Bainbridge (eds.), *Societal Implications of Nanoscience and Nanotechnology*, Dordrecht: Kluwer, pp. 45-54.

Decker, M.; Fiedeler, U.; Fleischer, Th.: 2004, 'Ich sehe was, was Du nicht siehst... zur Definition von Nanotechnologie', *Technikfolgenabschätzung ' Theorie und Praxis*, 13 (2), 10-16.

Dresselhaus, M., and G. Dresselhaus. 1995. *Ann. Rev. Mat. Sci.* 25:487.

Dresselhaus, M.S. 1998. Carbon-based nanostructures. In *R&D status and trends*, ed. Siegel et al.

Dresselhaus, M.S., G. Dresselhaus, and P. Eklund. 1996. *Science of fullerenes and carbon nanotubes*. San Diego: Academic Press.

Drexler, K.E.: 1981, 'Molecular Engineering – an Approach to the Development of General Capabilities for Molecular Manipulation', *Proceedings of the National Academy of Sciences of the United States of America-Physical Sciences*, 78, 5275-5278.

Drexler, K.E.: 1992, *Nanosystems: Molecular Machinery, Manufacturing, and Computation*, Wiley, New York.

Drexler, K.E.: 2004, 'Nanotechnology: From Feynman to Funding', *Bulletin of Science, Technology & Society*, 24, 21-27.

Dupuy, J.-P.: 2004, 'Complexity and Uncertainty: A Prudential Approach To Nanotechnology', in: European Commission (Community Health and Consumer Protection): *Nanotechnologies: A Preliminary Risk Analysis on the Basis of a Workshop, Brussels,* 1-2 March 2004, pp. 71-93.

Dupuy, J.P.: 2004, 'Pour une évaluation normative du programme nanotechnologique', *Annales des Mines*, (February 2004), 27-32.

Eigler, D.M.; Schweizer, E.K.: 1990, 'Positioning Single Atoms With a Scanning Tunneling Microscope', *Nature*, 344, 524-526.

Einsiedel, E.F.; Goldenberg, L.: 2004, 'Dwarfing the Social? Nanotechnology Lessons from the Biotechnology Front', *Bulletin of Science, Technology & Society*, 24, 28-33.

ETC: 2003a, *The Big Down: Atom Tech ' Technologies Converging at the Atomic Scale*. Winnipeg, Canada: Action Group on Erosion, Technology and Concentration [www.etcgroup.org/documents/TheBigDown.pdf].

ETC: 2003b, 'No Small Matter II: The Case for a Global Moratorium ' Size Matters!' *Occasional Paper Series* 7(1) [www.etcgroup.org/documents/Occ.Paper_Nanosafety.pdf].

Etzkowitz, H., 2001, 'Nano-Science and Society: Finding a Social Basis for Science Policy', in: M.C. Roco and W.S. Bainbridge (eds.), *Societal Implications of Nanoscience and Nanotechnology*, Dordrecht: Kluwer, pp. 121-128.

European Comission: *European Workshop on Social and Economic Research on Nanotechnologies and Nanosciences, Brussels, 14-15 April 2004* [www.stage-research.net/STAGE/PAGES/Nano.html].

European Commission (Community Health and Consumer Protection): 2004, *Nanotechnologies: A Preliminary Risk Analysis on theBasis of a Workshop, Brussels, 1-2 March 2004.*

European Commission: 2004, *Converging Technologies: Shaping the Future of European Societies (Report of the High Level Expert Group "Foresighting the New Technology Wave")*, Brussels: European Commission Research [http://europa.eu.int/comm/research/conferences/2004/ntw/index_en.html].

F. Buot, "Mesoscopic Physics and Nanoelectronics: Nanoscience and Nanotechnology," *Physics Reports*, pp.73-174, 1993.

F. Capasso and S. Datta, "Quantum Electron Devices," *Physics Today*, pp.74-82, February 1990.

Feynman, R., "There's Plenty of Room at the Bottom: An invitation to Enter a New Field of Physics," Talk at the Annual Meeting of the American Physical Society, 29 December 1959.

Feynman, R.: 1999 [1959], 'There's Plenty of Room at the Bottom: An Invitation to Enter a New Field of Physics', in: *The Pleasure of Finding Things Out*, Perseus, Cambridge, MA, pp. 117-139.

Fielder, F.A.; Reynolds, G.H.: 1994, 'Legal Problems of Nanotechnology: An Overview', *Southern California Interdisciplinary Law Journal*, 3, 593-629.

Fleischer, Th.; Decker, M. & Fiedeler, U. (eds.): 2004, *Große Aufmerksamkeit für kleine Welten ' Nanotechnologie und ihre Folgen*, special issue of *Technikfolgenabschätzung ' Theorie und Praxis*, 13 (2), 5-85 [www.itas.fzk.de/tatup/042/inhalt.htm].

Fogelberg, H. & Glimell, H.: 2003, 'Molecular Matters: In Search of the Real Stuff', in: H. Fogelberg & H. Glimell, *Bringing Visibility to the Invisible: Towards A Social Understanding of Nanotechnology*, Göteborg: Göteborg University, pp. 5-32 [www.sts.gu.se/publications/STS_report_6.pdf].

Fogelberg, H.: 2003, 'The Material Culture of Nanotechnology', in: H. Fogelberg & H. Glimell, *Bringing Visibility to the Invisible: Towards A Social Understanding of Nanotechnology*, Göteborg: Göteborg University, pp. 99-114.

Francoeur, E.: 1997, 'The Forgotten Tool: The Design and Use of Molecular Models', *Social Studies of Science*, 27, 7-40.

Frazier, G., "An Ideology For Nanoelectronics," in *Concurrent Computations: Algorithms, Architecture, and Technology*, Plenum Press, New York, 1988.

Galison, P.: 1996, 'Computer Simulations and the Trading Zone', in: P. Galison, D.J. Stump (eds.), *The Disunity of Science: Boundaries, Contexts, and Power*, Stanford University Press, Stanford, pp. 118-157.

Galison, P.: 1997, *Image and Logic: A Material Culture of Microphysics*, University of Chicago Press, Chicago.

Galison, P.; Stump, D.J. (eds.): 1996, *The Disunity of Science: Boundaries, Contexts, and Power*, Stanford University Press, Stanford.

Gieryn, T.F.: 1999, *Cultural Boundaries of Science: Credibility on the Line*, University of Chicago Press, Chicago.

Gieryn, T.F.; Figert, A.E.: 1986, 'Scientists Protect Their Cognitive Authority: The Status Degradation Ceremony of Sir Cyril Burt', in: G. Böhme, N. Stehr (eds.), *The Knowledge Society*, Reidel, Dordrecht, pp. 67-89.

Glimell, H.: 2001, 'Challenging Limits' Excerpts from an Emerging Ethnography of Nano Physicists', in: H. Glimell & O. Johlin (eds.), *The Social Production of Technology: On the Everyday Life with Things*, Göteburg, SE: BAS Publisher, chapter 7, pp. 111-131 (reprinted in: H. Fogelberg & H. Glimell, *Bringing Visibility to the Invisible: Towards A Social Understanding of Nanotechnology*, Göteborg: Göteborg University, pp. 115-139.

Glimell, H.: 2001, 'Dynamics of the Emerging Field of Nanoscience', in: M.C. Roco and W.S. Bainbridge (eds.), *Societal Implications of Nanoscience and Nanotechnology*, Dordrecht: Kluwer, pp. 156-160.

Glimell, H.: 2003, 'Dynamics of the Emerging Field of Nanoscience', in: H. Fogelberg & H. Glimell, *Bringing Visibility to the Invisible: Towards A Social Understanding of Nanotechnology*, Göteborg: Göteborg University, pp. 79-85 [www.sts.gu.se/publications/STS_report_6.pdf].

Glimell, H.: 2004, 'Grand Visions and Lilliput Politics: Staging the Exploration of the 'Endless Frontier'', in: D. Baird, A. Nordmann & J. Schummer (eds.), *Discovering the Nanoscale*, Amsterdam: IOS Press, pp. 231-246.

Goldhaber-Gordon, D., Montemerlo, M.S., Love, J.C., Opiteck, G.J., and Ellenbogen, J.C., "Overview of Nanoelectronic Devices," submitted to the *Proceedings of the IEEE*, February 1997. For more information, please send e-mail to nanotech@mitre.org.

Gorman, M.E.; Groves, J.F. & Shrager, J.: 2004, 'Societal Dimensions of Nanotechnology as a Trading Zone: Results from a Pilot Project', in: D. Baird, A. Nordmann & J. Schummer (eds.), *Discovering the Nanoscale*, Amsterdam: IOS Press, pp. 63-73.

Grinbaum, A. & Dupuy, J.-P.: 2004, 'Living with Uncertainty: Toward the Ongoing Normative Assessment of Nanotechnology', *Techne: Research in Philosophy and Technology*, 8(3) (forthcoming).

Grinbaum, A.: 2004, 'La condition de l'homme moderne et les nanotechnologies', in: G. Nivat (ed.), *Les limites de l´humain. 39èmes Rencontres Internationales de Genève*, L´Age d'Homme, Genève, p. 141.

Gross, M.: 1999, *Travels to the Nanoworld: Miniature Machinery in Nature and Technology*, Plenum, New York.

Grunwald, A.: 2004, 'Ethische Aspekte der Nanotechnologie. Eine Felderkundung', *Technikfolgenabschätzung ' Theorie und Praxis*, 13 (2), 71-78.

Gupta, V.K. & Pangannaya, N.B.: 2000, 'Carbon nanotubes: bibliometric analysis of patents', *World Patent Information*, 22, 185-189.

Guthold, M.; Falvo, M.R.; Matthews, W.G.; Paulson, S.; Washburn, S.; Erie, D.A.; Superfine, R.; Brooks, F.P.; Taylor, R.M.: 2000, 'Controlled Manipulation of Molecular Samples with the NanoManipulator', *IEEE-ASME Transactions on Mechatronics*, 5, 189-198.

Hacking, I.: 1983, *Representing and Intervening: Introductory Topics in the Philosophy of Natural Science*, Cambridge University Press, Cambridge, UK.

Hacking, I.: 1992, 'The Self-Vindication of the Laboratory Sciences', in: A. Pickering (ed.), *Science as Practice and Culture*, University of Chicago Press, Chicago, pp. 29-64.

Hansson, S.O.: 2004, 'Great Uncertainty about Small Things', *Techne: Research in Philosophy and Technology*, 8(3) (forthcoming).

Haruta, M. 1997. *Catalyst surveys of Japan* 1:61 and references therein.

Hayles, N.K. (ed.): 2004, *Nanoculture: Implications of the New Technoscience*, Bristol, UK: Intellect Books.

Hayles, N.K.: 2004, 'Connecting the Quantum Dots: Nanotechscience and Culture', in: N.K. Hayles (ed.), *Nanoculture: Implications of the New Technoscience*, Bristol, UK: Intellect Books, pp. 11-26.

Ihde, D.: 1991, *Instrumental Realism: The Interface between Philosophy of Science and Philosophy of Technology*, Indiana University Press, Bloomington.

Jena, P., S.N. Khanna, and B.K. Rao. 1996. In *Science and technology of atomically engineered materials*, ed. P. Jena. River Edge, NJ: World Scientific.

Johansson, M.: 2003, 'Plenty of room at the bottom: Towards an anthropology of nanoscience', *Anthropology Today*, 19 (No. 6), 3-6.

Johnson, A.: 2004, 'The End of Pure Science: Science Policy from Bayh-Dole to the NNI', in: D. Baird, A. Nordmann & J. Schummer (eds.), *Discovering the Nanoscale*, Amsterdam: IOS Press, pp. 217-230.

Jones, C.W., and W.J. Koros. 1994. *Carbon* 32:1419.

Jounet, C., W.K. Maser, P. Bernier, A. Loiseau, M. Lamy de la Chapelle, S. Lefrant, P. Deniard, R. Lee, and J.E. Fischer. 1997. *Nature* 388:756.

K.E. Drexler, *Nanosystems: Molecular Machinery, Manufacturing, and Computation*, Wiley, New York, 1992.

Karger, J., and D.M. Ruthven. 1992. *Diffusion in zeolites*. New York: J. Wiley.

Kay, L.: 2000, *Who Wrote the Book of Life? A History of the Genetic Code*, Stanford University Press, Stanford.

Keiper, A.: 2003, 'The Nanotechnology Revolution', *The New Atlantis*, 2, 17-34.

Khushf, G.: 2004, 'A Hierarchical Architecture for Nano-scale Science and Technology: Taking Stock of the Claims About Science Made By Advocates of NBIC Convergence', in: D. Baird, A. Nordmann & J. Schummer (eds.), *Discovering the Nanoscale*, Amsterdam: IOS Press, pp. 21-33.

Khushf, G.: 2004, 'Systems Theory and the Ethics of Human Enhancement: A Framework for NBIC Convergence', *Annals of the New York Academy of Sciences*, 1013, 124-149.

Kline, R.: 1992, *Steinmetz: Engineer and Socialist*, Johns Hopkins University Press, Baltimore.

Kline, R.: 1995, 'Construing 'Technology' as 'Applied Science'': Public Rhetoric of Scientists and Engineers in the United States, *Isis*, 86, 194-221.

Kline, R.: 2000, 'The Paradox of 'Engineering Science': A Cold War Debate about Education in the United States', *IEEE Technology and Society Magazine*, 19, 19-25.

Kline, R.; Pinch, T.: 1996, 'Users as Agents of Technological Change: The Social Construction of the Automobile in the Rural United States', *Technology and Culture*, 37, 763-95.

Knorr-Cetina, K.: 1992, 'The Couch, the Cathedral, and the Laboratory: On the Relationship between Experiment and Laboratory in Science', in: A. Pickering (ed.), *Science as Practice and Culture*, University of Chicago Press, Chicago, pp. 113-138.

Knorr-Cetina, K.: 1999, *Epistemic Cultures: How the Sciences Make Knowledge*, Harvard University Press, Cambridge, MA.

Kranakis, E.: 1997, *Constructing a Bridge: An Exploration of Engineering Culture, Design, and Research in Nineteenth-Century France and America*, MIT Press, Cambridge, MA.

Krätschmer, W., L.D. Lamb, K. Fostiropoulos, and D.R. Huffman. 1990. *Nature* 347:354.

Kresge, C.T., M.E. Leonowicz, W.J. Roth, J.C. Vartuli, and J.S. Beck. 1992. *Nature* 359:710.

Kuhn, T.S.: 1996, *The Structure of Scientific Revolutions*, University of Chicago Press, Chicago.

Latour, B.: 1983, 'Give Me a Laboratory and I Will Raise the World', in: K. Knorr-Cetina, M. Mulkay (eds.), *Science Observed*, Sage, London, pp. 141-170.

Latour, B.: 1996, *Aramis or The Love of Technology*, Harvard University Press, Cambridge, MA.

Layton, E.T., Jr.: 1971, 'Mirror-Image Twins: The Communities of Science and Technology in 19th-Century America', *Technology and Culture*, 12, 562-580.

López, J.: 2004, 'Bridging the Gaps: Science Fiction in Nanotechnology', *Hyle: International Journal for Philosophy of Chemistry*, 10(2), 129-152.

Lösch, A.: 2004, 'Nanomedicine and Space: Discursive Orders of Mediating Innovations', in: D. Baird, A. Nordmann & J. Schummer (eds.), *Discovering the Nanoscale*, Amsterdam: IOS Press, pp. 193-202.

Mackenzie, D.: 1996a, 'Economic and Sociological Explanations of Technological Change', in: *Knowing Machines: Essays on Technical Change*, MIT Press, Cambridge, MA, pp. 49-65.

Mackenzie, D.: 1996b, 'Marx and the Machine', in: *Knowing Machines: Essays on Technical Change*, MIT Press, Cambridge, MA, pp. 23-47.

Marshall, K.: 2004, 'Atomizing the Risk Technology', in: N.K. Hayles (ed.), *Nanoculture: Implications of the New Technoscience*, Bristol, UK: Intellect Books, pp. 147-160.

Martin, T.P., N. Malinowski, U. Zimmerman, U. Naher, and H. Schaber. 1993. *J. Chem. Phys.* 99:4210.

Martin, T.P., U. Naher, H. Schaber, U. Zimmerman. 1993. *Phys. Rev. Lett.* 70:3079.

Marty, A.: 2003, *Testimony before the Committee on Science on HR 766, Nanotechnology Research and Development Act*, US Government Printing Office, Washington, DC.

Marx, K.: 1963 [1847], *The Poverty of Philosophy*, International Publishers, New York.

Mayer, S.: 2002, 'From genetic modification to nanotechnology: the dangers of 'sound science'', in: T. Gilland (ed.), *Science: Can We Trust the Experts?*, London: Hodder and Stoughton, pp. 1-15.

Mehta, M.D.: 2002, 'Nanoscience and Nanotechnology: Assessing the Nature of Innovation in These Fields', *Bulletin of Science, Technology & Society*, 22(4), 269-273.

Mehta, M.D.: 2004, 'From Biotechnology to Nanotechnology: What Can We Learn from Earlier Technologies?, *Bulletin of Science, Technology & Society*, 24, 34-39.

Meyer, M. & Kuusi, O.: 2004, 'Nanotechnology: Generalizations in an Interdisciplinary Field of Science and Technology', *Hyle: International Journal for Philosophy of Chemistry*, 10(2), 153-168.

Meyer, M.: 2000a, 'Does science push technology? Patents citing scientific literature', *Research Policy*, 29, 409-434.

Meyer, M.: 2000b, 'Patent citations in a novel field of technology: What can they tell about interactions of emerging communities of science and technology?', *Scientometrics*, 48, 151-178.

Meyer, M.: 2001a, 'Patent citations in a novel field of technology: An exploration of nano-science and nano-technology', *Scientometrics*, 51, 163-183.

Meyer, M.: 2001b, 'Socio-economic Research on Nanoscale Science and Technology: A European Overview and Illustration', in: M.C. Roco and W.S. Bainbridge (eds.), *Societal Implications of Nanoscience and Nanotechnology*, Dordrecht: Kluwer, pp. 217-241.

Meyer, M.; Persson, O.: 1998, 'Nanotechnology ' interdisciplinarity, patterns of collaboration and differences in application', *Scientometrics*, 47, 195'205.

Milburn, C.: 2002, 'Nanotechnology in the Age of Posthuman Engineering: Science Fiction as Science', *Configurations*, 10, 261-295.

Milburn, C.: 2004, 'Nano/Splatter: Disintegrating the Postbiological Body', *New Literary History*, 35 (in print).

Misa, T.: 1988, 'How Machines Make History and How Historians (and Others) Help Them to Do So', *Science, Technology, and Human Values*, 13, 308-331.

Mnyusiwalla, A.; Abdallah, S.D.; Singer, P.A.: 2003, 'Mind the gap: science and ethics in nanotechnology', *Nanotechnology*, 14, R9-R13.

Mody, C.C.M.: 2000, "A New Way of Flying': Différance, Rhetoric, and the Autogiro in Interwar Aviation', *Social Studies of Science*, 30, 513-543.

Mody, C.C.M.: 2001, 'A Little Dirt Never Hurt Anyone: Knowledge-Making and Contamination in Materials Science', *Social Studies of Science*, 31, 7-36.

Mody, C.C.M.: 2004, 'How Probe Microscopists Became Nanotechnologists', in: D. Baird, A. Nordmann & J. Schummer (eds.), *Discovering the Nanoscale*, Amsterdam: IOS Press, pp. 119-133.

Mody, C.C.M.: 2004, 'How Probe Microscopists Became Nanotechnologists', in: D. Baird, A. Nordmann, J. Schummer (eds.), *Discovering the Nanoscale*, IOS Press, Amsterdam, pp. 119-133.

Mody, C.C.M.: 2004, 'Instruments in Training: The Growth of American Probe Microscopy in the 1980s', in: D. Kaiser (ed.), *Pedagogy and the Practice of Science: Producing Physical Scientists, 1800-2000*, Cambridge, MA: MIT Press (forthcoming).

Mody, C.C.M.: 2004, 'Small, but Determined: Technological Determinism in Nanoscience', *Hyle: International Journal for Philosophy of Chemistry*, 10(2), 99-128.

Mody, C.C.M.: 2004, *Crafting the Tools of Knowledge: The Invention, Spread, and Commercialization of Probe Microscopy, 1960-2000*, Ph.D. dissertation, Cornell University.

Montemerlo, M.S., Love, J.C., Opiteck, G.J., Goldhaber, D. J., and Ellenbogen, J.C., "Technologies and Designs for Electronic Nanocomputers," MITRE Technical Report 96W0000044, The MITRE Corporation, McLean, VA, July 1996. For more information, send e-mail to nanotech@mitre.org.

Moor, J.H. & Weckert, J.: 2004, 'Nanoethics: Assessing the Nanoscale From an Ethical Point of View', in: D. Baird, A. Nordmann & J. Schummer (eds.), *Discovering the Nanoscale*, Amsterdam: IOS Press, pp. 301-310.

Munn Sanchez, E.: 2004, 'The Expert's Role in Nanoscience and Technology', in: D. Baird, A. Nordmann & J. Schummer (eds.), *Discovering the Nanoscale*, Amsterdam: IOS Press, pp. 257-266.

Nordmann, A.: 2003, 'Shaping the World Atom by Atom: Eine nanowissenschaftliche WeltBildanalyse', in: A. Grunwald (ed.), *Technikgestaltung zwischen Wunsch und Wirklichkeit*, Berlin: Springer, pp. 191-199.

Nordmann, A.: 2004, 'Molecular Disjunctions: Staking Claims at the Nanoscale', in: D. Baird, A. Nordmann & J. Schummer (eds.), *Discovering the Nanoscale*, Amsterdam: IOS Press, pp. 51-62.

Nordmann, A.: 2004, 'Molecular Disjunctions', in: D. Baird, A. Nordmann, J. Schummer (eds.), *Discovering the Nanoscale*, IOS Press, Amsterdam, pp. 51-62.

Nordmann, A.: 2004, 'Nanotechnology: Convergence and Integration', Presentation at *EuroNanoForum, Trieste, December 10, 2003* (Proceedings in preparation).

Nordmann, A.: 2004, 'Nanotechnology's WorldView: New Space for Old Cosmologies', *IEEE Technology and Society Magazine*, 23 (forthcoming).

Nordmann, A.: 2004, 'Social Imagination for Nanotechnology', in: European Commission (Community Health and Consumer Protection): *Nanotechnologies: A Preliminary Risk Analysis on the Basis of a Workshop, Brussels, 1-2 March 2004*, pp. 111-113 [www.europa.eu.int/comm/health/ph_risk/documents/ev_20040301_en.pdf].

Nordmann, A.: 2004, 'Was ist TechnoWissenschaft? ' Zum Wandel der Wissenschaftskultur am Beispiel von Nanoforschung und Bionik', in: T. Rossmann & C. Tropea (eds.), *Bionik ' Neue Forschungsergebnisse aus Natur-, Ingenieur- und Geisteswissenschaften*, Berlin: Springer, 2004 (forthcoming)

Nye, D.E.: 1994, *American Technological Sublime*, MIT Press, Cambridge, MA.

Nye, D.E.: 2003, *America as Second Creation: Technology and Narratives of New Beginnings*, MIT Press, Cambridge, MA.

Paschen, H.; Coenen, C.; Fleischer, T.; Grünwald, R.; Oertel, D. & Revermann, C.: 2004, *Nanotechnologie: Forschung, Entwicklung, Anwendung*, Berlin: Springer (*Nanotechnologie, TAB-Arbeitsbericht 92*, Berlin: Büro für Technikfolgen-Abschätzung beim Deutschen Bundestag).

Regis, E.: 1995, *Nano: The Emerging Science of Nanotechnology*, Little, Brown, Boston.

Bibliography

Akis, R., D. Ferry, C. Musgrave, "Kinetic Lattice Monte Carlo Simulations of Processes on the Silicon (100) Surface," *Physica E-Low Dimensional Systems and Nanostructures* 19, 183-187 (2003).

Aksay, Ilhan. 1998. Nanostructured ceramics through self-assembly. In *R&D status and trends*, ed. Siegel et al.

Al Globus, Charles W. Bauschlicher Jr., Jie Han, Richard L. Jaffe, Creon Levit, Deepack Srivastava, "Machine phase fullerene nanotechnology," *Nanotechnology* 9(1998):192-199.

Alivisatos, A.P. 1996. Perspectives on the physical chemistry of semiconductor nanocrystals. *Journal of Physical Chemistry* 100:13226-13239.

Allara, D.L. 1996. Nanoscale structures engineered by molecular self-assembly of functionalized monolayers. In *Nanofabrication and biosystems*, ed. Hoch et al.

Birge, R.R. 1995. Protein based computers. *Sci. Am.* (Mar.):90-95.

Altmann, J. & Gubrud, M.A.: 2004, 'Military, Arms Control, and Security Aspects of Nanotechnology', in: D. Baird, A. Nordmann & J. Schummer (eds.), *Discovering the Nanoscale*, Amsterdam: IOS Press, pp. 269-277.

Altmann, J.: 2004, 'Military Uses of Nanotechnology: Perspectives and Concerns', *Security Dialogue*, 35 (1): 61-79.

Andres, R.P., S. Datta, D.B. Janes, C.P. Kubiak, and R. Reifenberger. 1998. The design, fabrication and electronic properties of self-assembled molecular nanostructures. In *The handbook of nanostructured materials and nanotechnology*, ed. H.S. Nalwa. San Diego: Academic Press.

Anonymous: 2002, *Small Wonders, Endless Frontiers: A Review of the National Nanotechnology Initiative*, National Academies Press, Washington, DC.

Antonio Regalado, "Nanotechnology Patents Surge as Companies Vie to Stake Claim," Wall Street Journal, June 18, 2004, p. 1.

Axelbaum, R.L. 1997. Developments in sodium/halide flame technology for the synthesis of unagglomerated non-oxide nanoparticles. In *Proc. of the Joint NSF-NIST Conference on Nanoparticles: Synthesis, Processing into Functional Nanostructures and Characterization* (May 12-13, Arlington, VA).

B. Liu, N.B. Leontis and N.C. Seeman, Bulged 3-arm DNA Branched Junctions as Components for Nanoconstruction, Nanobiology 3, 177-188 (1994). N.C. Seeman, Molecular Craftwork with DNA, The Chemical Intelligencer 1(3), 38-47 (1995).

Bainbridge, W.S.: 2002, 'Public Attitudes Toward Nanotechnology', *Journal of Nanoparticle Research*, 4, 561-570.

Baird, D. & Shew, A.: 2004, 'Probing the History of Scanning Tunneling Microscopy', in: D. Baird, A. Nordmann & J. Schummer (eds.), *Discovering the Nanoscale*, Amsterdam: IOS Press, pp. 145-156.

Baird, D.; Nordmann, A. & Schummer, J. (eds.): 2004, *Discovering the Nanoscale*, Amsterdam: IOS Press.

Baker, R.T.K. Synthesis, properties and applications of graphite nanofibers. In *R&D status and trends*, ed. Siegel et al.

Ball, P.: 2002, 'Natural Strategies for the molecular engineer', *Nanotechnology*, 13, 15-28.

Ball, P.: 2003, 'Nanotechnology in the firing line', *Nanotechweb.org,* 23 December [http://www.nanotechweb.org/articles/society/2/12/1/1].

Bate, R., Frazier, G., Frensley, W., Reed., M., "An Overview of Nanoelectronics," *Texas Instruments Technical Journal*, July-August 1989, pp. 13-20.

Baum, R.; Drexler, K.E.; Smalley, R.: 2003, 'Nanotechnology: Drexler and Smalley Make the Case For and Against 'Molecular Assemblers', *Chemical and Engineering News*, 81, 37-42.

Becker, M.F., J.R. Brock, H. Cai, N. Chaudhary, D. Henneke, L. Hilsz, J.W. Keto, J. Lee, W.T. Nichols, and H.D. Glicksman. 1997. Nanoparticles generated by laser ablation. In *Proc. of the Joint NSF-NIST Conf. on Nanoparticles*.

Bensaude-Vincent, B.: 2004, 'Two Cultures of Nanotechnology?', *Hyle: International Journal for Philosophy of Chemistry*, 10(2), 65-82.

Bensaude-Vincent, B.; Arribart, B.Y.; Sanchez, C.: 2002, 'Chemists at the School of Nature', *New Journal of Chemistry*, 26, 1-5.

Berndt, C.C., J. Karthikeyan, T. Chraska, and A.H. King. 1997. Plasma spray synthesis of nanozirconia powder. In *Proc. of the Joint NSF-NIST Conf. on Nanoparticles*.

Berube, D.M.: 2004, 'The Rhetoric of Nanotechnology', in: D. Baird, A. Nordmann & J. Schummer (eds.), *Discovering the Nanoscale*, Amsterdam: IOS Press, pp. 173-192.

Bishop, J. et al. 1990. *Surface modified drug nanoparticles*. U.S. Patent application. Docket 61894 Filed 9/17/90.

Blackman, M. 1994. *An evaluation of the Link Nanotechnology Program and the National Initiative on Nanotechnology*. Cambridge.

Brotzman, R. 1998. Nanoparticle dispersions. In *R&D status and trends*, ed. Siegel et al. Brus, L.E. 1996. Theoretical metastability of semiconductor crystallites in high pressure phases with applications to beta tin structures of silicon. *J. Am. Chem. Soc.* 118:4834-38.

Brown, J. S. and Duguid, P. Don't Count Society Out: A Response to Bill Joy. Paper prepared for the Societal Implications of Nanotechnology Conference, 2000.

Bueno, O.: 2004, 'The Drexler-Smalley Debated on Nanotechnology: Incommensurability at Work?', *Hyle: International Journal for Philosophy of Chemistry*, 10(2), 83-98.

Bueno, O.: 2004, 'Von Neumann, Self-Reproduction and the Constitution of Nanophenomena', in: D. Baird, A. Nordmann & J. Schummer (eds.), *Discovering the Nanoscale*, Amsterdam: IOS Press, pp. 101-115.

C. Mao, W. Sun, Z. Shen and N.C. Seeman, A DNA Nanomechanical Device Based on the B-Z Transition, Nature 397, 144-146 (1999).

C.D. Montemagno, G.D. Bachand, "Constructing nanomechanical devices powered by biomolecular motors," *Nanotechnology* 10(1999):225-231; G.D. Bachand, C.D. Montemagno, "Constructing organic/inorganic NEMS devices powered by biomolecular motors," *Biomedical Microdevices* 2(2000):179-184.

Calcote, H.F., and D.G. Keil. 1997. Combustion synthesis of silicon carbide powder. In *Proc. of the Joint NSF-NIST Conf. on Nanoparticles*.

Carsley, J.E., R. Shaik, W.W. Milligan, and E.C. Aifantis. 1997. In *Chemistry and physics of nanostructures and related non-equilibrium materials*. ed. E. Ma, B. Fultz, R. Shull, J. Morral, and P. Nash. Warrendale, PA: TMS.

Chantrell. 1994. *J. Appl. Phys.* 76:6811. Erb, U., G. Palumbo, R. Zugic, and K.T. Aust. 1996. In *Processing and properties of nanocrystalline materials*, ed. C. Suryanarayana, J. Singh, and F.H. Froes. Warrendale, PA: TMS. Gertsman, V.Y., M. Hoffman, H. Gleiter and R. Birringer. 1994. *Acta Metall. Mater.* 42:3539-3544.

Charles B. Musgrave, Jason K. Perry, Ralph C. Merkle, William A. Goddard III, "Theoretical studies of a hydrogen abstraction tool for nanotechnology," *Nanotechnology* 2(1991):187-195.

Chianelli, R.R. 1998. Synthesis, fundamental properties and applications of nanocrystals, sheets, and fullerenes based on layered transition metal chalcogenides. In *R&D status and trends*, ed. Siegel et al.

Chondroudis, K. and D.B. Mitzi. 1999. Electroluminescence from an organic-inorganic perovskite incorporating a quaterthiophene dye within lead halide perovskite layers. *Chemistry of Materials.* Vol 11. p. 3028.

Coenen, C.: 2004, 'Nanofuturismus: Anmerkungen zu seiner Relevanz, Analyse und Bewertung', *Technikfolgenabschätzung ' Theorie und Praxis*, 13 (2), 78-85.

Colbert, D.T. and R.E. Smalley. 1999. Fullerene nanotubes for molecular electronics. *Trends in Biotechnology*. Vol 17. 46-50.

Crandall, B.: 1999, 'Molecular Engineering', in: B. Crandall (ed.), *Nanotechnology: Molecular Speculations on Global Abundance*, MIT Press, Cambridge, MA, pp. 1-46.

Crandall, B.C. and Lewis, J., *Nanotechnology: Research and Perspectives*, MIT Press, Cambridge, 1992.

Czekai, D., et al. 1994. *Use of smaller milling media to prepare nanoparticulate dispersions*. U.S. Patent application. Docket 69802(02) Filed 2/25/94.

D.W. Brenner, S.B. Sinnott, J.A. Harrison, O.A. Shenderova, "Simulated engineering of nanostructures," *Nanotechnology* 7(1996):161-167; http://www.zyvex.com/nanotech/nano4/brennerPaper.pdf

Dai, H. J. 2000. "Controlling nanotube growth," *Physics World*, 13, 43-47.

Decker, M.; Fiedeler, U.; Fleischer, Th.: 2004, 'Ich sehe was, was Du nicht siehst... zur Definition von Nanotechnologie', *Technikfolgenabschätzung ' Theorie und Praxis*, 13 (2), 10-16.

Dresselhaus, M.S. 1998. Carbon-based nanostructures. In *R&D status and trends*, ed. Siegel et al.

Drexler K.E.:. 1995, 'Introduction to nanotechnology', in: Krummenacker, M. & Lewis, J. (eds.), *Prospects in Nanotechology. Proceedings of the 1st general conference on nanotechnology: developments, applications, and opportunities, November 11-14, 1992, Palo-Alto*, John Wiley & Sons, New York, pp. 1-20.

Drexler, E., Phoenix, C.: 2004, 'Self-Replication in Nanotechnology: Feasible, Potentially Safe, and Unnecessary' (unpublished paper, in preparation).

Drexler, E.: 1986, *Engines of Creation: The Coming Era of Nanotechnology*, Anchor Books, New York (expanded edition with a new afterword, 1990).

Drexler, E.: 1992, *Nanosystems: Molecular Machinery, Manufacturing, and Computation*, John Wiley & Sons, New York.

Drexler, E.; Smalley, R.E.: 2003, 'Point-Counterpoint: Nanotechnology', *Chemical and Engineering News*, 81(48), 37-42.

Drexler, K.E.: 1990, *Engines of Creation: The Coming Era of Nanotechnology*, Anchor Books, New York.

Drexler, K.E.: 1992, *Nanosystems. Molecular machinery, manufacturing and computation*, John Wiley & Sons, New York.

Drexler, K.E.: 2001, 'Machine-Phase nanotechnology', *Scientific American*, (Sept.), 66-67.

Dupuy, J.-P.: 2004, 'Complexity and Uncertainty: A Prudential Approach To Nanotechnology', in: European Commission (Community Health and Consumer Protection): *Nanotechnologies: A Preliminary Risk Analysis on the Basis of a Workshop, Brussels, 1-2 March 2004*, pp. 71-93 (www.europa.eu.int/comm/health/ph_risk/documents/ev_20040301_en.pdf).

Dupuy, J.P.: 2004, 'Pour une évaluation normative du programme nanotechnologique', *Annales des Mines*, (February 2004), 27-32.

Empedocles, S.; Bawendi, M. 1999. "Spectroscopy of single CdSe nanocrystallites," *Accounts of Chemical Research*, 32, 389-396.

ETC: 2003b, 'No Small Matter II: The Case for a Global Moratorium ' Size Matters!' *Occasional Paper Series* 7(1) [www.etcgroup.org/documents/Occ.Paper_Nanosafety.pdf].

European Commission (Community Health and Consumer Protection): 2004, *Nanotechnologies: A Preliminary Risk Analysis on theBasis of a Workshop, Brussels, 1-2 March 2004* [www.europa.eu.int/comm/health/ ph_risk/documents/ev_20040301_en.pdf].

Evans, D.F.; Wennerstrom, H.: 1999, *The Colloidal Domain: Where Physics, Chemistry, Biology and Technology Meet*, John Wiley & Sons, New York.

F. Liu, H. Wang and N.C. Seeman, Short Extensions to Sticky Ends for DNA Nanotechnology and DNA-Based Computation, Nanobiology 4, 257-262 (1999).

F. Mathieu, S. Liao, C. Mao, J. Kopatsch, T. Wang, and N.C. Seeman, Six-Helix Bundles Designed from DNA, NanoLetters 5, 661-665 (2005).

Fedor N. Dzegilenko, Deepak Srivastava, Subhash Saini, "Simulations of carbon nanotube tip assisted mechano-chemical reactions on a diamond surface," *Nanotechnology* 9(December 1998):325-330.

Fedor N. Dzegilenko, Deepak Srivastava, Subhash Saini, "Simulations of carbon nanotube tip assisted mechano-chemical reactions on a diamond surface," *Nanotechnology* 9(December 1998):325-330.

Fielder, F.A.; Reynolds, G.H.: 1994, 'Legal Problems of Nanotechnology: An Overview', *Southern California Interdisciplinary Law Journal*, 3, 593-629.

Fogelberg, H. & Glimell, H.: 2003, 'Molecular Matters: In Search of the Real Stuff', in: H. Fogelberg & H. Glimell, *Bringing Visibility to the Invisible: Towards A Social Understanding of Nanotechnology*, Göteborg: Göteborg University, pp. 5-32 [www.sts.gu.se/publications/STS_report_6.pdf].

Fogelberg, H. & Glimell, H.: 2003, *Bringing Visibility to the Invisible: Towards A Social Understanding of Nanotechnology*, Göteborg: Göteborg University [www.sts.gu.se/publications/STS_report_6.pdf].

Fogelberg, H.: 2003, 'The Material Culture of Nanotechnology', in: H. Fogelberg & H. Glimell, *Bringing Visibility to the Invisible: Towards A Social Understanding of Nanotechnology*, Göteborg: Göteborg University, pp. 99-114.

Fogelberg, H.; Glimell, H.: 2003, *Bringing Visibility to the Invisible: Towards a Social Understanding of Nanotechnology*, Göteborg University, Göteborg.

Franklin, N.R.; Dai, H.J. 2000. "An enhanced CVD approach to extensive nanotube networks with directionality", *Advanced Materials*, 12, 890-894.

Frazier, G., "An Ideology For Nanoelectronics," in *Concurrent Computations: Algorithms, Architecture, and Technology*, Plenum Press, New York, 1988.

Frenken, J.W.M. 1998. "Scanning Tunneling Microscopy", section 5.2, A. ten Wolde (ed.), *Nanotechnology: Towards a molecular construction kit*. STT Report #60, The Hague, 289-299.

Friedlander, S.K. 1993. Controlled synthesis of nanosized particles by aerosol processes. *Aerosol Sci. Technol*. 19:527.

Friedlander, S.K. 1998. Synthesis of nanoparticles and their agglomerates: aerosol reactors. In *R&D status and trends*, ed. Siegel et al.

G. Serrano & N.C. Seeman, Nanotecnologia basada en ADN, Revista de Química 19, 11-20 (2004), in Spanish.

George M. Whitesides, "The Once and Future Nanomachine," *Scientific American* 285(September 2001):78-83.

Gleiter, H. 1989. *Prog. Mater. Sci.* 33:223. Goddard, W.A. 1998. Nanoscale theory and simulation. In *R&D status and trends*, ed. Siegel et al.

Glimell, H.: 2003, 'A Nano Narrative: The Mircopolitics of a New Generic Enabling Technology', in: H. Fogelberg & H. Glimell, *Bringing Visibility to the Invisible: Towards A Social Understanding of Nanotechnology*, Göteborg: Göteborg University, pp. 55-77.

Glimell, H.: 2004, 'Grand Visions and Lilliput Politics: Staging the Exploration of the 'Endless Frontier", in: D. Baird, A. Nordmann & J. Schummer (eds.), *Discovering the Nanoscale*, Amsterdam: IOS Press, pp. 231-246.

Goddard, W.A. 1998. Nanoscale theory and simulation. In *R&D status and trends*, ed. R. Siegel et al.

Godsell, D.: 2003, *Living Machinery. Bionanotechnology: Lessons from Nature*, Wiley-Liss, New-York.

Goldhaber-Gordon, D., Montemerlo, M.S., Love, J.C., Opiteck, G.J., and Ellenbogen,

J.C., “Overview of Nanoelectronic Devices,” submitted to the *Proceedings of the IEEE*, February 1997. For more information, please send e-mail to nanotech@mitre.org.

Gorman, M.E.; Groves, J.F. & Shrager, J.: 2004, ‘Societal Dimensions of Nanotechnology as a Trading Zone: Results from a Pilot Project’, in: D. Baird, A. Nordmann & J. Schummer (eds.), *Discovering the Nanoscale*, Amsterdam: IOS Press, pp. 63-73.

Grinbaum, A. & Dupuy, J.-P.: 2004, ‘Living with Uncertainty: Toward the Ongoing Normative Assessment of Nanotechnology’, *Techne: Research in Philosophy and Technology*, 8(3) (forthcoming).

Grinbaum, A.: 2004, ‘La condition de l’homme moderne et les nanotechnologies’, in: G. Nivat (ed.), *Les limites de l´humain. 39èmes Rencontres Internationales de Genève*, L´Age d´Homme, Genève, p. 141.

Gross, M.: 1999, *Travels to the Nanoworld: Miniature Machinery in Nature and Technology*, Plenum, New York.

Grunwald, A.: 2004, ‘Ethische Aspekte der Nanotechnologie. Eine Felderkundung’, *Technikfolgenabschätzung ’ Theorie und Praxis*, 13 (2), 71-78.

Gubrud, M. A., *Nanotechnology and International Security*, Fifth Foresight Conference on Molecular Nanotechnology, November 5-8, 1997. Ref: http://www.foresight.org/ Conferences/MNT05/Papers/Gubrud/.

Gupta, V.K. & Pangannaya, N.B.: 2000, ‘Carbon nanotubes: bibliometric analysis of patents’, *World Patent Information*, 22, 185-189.

Guthold, M.; Falvo, M.R.; Matthews, W.G.; Paulson, S.; Washburn, S.; Erie, D.A.; Superfine, R.; Brooks, F.P.; Taylor, R.M.: 2000, ‘Controlled Manipulation of Molecular Samples with the NanoManipulator’, *IEEE-ASME Transactions on Mechatronics*, 5, 189-198.

H. Yan and N.C. Seeman, Edge-Sharing Motifs in DNA Nanotechnology. Journal of Supramolecular Chemistry 1, 229-237 (2003).

Hadjipanayis, C.G. 1998. Nanostructured magnetic materials. In *R&D Status and Trends*, ed. Siegel et al.

Hemerén, G.: 2004, ‘Nano Ethics Primer’, in: European Commission (Community Health and Consumer Protection): *Nanotechnologies: A Preliminary Risk Analysis on theBasis of a Workshop, Brussels, 1-2 March 2004* (www.europa.eu.int/ comm/health/ ph_risk/documents/ev_20040301_en.pdf), pp. 95-102.

Hessenbruch, A.: 2004, ‘Nanotechnology and the Negotiation of Novelty’, in: D. Baird, A. Nordmann & J. Schummer (eds.), *Discovering the Nanoscale*, Amsterdam: IOS Press, pp. 135-144.

Hietpas, P.B., S.D. Gilman, R.A. Lee, M.R. Wood, N. Winograd, and A.G. Ewing.

1996. Development of votammetric methods, capillary electrophoresis and tof sims imaging for constituent analysis of single cells. In *Nanofabrication and biosystems*, ed.

Higgins, R.J. 1997. An economical process for manufacturing of nano-sized powders based on microemulsion-mediated synthesis. In *Proc. of the Joint NSF-NIST Conf. on Nanoparticles.*

Hoch, H.C., L.W. Jelinski, and H.G. Craighead, eds. 1996. *Nanofabrication and biosystems*. New York: Cambridge University Press.

Hullmann, A. & Meyer, M.: 2003, 'Publications and Patents in Nanotechnology. An overview of previous studies and the state of the art', *Scientometrics*, 58 (3) 507-527.

J. D. Le, Y. Pinto, N.C. Seeman, K. Musier-Forsyth, T. A. Taton and R.A. Kiehl, Self-Assembly of Nanoelectronic Component Arrays by In Situ Hybridization to 2D DNA Scaffolding, NanoLetters 4, 2343-2347 (2004).

J. Storrs Hall, "Architectural considerations for self-replicating manufacturing systems," *Nanotechnology* 10(1999):323-330; http://www.foresight.org/Conferences/MNT6/Papers/Hall/index.html

J. Storrs Hall, "Architectural considerations for self-replicating manufacturing systems," *Nanotechnology* 10(1999):323-330; http://www.foresight.org/Conferences/MNT6/Papers/Hall/index.html

J.W. Lyding, K. Hess, G.C. Abeln, D.S. Thompson, J.S. Moore, M.C. Hersam, E.T. Foley, J. Lee, Z. Chen, S.T. Hwang, H. Choi, P.H. Avouris, I.C. Kizilyalli, "UHV-STM nanofabrication and hydrogen/deuterium desorption from silicon surfaces: implications for CMOS technology," *Appl. Surf. Sci.* 130(1998):221-230.

Junno, T., S.-B. Carlsson, H. Xu, L. Montelius, and L. Samuelson. 1998. Fabrication of quantum devices by angstrom-level manipulation of nanoparticles with an atomic force microscope. *Appl. Phys. Lett.* 72:548-550.

K. Eric Drexler, "Molecular nanomachines: Physical principles and implementation strategies," *Annu. Rev. Biophys. Biomol. Struct.* 23(1994):377-405.

K. Eric Drexler, "Nanomachinery: Atomically precise gears and bearings," *IEEE Micro Robots and Teleoperators Workshop*, Hyannis, Cape Cod, November 1987.

K. Eric Drexler, David Forrest, Robert A. Freitas Jr., J. Storrs Hall, Neil Jacobstein, Ralph Merkle, Christine Peterson, "On Physics, Fundamentals, and Nanorobots: A Rebuttal to Smalley's Assertion that Self-Replicating Mechanical Nanorobots Are Simply Not Possible"; http://www.imm.org/SciAmDebate2/smalley.html

K. Eric Drexler, *Engines of Creation: The Coming Era of Nanotechnology*, Anchor Press/Doubleday, New York, 1986; http://www.foresight.org/EOC/

K. Eric Drexler, *Nanosystems: Molecular Machinery, Manufacturing, and Computation*, John Wiley & Sons, New York, 1992.

Kang, J. and C. Musgrave, "A Quantum Chemical Study of the Self-Directed Growth Mechanism of Styrene and Propylene Molecular Nanowires on the Silicon (100) 2x1 Surface," *Journal of Chemical Physics* 116, 9907-9913 (2002).

Ke, M., S.A. Hackney, W.W. Milligan, and E.C. Aifantis. 1995. *Nanostructured Mater*. 5:689.

Kear, B., and G. Skandan. 1998. Nanostructural bulk materials: Synthesis, processing, properties and performance. *In R&D status and trends*, ed.

Kear, B.H., R.K. Sadangi, and S.C. Liao. 1997. Synthesis of WC/Co/diamond nanocomposites. In *Proc. of the Joint NSF-NIST Conf. on Nanoparticles*.

Keiper, A.: 2003, 'The Nanotechnology Revolution', *The New Atlantis*, 2, 17-34.

Laszlo, P.: 2004, 'Is There Life After Partington?', *Hyle: International Journal for Philosophy of Chemistry*, 10(2), 169-178.

Lee, M.S. and M. Head-Gordon. 2000. Absolute and relative energies from polarized atomic orbital selfconsistent field calculations and a second order convergence with size and composition of the secondary basis. *Computers and Chemistry*. Vol. 24(3,4). 295-301.

Lehn, J.M.: 1995, *Supramolecular Chemistry*, VCH, Weinheim.

Lent, C.S., Tougaw, P.D., Porod, W., Bernstein, G.H., "Quantum Cellular Automata," *Nanotechnology*, Vol. 4, p. 49, 1993.

Lide, D.R., ed. 1993-1994. *CRC Handbook of Chemistry and Physics*, 74th ed.

Lin-Easton, P.C.: 2001, 'It's Time for Environmentalists to Think Small ' Real Small: A Call for the Involvement of Environmental Lawyers in Developing Precautionary Policies Molecular Nanotechnology', *Georgetown International Law Review*, 14, 106-134.

Lösch, A.: 2004, 'Nanomedicine and Space: Discursive Orders of Mediating Innovations', in: D. Baird, A. Nordmann & J. Schummer (eds.), *Discovering the Nanoscale*, Amsterdam: IOS Press, pp. 193-202.

Lukens, R.: 2004, 'Historic Nanotechnology', *Chemical Heritage*, 22, 29.

M.C. Hersam, G.C. Abeln, J.W. Lyding, "An approach for efficiently locating and electrically contacting nanostructures fabricated via UHV-STM lithography on Si(100)," *Microelectronic Engineering* 47(1999):235-.

M.C. Hersam, G.C. Abeln, J.W. Lyding, "An approach for efficiently locating and electrically contacting nanostructures fabricated via UHV-STM lithography on Si(100)," *Microelectronic Engineering* 47(1999):235-.

M.C. Roco, "National Nanotechnology Initiative: Overview," September 20, 2004.

Marshall, K.: 2004, 'Atomizing Risk Technology', in: K. Hayles (ed.), *NanoCulture*, Intellect Books, Bristol, pp. 79-102.

McConnell, H.M. 1996. Light-addressable potentiometric sensor: Applications to drug discovery. In *Nanofabrication and biosystems*, ed. Hoch et al.

Mehl, R.F., and R.W. Cahn. 1983. Historical development. In *Physical metallurgy*. North Holland. Milligan, W.W., S.A. Hackney, M. Ke, and E.C. Aifantis. 1993. *Nanostructured Materials* 2:267.

Meyer, M. and O. Persson. 1998. "Nanotechnology - interdisciplinarity, patterns of collaboration and differences in application," *Scientometrics*, XXXXII (2), 195-205.

Meyer, M.; Persson, O.: 1998, 'Nanotechnology ' interdisciplinarity, patterns of collaboration and differences in application', *Scientometrics*, 47, 195'205.

Milburn, C.: 2004, 'Nano/Splatter: Disintegrating the Postbiological Body', *New Literary History*, 35 (in print).

Minsky, M.: 1995, 'Virtual Molecular Reality', in: Krummenacker, M. & Lewis, J. (eds.), *Prospects in Nanotechology. Proceedings of the 1st general conference on nanotechnology: developments, applications, and opportunities, November 11-14, 1992, Palo-Alto*, John Wiley & Sons, New York, pp. 187-205.

Mishra, R.S., and A.K. Mukherjee. 1997. Oral presentation at TMS meeting, Indianapolis, Indiana, 16-18 September 1997, to be published in proceedings of Symp. "Mechanical Behavior of Bulk Nano-Materials."

Mishra, R.S., R.Z. Valiev, and A.K. Mukherjee. 1997. *Nanostructured Materials* 9:473. Morris, D.G., and M.A. Morris. 1991. *Acta Metall. Mater*. 39:1763-1779.

Mody, C.C.M.: 2004, 'How Probe Microscopists Became Nanotechnologists', in: D. Baird, A. Nordmann & J. Schummer (eds.), *Discovering the Nanoscale*, Amsterdam: IOS Press, pp. 119-133.

Montemerlo, M.S., Love, J.C., Opiteck, G.J., Goldhaber, D. J., and Ellenbogen, J.C., "Technologies and Designs for Electronic Nanocomputers," MITRE Technical Report 96W0000044, The MITRE Corporation, McLean, VA, July 1996. For more information, send e-mail to nanotech@mitre.org.

Moor, J.H. & Weckert, J.: 2004, 'Nanoethics: Assessing the Nanoscale From an Ethical Point of View', in: D. Baird, A. Nordmann & J. Schummer (eds.), *Discovering the Nanoscale*, Amsterdam: IOS Press, pp. 301-310.

Moronne, M.M.; Ben Dahan, M.; Bruchez, M.; Hamamoto, D.J.; Weiss, S.; Alivisatos, A.P. 1999. "Nanocrystal phosphors as biological probes for fluorescence microscopy", *Biophysical Journal*, 76, A450-A450.

Mui, C., S. Bent and C. Musgrave, "A Quantum Chemistry Based Statistical Mechanical

Model of Hydrogen Deposition from Si(100)-2x1, Ge(100)-2x1, and SiGe Alloy Surfaces," *Journal of Physical Chemistry B* , 108, 18243-18253, (2004).

N.C. Seeman and P.S. Lukeman, Nucleic Acid Nanostructures, Reports on Progress in Physics 68, 237-270 (2005).

N.C. Seeman, At the Crossroads of Chemistry, Biology and Materials: Structural DNA Nanotechnology, Chemistry & Biology 10, 1151-1159 (2003).

N.C. Seeman, B. Ding , S. Liao , T. Wang, W.B. Sherman, P.E. Constantinou, J. Kopatsch, C. Mao, R. Sha, F. Liu, H. Yan & P.S. Lukeman, Experiments in Structural DNA Nanotechnology: Arrays and Devices , Proc. SPIE; Nanofabrication: Technologies, Devices and Applications 5592, 71-81 (2005).

N.C. Seeman, Biochemistry and Structural DNA Nanotechnology: An Evolving Symbiotic Relationship, Biochemistry 42, 7259-7269 (2003).

N.C. Seeman, C. Mao, F. Liu, R. Sha, X. Yang, L. Wenzler, X. Li, Z. Shen, H. Yan, P. Sa-Ardyen, X. Zhang, W. Shen, J. Birac, P. Lukeman, Y. Pinto, J. Qi, B. Liu, H. Qiu, S.M. Du, H. Wang, W. Sun, Y. Wang, T.-J. Fu, Y. Zhang, J.E. Mueller and J. Chen, Nicks, Nodes, and New Motifs for DNA Nanotechnology, Frontiers of Nano-Optoelectronic Systems, ed. by L. Pavesi & E. Buzanova, Kluwer, Dordrecht, 177-198 (2000).

N.C. Seeman, Design and Engineering of Nucleic Acid Nanoscale Assemblies, Current Opinion in Structural Biology, 6, 519-526 (1996).

N.C. Seeman, Directing the Structure of Matter through Nanotechnology, Proceedings of IEEE International Joint Symposia on Intelligence and Systems IEEE Computer Society, Los Alamitos, CA, 146-150 (1998).

N.C. Seeman, DNA Engineering and its Application to Nanotechnology, Trends in Biotechnology 17, 437-443 (1999).

N.C. Seeman, DNA Nanostructures for Mechanics and Computing: Nonlinear Thinking With Life's Central Molecule. NanoBiotechnology. Editors, Chad Mirkin and Christof Niemeyer, Wiley-VCH Verlag GmbH & Co., KgaA, Weinheim, Chapter 20, pp 308-318 (2004).

N.C. Seeman, DNA Nanostructures for Mechanics and Computing: Nonlinear Thinking With Life's Central Molecule. NanoBiotechnology. Editors, Chad Mirkin and Christof Niemeyer, Wiley-VCH Verlag GmbH & Co., Weinheim, Chapter 20, pp 308-318 (2004).

N.C. Seeman, DNA Nanotechnology, *Encyclopedia of Supramolecular Chemistry*, Marcel Dekker, New York, 475-483 (2004).

N.C. Seeman, DNA Nanotechnology: From Topological Control to Structural Control, in Pattern Formation in Biology, Vision and Dynamics, ed. by A.Carbone,

M.Gromov, P.Pruzinkiewicz, World Scientific Publishing Company, Singapore, 271-309 (2000).

N.C. Seeman, DNA Nanotechnology: Life's Central Performer in a New Role, Biological Physics Newsletter 2 (1) 2-6 (2002).

N.C. Seeman, DNA Nanotechnology: Novel DNA Constructions. Annual Review of Biophysics and Biomolecular Structure 27, 225-248 (1998).

N.C. Seeman, DNA Nicks and Nodes and Nanotechnology, *NanoLetters* 1, 22-26 (2001).

N.C. Seeman, F. Liu, C. Mao, X. Yang, L. A. Wenzler, R. Sha, W. Sun, Z. Shen, X. Li, J. Qi, Y. Zhang, T.-J. Fu, J. Chen and E. Winfree, Two Dimensions and Two States in DNA Nanotechnology, Proceedings of the 11th Conversation in Biomolecular Stereodynamics, ed. by R.H. Sarma and M.H. Sarma, Adenine Press, New York, 253-262 (2000).

N.C. Seeman, H. Wang, J. Qi., X.J. Li, X.P. Yang, Y. Wang, H. Qiu, B. Liu, Z. Shen, W. Sun, F. Liu, J.J. Molenda, S.M. Du, J. Chen, J.E. Mueller., Y. Zhang, T.-J. Fu, and S. Zhang, DNA Nanotechnology and Topology. In: Biological Structure and Dyinamics, ed. by R.H. Sarma and M.H. Sarma, Adenine Press, New York, vol. 2, pp. 319-341 (1996).

N.C. Seeman, H. Wang, X. Yang, F. Liu, C. Mao, W. Sun, L. Wenzler, Z. Shen, R. Sha, H. Yan, M.H. Wong, P. Sa-Ardyen, B. Liu, H. Qiu, X. Li, J. Qi, S.M. Du, Y. Zhang, J.E. Mueller, T.-J. Fu, Y. Wang, and J. Chen, New Motifs in DNA Nanotechnology, Nanotechnology 9, 257-273 (1998).

N.C. Seeman, In the Nick of Space: Generalized Nucleic Acid Complementarity and the Development of DNA Nanotechnology, Synlett 2000, 1536-1548 (2000).

N.C. Seeman, J. Chen, S.M. Du, J.E. Mueller, Y. Zhang, T.-J. Fu, H. Wang, Y. Wang, and S. Zhang, Synthetic DNA Knots and Catenanes, New Journal of Chemistry 17, 739-755 (1993).

N.C. Seeman, J. Chen, Y. Zhang, B. Liu, H. Qiu, T.-J. Fu, Y. Wang, X. Li, X. Yang, J. Qi, F. Liu, L.A. Wenzler, S, Du, J.E. Mueller, H. Wang, C. Mao, W. Sun, Z. Shen, M.H. Wong, H, Yan and R. Sha, A Bottom-Up Approach to Nanotechnology Using DNA, In: Biological Molecules in Nanotechnology, ed. by S.C. Lee, IBC Libraries, Southborough, MA, Chapter 8 (1998).

N.C. Seeman, J. Qi, X. Li, X. Yang, N.B. Leontis, B. Liu, Y.Zhang, S.M. Du and J. Chen, The Control of DNA Structure: From Topological Modules to Geometrical Modules. In: Modular Chemistry, ed. by J. Michl, Kluwer, 95-104 (1997).

N.C. Seeman, Key Experimental Approaches in DNA Nanotechnology, Current Protocols in Nucleic Acid Chemistry, Unit 12.1, John Wiley & Sons, New York (2002).

N.C. Seeman, Nanoengineering with DNA, Biomolecular Materials: Materials Research Society Symposium Proceedings 292, 123-134 (1993).

N.C. Seeman, Nanoscale Assembly and Manipulation of Branched DNA: A Biological Starting Point for Nanotechnology, NANOCON Proceedings, ed. by J. Lewis and J.L. Quel, NANOCON, P.O. Box 40176, Bellevue, WA 98004, pp. 101-123; transcript of oral presentation, pp. 30-36 (1989).

N.C. Seeman, Nucleic Acid Nanostructures and Topology. Angewandte Chemie. 110, 3408-3428 (1998); Angewandte Chemie International Edition 37, 3220-3238 (1998).

N.C. Seeman, Structural DNA Nanotechnology: A New Organizing Principle for Advanced Nanomaterials, Materials Today 6 (7), 24-29 (2003).

N.C. Seeman, The Use of Branched DNA for Nanoscale Fabrication, Nanotechnology 2, 149-159 (1991).

N.C. Seeman, Y. Zhang, S.M. Du and J. Chen, Construction of DNA Polyhedra and Knots Through Symmetry Minimization, In: Supramolecular Stereochemistry, ed. by J. Siegel, NATO Advanced Research Workshops Series, Kluwer, Dordrecht, 27-32 (1995).

N.C. Seeman, Y. Zhang, T.-J. Fu, S. Zhang, Y. Wang and J. Chen, Chemical Synthesis of Nanostructures, Biomolecular Materials by Design: Materials Research Society Symposium Proceedings 330, 45-56 (1994).

N.C. Seeman, *Nanotechnology and the Double Helix*, Scientific American., 290, (6) 64-75 (2004).

N.C. Seeman. Structural DNA Nanotechnology: An Overview. Methods in Molecular Biology 303: Bionanotechnology Protocols, Editors, Sandra J. Rosenthal and David W. Wright, Humana Press, Totowa, NJ, pp. 143-166 (2005).

N.C.Seeman, J. Chen, Y. Zhang, T.-J. Fu, X. Li, X. Yang, Y. Wang, J. Qi, B. Liu and F. Liu, Control of Structure and Topology in DNA Nanotechnology. In: Molecular Nanotechnology: Biological Approaches and Novel Applications, ed. by S.A. Minden, IBC Libraries, Southborough, MA, Chapter 2.2 (31 pages) (1997).

Nordmann, A.: 2003, 'Shaping the World Atom by Atom: Eine nanowissenschaftliche WeltBildanalyse', in: A. Grunwald (ed.), *Technikgestaltung zwischen Wunsch und Wirklichkeit*, Berlin: Springer, pp. 191-199.

Nordmann, A.: 2004, 'Molecular Disjunctions: Staking Claims at the Nanoscale', in: D. Baird, A. Nordmann & J. Schummer (eds.), *Discovering the Nanoscale*, Amsterdam: IOS Press, pp. 51-62.

Nordmann, A.: 2004, 'Molecular Disjunctions', in: D. Baird, A. Nordmann, J. Schummer (eds.), *Discovering the Nanoscale*, IOS Press, Amsterdam, pp. 51-62.

Nordmann, A.: 2004, 'Nanotechnology: Convergence and Integration', Presentation at *EuroNanoForum, Trieste, December 10, 2003* (Proceedings in preparation).

Nordmann, A.: 2004, 'Nanotechnology's WorldView: New Space for Old Cosmologies', *IEEE Technology and Society Magazine*, 23 (forthcoming).

Nordmann, A.: 2004, 'Social Imagination for Nanotechnology', in: European Commission (Community Health and Consumer Protection): *Nanotechnologies: A Preliminary Risk Analysis on the Basis of a Workshop, Brussels, 1-2 March 2004*, pp. 111-113

Nordmann, A.: 2004, 'Was ist TechnoWissenschaft? ' Zum Wandel der Wissenschaftskultur am Beispiel von Nanoforschung und Bionik', in: T. Rossmann & C. Tropea (eds.), *Bionik ' Neue Forschungsergebnisse aus Natur-, Ingenieur- und Geisteswissenschaften*, Berlin: Springer, 2004 (forthcoming)

Paschen, H.; Coenen, C.; Fleischer, T.; Grünwald, R.; Oertel, D. & Revermann, C.: 2004, *Nanotechnologie: Forschung, Entwicklung, Anwendung*, Berlin: Springer (*Nanotechnologie, TAB-Arbeitsbericht 92*, Berlin: Büro für Technikfolgen-Abschätzung beim Deutschen Bundestag).

Peng, X.G.; Manna, L.; Yang, W.D.; Wickham, J.; Scher, E.; Kadavanich, A.; Alivisatos, A.P. 2000. "Shape control of CdSe nanocrystals", *Nature*, 404, 59-61.

Philip G. Collins, Phaedon Avouris, "Nanotubes for electronics," *Scientific American* 283(December 2000):62-69.

Philipse, A.P. 1998. "Colloidal Dispersions," section 3.4, A. ten Wolde (ed.), *Nanotechnology: Towards a molecular construction kit*. STT Report #60, The Hague, 171-178.

Phoenix, C.; Drexler, E.: 2004, 'Safe Exponential Manufacturing', *Nanotechnology*, 15, 869-72.

Pitt, J.C.: 2004, 'The Epistemology of the Very Small', in: D. Baird, A. Nordmann & J. Schummer (eds.), *Discovering the Nanoscale*, Amsterdam: IOS Press, pp. 157-163.

R. Sha, F. Liu and N.C. Seeman, Atomic Force Measurement of the Interdomain Angle in Symmetric Holliday Junctions, Biochemistry 41, 5950-5955 (2002).

R. Sha, F. Liu, D.P. Millar N.C. Seeman, Atomic Force Microscopy of Parallel DNA Branched Junction Arrays, Chemistry & Biology 7, 743-751 (2000).

R. Sha, F. Liu, M. Bruist and N.C. Seeman, Parallel Helical Domains in DNA Branched Junctions Containing 5', 5' and 3', 3' Linkages, Biochemistry 38, 2832-2841 (1999).

R. T. Bate, "Nanoelectronics," *Nanotechnology*, Vol. 1, pp. 1-7, 1990.

Ralph C. Merkle, "A new family of six degree of freedom positional devices," *Nanotechnology* 8(1997):47-52; http://www.zyvex.com/nanotech/6dof.html

Ralph C. Merkle, "A Proof About Molecular Bearings," *Nanotechnology* 4(1993):86-90; http://www.zyvex.com/nanotech/bearingProof.html

Ralph C. Merkle, "Binding sites for use in a simple assembler," *Nanotechnology* 8(1997):23-28; http://www.zyvex.com/nanotech/bindingSites.html

Ralph C. Merkle, "Casing an assembler," Nanotechnology 10(1999):315-322; http://www.zyvex.com/nanotech/casing

Ralph C. Merkle, "Convergent assembly," *Nanotechnology* 8(1997):18-22; http://www.zyvex.com/nanotech/convergent.html

Ralph C. Merkle, "Design considerations for an assembler," *Nanotechnology* 7(1996):210-215; http://www.zyvex.com/nanotech/nano4/merklePaper.html

Ralph C. Merkle, "Design-ahead for nanotechnology," in Markus Krummenacker, James Lewis, eds., *Prospects in Nanotechnology: Toward Molecular Manufacturing*, John Wiley & Sons, New York, 1995, pp. 23-52.

Ralph C. Merkle, "Molecular building blocks and development strategies for molecular nanotechnology," *Nanotechnology* 11(2000):89-99; http://www.zyvex.com/nanotech/mbb/mbb.html

Ralph C. Merkle, "Self replicating systems and molecular manufacturing," *J. Brit. Interplanet. Soc.* 45(1992):407-413; http://www.zyvex.com/nanotech/selfRepJBIS.html

Ramanan, V.R. 1998. Nanocrystalline soft magnetic alloys for applications in electrical and electronic devices. In *R&D Status and Trends*, ed. Siegel et al.

RAND Corporation. *The Potential of Nanotechnology for Molecular Manufacturing*. 1995.

Ratner, M.; Ratner, D.: 2003, *Nanotechnology: A Gentle Introduction to the Next Big Idea,* Prentice Hall, Upper Saddle River, NJ.

Richard E. Smalley, "Of chemistry, love, and nanobots," *Scientific American* 285(September 2001):76-77.

Rietman, E.A.: 2001, 'Drexler hypothesis of a universal assembler is supported not by theoretical arguments alone but by existence proof in the form of biological life', in: *Molecular Engineering of Nanosystems*, Springer, New York & Berlin.

Robert A. Freitas Jr., "Exploratory design in medical nanotechnology: A mechanical artificial red cell," *Artificial Cells, Blood Substitutes, and Immobil. Biotech.* 26(1998):411-430; http://www.foresight.org/Nanomedicine/Respirocytes.html

Robert A. Freitas Jr., "Nanodentistry," *J. Amer. Dent. Assoc.* 131(November 2000):1559-1566.

Robert A. Freitas Jr., "Some Limits to Global Ecophagy by Biovorous Nanoreplicators, with Public Policy Recommendations," Zyvex preprint, April 2000.

Roberts, J.A.: 2004, 'Deciding the Future of Nanotechnologies: Legal Perspectives on Issues of Democracy and Technology', in: D. Baird, A. Nordmann & J. Schummer (eds.), *Discovering the Nanoscale*, Amsterdam: IOS Press, pp. 247-255.

Robison, W.L.: 2004, 'Nano-Ethics', in: D. Baird, A. Nordmann & J. Schummer (eds.), *Discovering the Nanoscale*, Amsterdam: IOS Press, pp. 285-300.

Roco, M. C., R. S. Williams, and P. Alivisatos, *Nanotechnology Research Directions: Vision for Nanotechnology in the Next Decade*, Kluwer Academic Publishers (2000).

Roco, M.C. & Tomellini, R. (eds.): 2002, *Nanotechnology: Revolutionary Opportunities and Societal Implications, Workshop, Lecce (Italy), 31 January - 1 February 2002*, Luxemburg: Office for Official Publications of the European Communities, [200 pp.]

Roco, M.C., R.S. Williams, and P. Alivisatos (eds). 1999. *Nanotechnology Research Directions: IWGN Workshop Report, Vision for Nanotechnology R&D in the Next Decade*. (Kluwer Academic Publishers.

Roco, M.C.: 2003a, 'Broader Societal Issues of Nanotechnology', *Journal of Nanoparticle Research*, 5, 181-189.

Roco, M.C.: 2004, 'The Emergence and Policy Implications of Converging Technologies' [www.nsf.gov/home/crssprgm/nano/nbic_roco_04_0422_@aaas_57sl.pdf, accessed May 2004].

Rohrer, H.: 1995, 'The Nanometer Age – Challenge and Chance', *Microelectronic Engineering*, 27, 3-15.

Rosen, A. 1998. A periodic table in three dimensions: A sightseeing tour in the nanometer world. In *Advances in quantum chemistry*. In press.

S. Xiao, F. Liu, A. Rosen, J.F. Hainfeld, N.C. Seeman, K.M. Musier-Forsyth & R.A. Kiehl, Self-Assembly of Nanoparticle Arrays by DNA Scaffolding, J. Nanoparticle Research 4, 313-317 (2002).

S. Zhang, C. Cordon-Cardo, H.S. Zhang, V.E. Reuter, S. Adluri, W.B. Hamilton, K.O. Lloyd, P.O. Livingston, "Selection of tumor antigens as targets for immune attack using immunohistochemistry: I. Focus on gangliosides," *Int. J. Cancer* 73(26 September 1997):42-49.

S.P. Walch, W.A. Goddard III, R.C. Merkle, "Theoretical studies of reactions on diamond surfaces," Fifth Foresight Conference on Molecular Nanotechnology, 1997; http://www.foresight.org/Conferences/MNT05/Abstracts/Walcabst.html

S.P. Walch, W.A. Goddard III, R.C: Merkle, "Theoretical studies of reactions on diamond surfaces," Fifth Foresight Conference on Molecular Nanotechnology, 1997; http://www.foresight.org/Conferences/MNT05/Abstracts/Walcabst.html

Sanders, P.G., J.A. Eastman, and J.R. Weertman. 1996. In *Processing and properties of nanocrystalline materials*, ed. Suryanarayana et al.

Sanders, P.G., M. Rittner, E. Kiedaisch, J.R. Weertman, H. Kung, and Y.C. Lu. 1997. *Nanostructured Mater*. 9:433.

Schiemann, G.: 2004, 'Dissolution of the Nature-Technology Dichotomy? Perspectives on Nanotechnology from an Everyday Understanding of Nature', in: D. Baird, A. Nordmann & J. Schummer (eds.), *Discovering the Nanoscale*, Amsterdam: IOS Press, pp. 209-213.

Schmidt, J.C.: 2004, 'Unbounded Technologies: Working Through Technological Reductionism of Nanotechnology', in: D. Baird, A. Nordmann & J. Schummer (eds.), *Discovering the Nanoscale*, Amsterdam: IOS Press, pp. 35-50.

Schuler, E.: 2004, 'Perception of Risks and Nanotechnology', in: D. Baird, A. Nordmann & J. Schummer (eds.), *Discovering the Nanoscale*, Amsterdam: IOS Press, pp. 279-284.

Schummer, J.: 2001, 'Ethics of Chemical Synthesis', *Hyle: International Journal for Philosophy of Chemistry*, 7, 103-124.

Schummer, J.: 2004, 'Interdisciplinary Issues in Nanoscale Research', in: D. Baird, A. Nordmann & J. Schummer (eds.), *Discovering the Nanoscale*, Amsterdam: IOS Press, pp. 9-20.

Schummer, J.: 2004, 'Interdisciplinary Issues in Nanoscale Research', in: D. Baird, A. Nordmann, J. Schummer (eds.), *Discovering the Nanoscale*, IOS Press, Amsterdam, pp. 9-20.

Schummer, J.: 2004, 'Why do Chemists Perform Experiments?', in: D. Sobczynska, P. Zeidler, E. Zielonacka-Lis (eds.), *Chemistry in the Philosophical Melting Pot*, Peter Lang (Frankfurt/M.), 2004, pp. 395-410.

Schummer, J.; Baird, D. (eds.): 2004-5, *Nanotech Challenges*, joint special issue of *Hyle: International Journal for Philosophy of Chemistry & Techne: Research in Philosophy and Technology* (forthcoming).

Schwarz, R.B. 1998. Storage of hydrogen in powders with nanosized crystalline domains. In *R&D status and trends*, ed. Siegel et al.

Shull, R.D. 1998. NIST activities in nanotechnology. In *R&D Status and Trends*, ed.

Shull, R.D., R.D. McMichael, and J.J. Ritter. 1993. *Nanostructured Mater*. 2:205.

Shultz, W.: 2000, 'Nanotechnology: The Next Big Thing', *Chemical and Engineering News*, 78(18), 41-47.

Siegel R.W. and G.E. Fougere. 1994. In *Nanophase materials*, ed. G.C. Hadjipanayis and R.W. Siegel. Netherlands: Kluwer Acad. Publ.

Siegel, R.W., E. Hu, and M.C. Roco, eds. 1998. *R&D status and trends in nanoparticles, nanostructured materials, and nanodevices in the United States*. Baltimore: Loyola College, International Technology Research Institute. NTIS #PB98-117914.

Smalley, R.: 2001, 'Of Chemistry, Love and Nanobots', *Scientific American*, 285, 76-77.

Snow, E.S., P.M. Campbell, and F.K. Perkins. 1997. Nanofabrication with proximal probes. *Proceedings of the IEEE* 85:601-11.

Stephen P. Walch, Ralph C. Merkle, "Theoretical studies of diamond mechanosynthesis reactions," *Nanotechnology* 9(1998):285-296.

Steve, Jurvetson, "Transcending Moore's Law with Molecular Electronics," Nanotechnology Law & Business Journal, Vol. 1, No. 1, article 9, p. 9.

Stix, G.: 1996, 'Trends in Nanotechnology: Waiting for Breakthroughs', *Scientific American*, 274(4), 94-99.

Storrs, J., Hall, *Flying Saucers?* http://www.nanocentral.com/NanoWorld/Perspectives/JoshHall/askjosh_aug96.html.

Stranick, S.J.; Atre, S.V.; Parikh, A.N.; Wood, M.C.; Allara, D.L.; Winograd, N.; Weiss, P.S. 1996. "Nanometer-scale phase separation in mixed composition self-assembled monolayers", *Nanotechnology*, 7, 438-442.

Sun, S.H.; Murray, C.B. 1999. "Synthesis of monodisperse cobalt nanocrystals and their assembly into magnetic superlattices", *Journal of Applied Physics*, 85, 4325-4330.

Surowiecki, J.: 2004, 'Bring on the Nanobubble', *The New Yorker*, March 15, p. 68.

Swiss Re: 2004, *Nanotechnology: Small Matter, Many Unknowns*, Zurich: Swiss Re [www.swissre.com].

Tighe, T.S., J.M. Worlock, and M.L. Roukes. 1997. Direct thermal conductance measurements on suspended monocrystalline nanostructures. *Appl. Phys. Lett*. 70:2687-9.

Tobio, M., R. Gref, A. Sanchez, R. Langer and M.J. Alonso. 1998. "Stealth PLA-PEG Nanoparticles as Protein Carriers for Nasal Administration", *Pharmaceutical Research*, Vol. 15, No 2, pp 270-275.

Toumey, C.: 2004, 'Anticipating Public Reactions to Nanotechnology', *Techne: Research in Philosophy and Technology*, 8(3) (forthcoming).

Tour, J.M.: 2003, *Molecular Electronics: Commercial Insights, Chemistry, Devices, Architecture and Programming*, World Scientific, Singapore.

Trudeau, M.L., and J.Y. Ying. 1996. *Nanostructured Materials* 7:245. Tschope, A.S., W. Liu, M. Flyzani-Stephanopoulos, and J.Y. Ying 1995. *J. Catal*. 157:42.

Vaia, R., D. Tomlin, M. Schulte, and T. Bunning. 2000. Two-phase nanoscale morphology of polymer/LC composites. *Polymer*. In press.

Vermaas, P.E.: 2004, 'Nanoscale Technology: A Two-Sided Challenge for Interpretations of Quantum Mechanics', in: D. Baird, A. Nordmann & J. Schummer (eds.), *Discovering the Nanoscale*, Amsterdam: IOS Press, pp. 77-91.

W. Shen, M. Bruist, S. Goodman & N.C. Seeman, A Nanomechanical Device for Measuring the Excess Binding Energy of Proteins that Distort DNA, Angew. Chem. Int. Ed. 43, 4750-4752 (2004); Angew.Chem. 116, 4854-4856 (2004).

W.B. Sherman and N.C. Seeman, A Precisely Controlled DNA Bipedal Walking Device, NanoLetters 4, 1203-1207 (2004); Erratum, 4, 1801-1801.

Wang, C.C., Z. Zhang, and J.Y. Ying. 1997. *Nanostructured Mater*. 9:583.

Weckert, J.: 2001, 'The Control of Scientific Research: The Case of Nanotechnology', *Australian Journal of Professional and Applied Ethics*, 3, 29-44.

Whitesides G. 2001 'The Once and Future nanomachines', *Scientific American*, (Sept.), 78-83.

Whitesides, G.: 1998, 'Nanotechnology: Art of the Possible', *Technology, MIT Magazine of Innovation*, (Nov.-Dec.), pp. 8-13.

Whitesides, G.M.: 2001, 'The Once and Future Nanomachine', *Scientific American*, 285, 78-83.

Wilbur, J.L.; Whitesides, G.M. 1999. *Self-Assembly and Self-Assembled Monolayers in Micro- and Nanofabrication*; Timp, G., Ed.; Springer-Verlag New York, Inc.: New York, pp 331-370.

Wildoer, J.W.G., L.C. Venema, A.G. Rinzler, R.E. Smalley, and C. Dekker. 1998. Electronic structure of atomically resolved carbon nanotubes. *Nature* 391:59-62.

Williams, R.S. 1998. Functional nanostructures. In *R&D status and trends*, ed. Siegel et al. Wiltzius, P. 1998.

Wolfson, J.R.: 2003, 'Social and Ethical Issues in Nanotechnology: Lessons from Biotechnology and Other High Technologies', *Biotechnology Law Report*, 22 (4), 376-96.

Wood, S., Jones, R.; Geldart, A.: 2003, *The Social and Economic Challenges of Nanotechnology*, UK Economic and Social Research Council [www.esrc.ac.uk/esrccontent/DownloadDocs/Nanotechnology.pdf, accessed: 3 Feb. 2004].

Woods, S.; Jones, R. & Geldart, A.: 2003, *The Social and Economic Challenges of Nanotechnology*, Report to the Economic and Social Research Council (ESRC), Swindon, UK [www.esrc.ac.uk/esrccontent/DownloadDocs/Nanotechnology.pdf].

Wu, M.K., R.S. Windeler, C.K. Steiner, T. Bors, and S.K. Friedlander. 1993. Controlled synthesis of nanosized particles by aerosol processes. *Aerosol Sci. Technol.* 19: 527.

X.J. Li, X.P. Yang, J. Qi, and N.C. Seeman, Antiparallel DNA Double Crossover Molecules as Components for Nanoconstruction, Journal of the American Chemical Society, 118, 6131-6140 (1996).

Xu, Y. and C. Musgrave, "A DFT Study of the Atomic Layer Deposition of Al 2 O 3 on SAMs: The Effect of SAM Termination," *Chemistry of Materials,* 16, 646 (2004).

Y. Wang, J.E. Mueller, B. Kemper , and N.C. Seeman, The Assembly and Characterization of 5-Arm and 6-Arm DNA Junctions, Biochemistry 30, 5667-5674 (1991).

Ying, J.Y. 1998. Nanostructure processing of advanced catalytic materials. In *R&D status and trends*, ed. Siegel et al.

Ying, J.Y., and T. Sun. 1997. Research needs assessment on nanostructured catalysts. *J. of Electroceramics* 1:219-238.

Index

Index